AMAZONS TO FIGHTER PILOTS

Contributing Editors

Gloria Allaire, University of Kentucky
Mary Allen, University of Toronto
Paul Davis, Texas Military Institute, San Antonio
Valerie Eads, Independent Scholar
Sherry Mou, Wellesley College, Massachusetts
Sarah Purcell, Central Michigan University
Bobbie Oliver, Curtin University of Technology, Washington

AMAZONS TO FIGHTER PILOTS

A Biographical Dictionary of Military Women

Volume One: A–Q

Reina Pennington, Editor
Robin Higham, Advisory Editor
Foreword by Gerhard Weinberg

Greenwood Press
Westport, Connecticut • London

Library of Congress Cataloging-in-Publication Data

Amazons to fighter pilots : a biographical dictionary of military women / Reina Pennington, editor ; Robin Higham, advisory editor ; foreword by Gerhard Weinberg.
    p. cm.
   Includes bibliographical references and index.
    ISBN 0-313-29197-7 (set : alk. paper)—ISBN 0-313-32707-6 (v. 1 : alk. paper)—
    ISBN 0-313-32708-4 (v. 2 : alk. paper)
   1. Women soldiers—Biography—Dictionaries. 2. Women sailors—Biography—Dictionaries. I. Pennington, Reina, 1956– II. Higham, Robin D. S.
U52.A44 2003
355'.0082—dc21   2002044777

British Library Cataloguing in Publication Data is available.

Copyright ©2003 by Reina Pennington

All rights reserved. No portion of this book may be reproduced, by any process or technique, without the express written consent of the publisher.

Library of Congress Catalog Card Number: 2002044777
ISBN: 0-313-29197-7 (set)
    0-313-32707-6 (v.1)
    0-313-32708-4 (v.2)

First published in 2003

Greenwood Press, 88 Post Road West, Westport, CT 06881
An imprint of Greenwood Publishing Group, Inc.
www.greenwood.com

Printed in the United States of America

The paper used in this book complies with the Permanent Paper Standard issued by the National Information Standards Organization (Z39.48–1984).

10 9 8 7 6 5 4 3 2 1

Contents

Foreword *by Gerhard Weinberg*	vii
Preface	ix
List of Entries by Geographic Region	xvii
List of Entries by Time Period/Conflict	xxv
List of Entries by Role/Branch of Service	xxxiii
List of Entries by Prisoner/POW Status	xxxix
List of Entries on Groups and Organizations	xli
Introduction: A Historical Overview of Women in the Military and in War	xliii
Alphabetical Entries	**1**
Numbered Entries	**511**
Timeline	525
Appendix A: Women as Prisoners of War: A Bibliographic Survey	651
Appendix B: Women, Medicine, and the Military: A Bibliographic Survey	667
Bibliography	691
Index	727
About the Contributors	749

Foreword

The United States has been changing in an important but barely noticed way. Two developments appear to be making for greater interest in and discussion of military issues. This was true even before the terrorist attacks on September 11, 2001 brought organized violence dramatically to public attention. One of these developments is the clear evidence of public interest in the wars in which the United States has been involved. The Civil War continues to attract the attention not only of scholars but also of large numbers of readers, reenactors, and battlefield tourists. World War II has perhaps drawn even more public interest, as the crowds who saw the movie *Saving Private Ryan*, the large sales of books dealing with World War II and Holocaust topics, and the programming of the History Channel all show. Students flock in disproportionate numbers into courses covering the country's wars.

The second development breaking into the public discourse, to a lesser extent but with a greater degree of acrimony, is a discussion of the role of women in the military. Without in any way rehearsing the arguments over this issue (and in particular the controversy over the assignment of women to combat roles), I will note one special characteristic of the discussion because of its astonishing absence: the role of women in the military, including combat, in World War II, especially in the armed forces of the Soviet Union. Since literally hundreds of thousands of women served in the Red Army and Red Air Force—in tanks, on guns, in planes—one might have expected some major efforts on the part of scholars at analyzing these experiences, not as models to be imitated but as events that could bring the light of reality to at least some of the aspects of the heated debates. The editor of this work has published a fine book that covers one aspect of those events—the role of women in entirely or predominantly female combat units of the Red Air Force—but the United States government, which during the Cold War poured research money into innumerable aspects of Soviet government and life, has not extended its interest to a topic of such obvious contemporary relevance.

Furthermore, the discussion of women in the military has been conducted largely in a vacuum. There has been no or very little attention to historical experience. If the role of women in the Soviet military in World War II was on an unprecedented scale, it was certainly not without precedent. Beyond those activities in the military that have come to be in some ways accepted as "normal"—nurses, cooks, and laundresses in the past, secretaries, drivers, and balloon handlers more recently—there have been women participating in combat for centuries; and it is that experience which has been absent from the discussion. In part this is a result of the way in which historians have come to specialize. The historians of the military pay little or no attention to female participants—whether disguised as men or not—while the development of women's history in recent decades has generally shied away from military topics. Moreover, with the steady dying of the surviving World War II and Korean War veterans, the general silence of the Vietnam veterans, the end of the draft, and the reduction of the military with the end of the Cold War, the percentage of the population with any personal or family ties to the military or to military affairs is steadily shrinking. This shrinkage is accentuated for the generation now coming into public life by the disappearance (for the above and additional reasons) of men and women with military experience or ties from the faculties of colleges and universities. Furthermore, the study of military affairs has become unfashionable in much of the scholarly community.

It is the special contribution of this work that it opens up the horizons of any who are interested in either the history of war or the history of women, as well as in the current debates, by offering a sample over time and space of women who have engaged in combat or other active military roles. Because of the newness of the endeavor, the paucity of sources in many cases, and the small number of specialists whose expertise can be tapped, the selection of persons and the lengths of entries are necessarily somewhat arbitrary. This characteristic of the collection does not, however, detract in any substantial way from its value. What is of great significance beyond the specific detail of the individual entries is the chronological and geographic range of coverage. By itself, that range should enable readers to recognize the importance of the subject to an understanding of the history of warfare and women's role in it and also stimulate further research. And perhaps most important, it should add a dimension of precedents and perspective to a discussion likely to continue for years, but hitherto conducted largely without either, in a society where, as indicated above, acquaintance with military affairs is steadily decreasing even as interest remains high.

Gerhard Weinberg

Preface

Until quite recently, scholars have had little to say about the military roles of women. For example, anthropologist Lionel Tiger contends that "almost universally war is an all-male enterprise."[1] Martin Van Creveld wrote in 1991 that "women have never taken a major part in combat–in any culture, in any country, in any period of history."[2] John Keegan claimed in 1993, "Warfare is . . . the one human activity from which women, with the most insignificant exceptions, have always and everywhere stood apart. . . . Women . . . do not fight . . . and they never, in any military sense, fight men."[3] Keegan and Van Creveld are in many ways brilliant historians. However, this does not mean that they are qualified to speak as experts on every aspect of military history—particularly not the history of military women. Their sweeping generalizations about women are deceptive and counterfactual. Such pronouncements fit neatly into society's myth that men fight while women do not, but the truth is rarely so simple. History reveals that both women and men played a variety of roles; not all men have been soldiers, while some women have. As Carol Tavris has pointed out, "all polarities of thinking, like all dichotomies of groups, are by nature artificial, misleading, and oversimplified."[4] Yet even when women's participation in combat has been acknowledged, it is generally dismissed as "historically insignificant," "anomalous," or "more at the symbolic level than the real."[5]

A study of military women must be positioned at the intersection of military history and women's history, and this is probably the reason it has rarely been undertaken. Specialists in military history rarely address "women's issues," while those who write the history of women seldom have an interest in military history. The overall neglect is so marked that the history of military women is not just a marginalized subject; it is a historiographical "no-man's land." History lags behind sociology, anthropology, and literature in the study of women in combat. Historian D'Ann Campbell observes that "the question of women in combat has generated a vast literature that draws from law, biology, and psychology, but seldom from history."[6]

There are many possible reasons for the failure of military historians to examine the role of women; a Western bias that dismisses the relevance of foreign experience is one probable explanation, while the gender bias of male historians who have written most military history has certainly been a factor. One reason women have been left out of military history is that the categories have been defined in traditional terms, by men. The category of "military service" is, by accepted social definition, something men did, and women did not. It was a category of activity that has been used in many societies to distinguish male and female roles. The arbitrary categorization of women's activities is often related to what M.C. Devilbiss has described as the "yes, but phenomenon," used to explain away women's military contributions: *Yes, but* that was not in modern times; *yes, but* that was combat support and not a direct combat role; *yes, but* they were Russian/African/South American women, not American citizens; *yes, but* that was not a real war.[7]

Military historians are increasingly interested in the behaviors, perceptions, and reactions of men in combat, but when women were present in combat, their presence has often been overlooked. Yet the recent recognition of the importance of the experience of the common soldier, the "history from below" approach (admirably accomplished by Keegan himself in *The Face of Battle* and other books) makes an examination of the experience of women combatants more imperative than ever. This has been recognized by some military historians, such as Richard Kohn (who suggests that "startling new material might emerge on such topics as the large and important role of women in military organization and war before the twentieth century"), but has not yet been integrated into mainstream military history.[8]

A few studies have recently begun to appear on military women. Works in English still focus heavily on the Western experience, primarily that of the twentieth century. Many studies still concentrate on military women as clerks, nurses, and drivers—traditional "women's" roles that do not threaten the social status quo or the military's self-definition. Women who fought have received very little attention indeed. For example, Roger J. Spiller's three-volume *Dictionary of American Military Biography* (1984) includes only two women. *Webster's American Military Biographies* (1978), with more than a thousand entries, fails to list such women as Anna Ella Carroll. Neither Spiller nor *Webster's* includes the topic of "women" as an index item. *The Harper Encyclopedia of Military Biography* (1992) has no index, but a scan of the entries reveals that even the best-known military women are missing. The 1937–1969 index for the *Journal of Military Affairs* contains no heading for "women." Reference works in both military and women's history most often cover American subjects. Yet hundreds of thousands of military women served in the Soviet Union, and many more fought as partisans in countries like Israel and Yugoslavia. Even the most prominent of these non-American military women have been sadly neglected. Almost any student of military history, in consulting existing secondary works, would come to the conclusion that women have made no significant military contribution.

The history of women in combat has also been neglected by those studying women's history. One often looks in vain for entries on "military," "soldiers," or even "war" in the indexes of biographical and bibliographical reference works that focus on women. Those that do include military women emphasize only a few legendary figures. However, the study of military women has much to contribute to women's history. This study examines an important historical arena in which women acted outside the private sphere.[9] Women who have played military roles exhibit many interesting characteristics that put them outside the framework of typical female experience. First, they demonstrate an acceptance of the concept of violence inherent in the military and often performed acts of violence themselves. Second, they exhibit historical agency—they have an active rather than a passive involvement in events. Third, they stretched or rejected the definitions of female identity accepted in their own time. Fourth, their desire and willingness to play a military role often attained at least a temporary priority over other roles (such as mother or wife).

Recent historiographical trends favor moving beyond women's history toward a broader approach of studying gender, or the creation of sex roles and the relations between women and men.[10] The study of military women offers the opportunity to examine the interaction and integration of women and men in a combat situation. A significant aspect of women's history is a stress on the importance of women's voices, emphasizing sources such as diaries, letters, and memoirs that reveal the women's own perception of their experience. This approach coincides with the trend in military history toward "history from below." The experiences of women in combat have sometimes been recorded in their own words, but, like the letters of Rosetta Wakeman, who fought in the Civil War, have been hidden or overlooked until recently.

Another recent trend in women's history is the examination of ways in which women have had power; women's experiences in combat provide one basis for assessing how and when women have been powerful in a wartime setting. Furthermore, some writers have suggested that the exclusion of women from ruling elites was due to and justified by their exclusion from the military.[11] If it can be demonstrated that when given the opportunity, women performed as well as men in military roles and even in combat, new avenues of historical questioning can be opened regarding the interconnection between gender roles, military service, and power.

However, women have rarely attained positions of power, either politically or militarily. Their military participation most often has been in the lower ranks and has often been informal or in disguise. Such women leave few if any documents, such as letters and diaries; when documents were created, they were rarely preserved. In short, we usually study what men say about women. From Herodotus and Tacitus forward, most of recorded history has been filtered through the eyes of men, and at least some of these historians were biased by the views of their societies on the "proper" roles of women.

But if historians have overlooked women's military roles, there has always been interest among other writers. The description of women's military service and experiences has been left largely to memoirists, biographers, and amateur enthusiasts. Popular works are unfortunately replete with errors in dates and names and are rarely cross-referenced with archival or other documentary sources. Academically trained historians have only begun to explore this vital field of history. They have brought to bear the tools of history that enable reevaluation and analysis of the true role of women in the military. For example, a variety of bibliographic aids are beginning to appear, of which Charlotte Seeley's *American Women and the U.S. Armed Forces,* a guide to the National Archives, is particularly useful.[12] One promising development is that the journal *Minerva: Quarterly Report on Women and the Military,* published since 1983 by Linda Grant De Pauw at George Washington University, is presently being upgraded to a peer-reviewed publication.

There are many possible approaches by which to examine the history of military women. New theoretical frameworks are being developed all the time. Two of the best are outlined in articles by Margaret Higonnet and Patrice Higonnet, "The Double Helix," and Mady Wechsler Segal, "Women's Military Roles Cross-Nationally: Past, Present, and Future."[13] Both these articles present excellent theories to explain why combat experience does not result in a linear expansion of women's military roles; why those roles expand and contract more as a result of cultural and social influences than because of military necessity; and why women's actual performance is rarely a factor in determining what roles are permitted them. In addition, "The Double Helix" suggests that whenever women's roles expand, they are still valued relatively less than men's roles.

We are all aware of the continuing debates over the role of women in combat. Most of the arguments about whether women can cope with combat duty are based on speculation rather than the actual experience of women and show little knowledge of history. This is a work of historical reference; it does not address the question of whether women *should* be part of the military or participate in combat. Our contributors seek only to document, with a trained and critical eye, the experience of military women in broad historical context and according to accepted standards of scholarship. The resulting entries do, naturally, address issues of capability. Historical information on questions of leadership, unit cohesion, small-group bonding, physical abilities, soldierly skills, and endurance of stress, injury, capture, and torture can be found in the pages of this book.

Preface

Other collections on military women have appeared, but none directly competes with this book. What makes *Amazons to Fighter Pilots* unique are (1) its international emphasis, (2) its consideration of women throughout history, rather than just in recent times, (3) its scholarly approach, and (4) its focus on women who fought. This collection adheres to a rather close definition of what constitutes a military woman. First, any woman who bore arms during hostile actions or otherwise filled what would normally be considered a combat role is a military woman; the primary emphasis is on women who fought. Second, a military woman is any woman who held military rank, including women who served in auxiliary units. Women who served in "traditional" women's roles are less stressed in this book. While nurses, clerks, and "women of the army" can be considered military auxiliaries, they have already received a great deal more attention from historians than have women who fought. Thus only selected, representative examples of women in these categories are included. An extensive bibliographic discussion of women in medical roles appears in an appendix to assist readers in pursuing this topic. Women who served in espionage are also generally omitted in this collection, as their actions are considered traditional for women and are not normally connected with direct military activity. However, guerrilla fighters and partisans did often engage in fighting and are among the entries in this collection. Several lists are provided that group entries by geographic region, by time period/conflict, and by role/branch of service to assist readers in identifying women in various categories. There is also a list of the entries that mention women who became prisoners, as well as a separate appendix providing a bibliographic survey on the topic of women as prisoners of war.

It is trendy in some circles to denigrate substantive (as opposed to analytical) history as "mere gap filling" and to praise those who "move beyond" documentary history. There is a common view in women's-history circles that "recovering" women's history has already been done (although the study of military women remains largely unrecovered). Taken out of context, the entries in this dictionary might be dismissed as a collection of gap-filling biographies. Therefore, the introduction provides a chronological overview of women's military experience, setting a context within which the contents of this book increase in meaning and importance. Additionally, an extensive timeline provides summaries of the entries in strict chronological order, plus other important information and examples of women's military activities, set off against a parallel listing of military history events. Sidebars throughout the book provide relevant quotations, statistics, and additional information on women and war, and photographs and illustrations add visual information to the text.

A biographical dictionary of this sort, if it focused only on individual women, would omit the masses of women who participated in combat but cannot be identified on an individual basis. Thus group entries and organizational entries are included to provide a more comprehensive picture of the

Preface

scope and diversity of women's military participation. Many interesting women and groups had to be omitted from this work because a qualified contributor could not be identified or because there was insufficient time and funding to permit primary-source research that would be required to determine the subject's suitability. The length of an entry is not necessarily related to its historical importance, but is more a reflection of the availability of sources, of the duration and extent of activities discussed, and of the subject's significance as being representative of broad historical trends. A master bibliography is provided, so in the entries, sources are listed in short form (author, short title, date), except for some sources used only in a single entry. Terms in boldface type are cross-references to other main entries in this volume.

Amazons to Fighter Pilots: A Biographical Dictionary of Military Women provides an extensive listing of compelling women who, throughout history and in all nations, have played military roles. As suggested earlier, this work is by no means comprehensive or the "last word" on military women. A book of this length reveals only the tip of the historical iceberg. A great deal more research and writing remain to be done on women throughout history and throughout the world. It is hoped that this book will serve as a springboard to more exhaustive work by historians. We hope and believe that this book will be a unique resource for both scholarly researchers and the general public.

NOTES

1. Lionel Tiger, *Men in Groups* (London: Nelson, 1969), 80.
2. Martin Van Creveld, "Women of Valor: Why Israel Doesn't Send Women into Combat," *Policy Review* (Fall 1991): 65–67.
3. John Keegan, *A History of Warfare* (New York: Knopf, 1993), 76.
4. Carol Tavris, *The Mismeasure of Woman: Why Women Are Not the Better Sex, the Inferior, or the Opposite Sex* (New York: Simon and Schuster, 1992), 288.
5. Edd D. Wheeler, "Women in Combat: A Demurrer," *Air University Review* 30.1 (1978): 62–68.
6. D'Ann Campbell, "Women in Combat: The World War II Experience in the United States, Great Britain, Germany, and the Soviet Union," *Journal of Military History* 57.2 (1993): 322.
7. M.C. Devilbiss, "Gender Integration and Unit Deployment: A Study of GI Jo," *Armed Forces and Society* 11.4 (1985): 526.
8. Richard E. Kohn, "The Social History of the American Soldier: A Review and Prospectus for Research," *American Historical Review* 86 (June 1981): 533–567.
9. Dorothy O. Helly and Susan M. Reverby, eds., *Gendered Domains: Rethinking Public and Private in Women's History* (Ithaca: Cornell University Press, 1991), xi, 6.
10. Pauline Schmitt Pantel, ed., *A History of Women in the West: I. From Ancient Goddesses to Christian Saints,* trans. Arthur Goldhammer (Cambridge, MA: Belknap, 1992), 465.
11. Elise Boulding, "Public Nurturance and the Man on Horseback," *Face to Face: Fathers, Mothers, Masters, Monsters: Essays for a Nonsexist Future,* ed. Meg Murray (Westport, CT: Greenwood Press, 1983), 273–292.

12. Charlotte Palmer Seeley, ed., *American Women and the U.S. Armed Forces: A Guide to the Records of Military Agencies in the National Archives Relating to American Women* (Washington, DC: National Archives and Records Administration, 1992).

13. Margaret Higonnet and Patrice Higonnet, "The Double Helix," *Behind the Lines: Gender and the Two World Wars*, ed. Margaret Higonnet et al. (New Haven: Yale University Press, 1987), 31–47, and Mady Wechsler Segal, "Women's Military Roles Cross-Nationally: Past, Present, and Future," *Gender & Society* 9.6 (1995): 757–775.

List of Entries by Geographic Region

Note: Some of the individuals listed may not have their own entry but appear in other entries (for example, within a group entry). Others may be mentioned in more than one entry. Check the index for precise locations.

Africa
Algerian-French War, women in
Arsinoë III
Berenice II Euergetes
Candace
Cleopatra VII
Dahomey, women in army of
Hatchepsut
Kāhinah
Libya, military women in
Masarico
Nehanda
Nhongo, Joyce Mugari
Zimbabwe, Women in the National Liberation War

African American
Bethune, Mary McLeod
Taylor, Susie King
Tubman, Harriet
WAC, African Americans in
Williams, Cathay

Australia
Australian Defence Forces, women in
Best, Kathleen
POWs, Australian women
WAAAF (Women's Auxiliary Australian Air Force)
Walker, Kath

Britain/Ireland/Scotland
Aethelburh
Aethelflaed of Mercia
Agnes of Dunbar
ATA (Air Transport Auxiliary)
ATS (Auxiliary Territorial Service)
Badlesmere, Lady Margaret
Bankes, Lady Mary

List of Entries by Geographic Region

Boudicca
Cartimandua
Catherine of Aragon
Cavanaugh, Kit
Dionisia de Grauntcourt
Elizabeth I
Emma FitzOsbern
FANY Corps (First Aid Nursing Yeomanry Corps)
Gower, Pauline
Gwynne-Vaughan, Helen
Harley, Lady Brilliana
Henrietta Maria
Isabel, countess of Buchan
Isabella of France
Knyvet, Alice
Lilliard of Ancrum
Margaret of Anjou
Mary I, queen of England
Matilda Augusta of England
Matilda of Boulogne
Nicolaa de le Haye
Philippa of Hainault
Sandes, Flora
Snell, Hannah
Stanley, Charlotte, countess of Derby
Talbot, Mary Anne
Thatcher, Margaret
WAAC (Women's Auxiliary Army Corps)
Women's Land Army
Women's Legion
Wulfbald, widow of

Canada

Brasseur, Deanna
Clay, Wendy
De La Tour, Françoise-Marie Jacquelin
Drucour, Seigneuress Anne-Marie
Erxleben, Heather
Foster, Jane
Hellstrom, Sheila
Secord, Laura Ingersoll
Verchères, Marie-Madeleine Jarret de

China

Aluzhen
Chi Zhaoping
Erketü Qatun
Gao (first name unknown)
Hua Mulan
Kang Keqing
Kong (first name unknown)
Li Xiu
Liang Hongyü
Liu (first name unknown)
Liu Shuying
Lu (first name unknown)
Mao, Empress
Meng, Consort
Pan (first name unknown)
Pingyang, Princess
Qin Liang-yü
Ren (first name unknown)
Shalizhi
Shao (first name unknown)
Shejie (first name unknown)
Shen Yunying
Wa (first name unknown)
Xi (first name unknown)
Xian (first name unknown)
Yan (first name unknown)
Yang Miaozhen

Europe

Beatrice of Lorraine
Chiomara
Christine de Pisan
Cimbrian Women
Cloelia
Crusades, women in
Eger, women in siege of

Eleanor of Aquitaine
Hungary, women in 1956 revolution
Hussite Wars, women in
Joanna of England
Johanna of Rožmitál
Lebstuck, Maria
Maria Theresa
Mary of Hungary
Resistance movements, Yugoslavian women
Teuta
Thirty Years' War, women in
Zrinyi, Countess Ilona

France

Albret, Jeanne d'
Alice de Montmorency
Barreau, Rose-Alexandrine
Bordereau, Renée
Bouillon, Rose
Brunehaut
Cezelly, Françoise de
Chapuy, Reine
Charpentier, Marie
Chevreuse, Marie de Rohan-Montbazon, duchesse de
Claude des Armoises
Clotild
Delaye, Margot
Duchemin, Marie-Angélique Josèphe
Emma, queen of France
Engle, Regula
Ermengard, viscountess of Narbonne
Figueur, Thérèse
Fredegund
French Revolution and Napoleonic era, women soldiers in
Ghesquière, Virginie
Gouges, Olympe de
Giralda de Laurac
Isabella de Lorraine
Jeanne de Montfort
Jeanne de Penthièvre
Joan of Arc
Julien, Sophie
Julienne du Guesclin
Khan, Noor Inayat
La Guette, Catherine de Meurdrac, Madame de
Lacombe, Rose
Léon, Pauline
Leubevére
Longueville, Anne-Geneviève de Bourbon-Condé, duchesse de
Mabel of Bellême
Mahaut d'Artois
Médici, Catherine de
Médici, Marie de
Michel, Louise
Montpensier, Anne Marie Louise d'Orléans, duchesse de
Mouron, Marie Magdelaine
Parent, Marie-Barbe
Paris Commune, women's military roles in
Petitjean, Madeleine
Pochetat, Catherine
Prémoy, Geneviève
Prothais, Geneviève
Quatsault, Anne
Resistance Movements, French women
Robin, Jeanne
Rochejaquelein, La Marquise de La
Rouget, Claudine
Saint-Baslemont, madame de
Schellinck, Marie-Jeanne
Théroigne de Méricourt
Valois, Marguerite de
Xaintrailles, Marie-Henriette Heiniken

Germany

Reitsch, Hanna
Stauffenberg, Melitta

List of Entries by Geographic Region

Greece/Macedon
Bouboulina, Laskarina
Cynane
Eurydice II
Mavrogenous, Manto
Olympias
Tomyris

India
Begum Samru
Bhai, Laxshmi
Holkar, Ahilyabhai
Raziya, Sultan

Israel
Ben-Yehuda, Netiva
Ben-Zvi, Rachel Yanait
CHEN
Deborah
Israeli War of Independence, women's military roles
Meir, Golda
Shochat, Mania

Italy
Adelaide of Turin
Alruda Frangipani
Attendolo, Margherita
Bianca di Rossi
Bianca Maria Visconti Sforza
Bona of Lombardy
Eleanora of Arborea
Forteguerri, Laudomia
Lanz, Katharina
Maria of Pozzuoli
Marzia degli Ubaldini
Matilda of Tuscany
Resistance movements, Italian women
Rodiana, Onorata
Rodolfi, Camilla
Sforza, Caterina
Sichelgaita of Salerno, duchess of Apulia
Siena, women in siege of
Stamira of Ancona
Triaria, Empress

Japan
Hōjō Masako
Tomoe Gozen

Latin America
Azurduy de Padilla, Juana
de Abreu e Lencastre, Doña Maria Ursula
Erauso, Catalina de
Feitosa, Jovita Alves
García Morales de Carrera, Petrona
Paraguayan women in the War of the Triple Alliance
Quitéria de Jesus, Maria

Mexico
Barrera, Maria de la Luz Espinosa
Bocanegra, Gertrudis
Herrera, Petra
Jiménex, Angela
Reachy, Ignacia
Ruiz, Petra
Soldaderas
Xochitl

Near East and Middle East
Amazons
Artemisia I of Halicarnassus
Artemisia II of Halicarnassus
Berenice Syra
Buhrayd, Jamilah
Hind bint 'Utba
Judith
Mavia
Semiramis
Tamar, queen of Georgia

Telesilla of Argos
United Arab Emirates, military women in
Yarmūk, women in Battle of
Zaynab "Rustam-ʿAli"
Zenobia, queen of Palmyra

Native America

Awashonks
Brant, Mary
Chief Earth Woman
Ehyophsta
Elk-Hollering-in-the-Water
Lozen
Native American women
Other Magpie, The
Running Eagle
Throwing Down
Ward, Nancy
Wetamoo
Woman Chief

Poland

Armia Krajowa
Emilia Plater Independent Women's Battalion
Gierczak, Emilia
Leska-Daab, Anna
Lewandowska, Janina
OLK (Ochotniza Legia Kobiet)
Plater, Emilia
PLSK (Pomocnicza Lotnicza Służba Kobiet)
Sosnowska-Karpik, Irena
Spiegel, Chaike Belchatowska
Zagórska, Aleksandra

Russia/Soviet Union

122nd Aviation Group
125th Guards Bomber Aviation Regiment
46th Guards Bomber Aviation Regiment
586th Fighter Aviation Regiment
Alekseeva, Ekaterina
Arzamasskaia, Alena
Bashkirova, Kira Aleksandrovna
Bochkareva, Mariia Leont'evna
Bogat, A.P.
Catherine II of Russia
Central Women's School for Sniper Training
Chaikina, Elizaveta Ivanova
Chechneva, Marina
Dadeshkeliani, Kati
Dolgorukaia, Princess Sofiia Aleksandrovna
Durova, Nadezhda
Egorova [Timofeeva], Anna Aleksandrovna
Fedutenko, Nadezhda (Natasha) Nikiforovna
Fomicheva-Levashova, Klavdia
Grizodubova, Valentina
Hero of the Soviet Union, women recipients
Ivanova, Rimma Mikhailovna
Khomiakova, Valeriia
Konstantinova, Tamara
Kosmodemianskaia, Zoia
Kovshova, Natal'ia/Polivanova, Mariia
Krasil'nikova, Anna
Levchenko, Irina
Litviak, Lidiia
Makarova, Tat'iana/Belik, Vera
Markina, Tat'iana
Mazanik, Elena
Oktiabrskaia, Mariia Vasil'evna
Onilova, Neonila Andreevna
Order of Glory, women recipients
Order of St. George, women recipients
Order of the Red Banner, women recipients
Osoaviakhim
Pal'shina, Antonia

List of Entries by Geographic Region

Pavlichenko, Liudmila Mikhailovna
Raskova, Marina
Reisner, Larisa
Resistance movements, Soviet women
Russian Civil War, women in
Samsonova, E.P.
Shakhovskaia, Evgeniia
Smirnova, Zoia
Tikhomirova, Aleksandra
Women's Battalions of Death
Women's Military Congress
Women's Military Unions
Zemliachka, Rozaliia Samoilovna
Zverova, Nina Ivanova

Spain
Aragón, Agustina
Ermessend of Carcassonne
Etchebéhère, Mika
Gidinild
Ibárruri, Dolores
Isabel de Conches
Malasaña, Manuela
Odena, Lina
Sánchez Mora, Rosario
Spanish Civil War, women in

Tibet
Ani Pachen

United States
Allen, Eliza
American Civil War, women in
Bailey, Ann
Bailey, Anne Hennis Trotter
Bean, Mollie
Blalock, Sarah Mjalinda
Brewer, Lucy
Brownell, Kate (Kady)
Budwin, Florena
Carroll, Anna Ella
Clark, Amy
Clarke, Mary
Cochran, Jacqueline
Corbin, Margaret Cochran
Cornum, Rhonda
Darragh, Lydia
Dix, Dorothea
Duerk, Arlene B.
Earley, Charity Adams
Edmonds, Sarah Emma Evelyn
Etheridge, Anna
Foote, Evelyn Patricia
Galloway, Mary
Hancock, Joy Bright
Hart, Nancy (Revolutionary War)
Hart, Nancy (American Civil War)
Hobby, Oveta Culp
Hodgers, Jennie
Holm, Jeanne
Hopper, Grace Murray
Lane, Anna Maria
Love, Nancy
Mariner, Rosemary
McAfee (Horton), Mildred
McCauley, Mary Ludwig Hays "Molly Pitcher"
Motte, Rebecca
"Nancy Harts"
Phipps, Anita
Rathburn-Nealy, Melissa
Rossi, Marie T.
Samson, Deborah
Scaberry, Mary
SPAR
St. Clair, Sally
Stratton, Dorothy Constance
Streeter, Ruth Cheney
Turchin, Nadine
Velazquez, Loreta

Vivandières
WAC (Women's Army Corps)
WAFS (Women's Auxiliary Ferrying Squadron)
Wakeman, Sarah Rosetta
Walker, Mary
Walker, Nancy Slaughter
WASP (Women's Airforce Service Pilots)
WAVES (Women Accepted for Voluntary Emergency Service)
West, Harriet W.
Whittle Tobiason, Reba
Women marines
Wright, Prudence

Vietnam
Dho Minde
Le Tunn Ding
Nguyen Thi Dinh
Phung Thi Chinh
Trung Trac and Trung Nhi
Vietnamese Revolution, women in

List of Entries by Time Period/Conflict

Note: Some of the individuals listed may not have their own entry but appear in other entries (for example, within a group entry). Others may be mentioned in more than one entry. Check the index for precise locations.

Ancient and Classical Times
Amazons
Arsinoë III
Artemisia I of Halicarnassus
Artemisia II of Halicarnassus
Berenice II Euergetes
Berenice Syra
Boudicca
Candace
Cartimandua
Chi Zhaoping
Chiomara
Cimbrian women
Cleopatra VII
Cloelia
Cynane
Deborah
Eurydice II
Hatchepsut
Judith
Olympias
Phung Thi Chinh
Semiramis
Telesilla of Argos
Teuta
Tomyris
Triaria, Empress
Trung Trac and Trung Nhi

Late Antiquity (3rd–7th Centuries)
Brunehaut
Clotild
Fredegund
Hind bint'Utba
Kāhina
Khawlah, Bint al-Azwar al-Kindiyyah
Leubevére
Li Xiu

List of Entries by Time Period/Conflict

Lu (first name unknown)
Mao, Empress
Mavia
Shao (first name unknown)
Yarmūk, women in Battle of
Zenobia, queen of Palmyra

Middle Ages (8th–15th Centuries)

Adelaide of Turin
Aethelburh
Aethelflaed of Mercia
Agnes of Dunbar
Alice de Montmorency
Alruda Frangipani
Aluzhen
Attendolo, Margherita
Badlesmere, Lady Margaret
Beatrice of Lorraine
Bianca di Rossi
Bianca Maria Visconti Sforza
Bona of Lombardy
Christine de Pisan
Claude des Armoises
Crusades, Women in
Dionisia de Grauntcourt
Eleanor of Aquitaine
Eleanora of Arborea
Emma FitzOsbern
Emma, queen of France
Ermengard, viscountess of Narbonne
Ermessend of Carcassonne
Gao (first name unknown)
Gidinild
Giralda de Laurac
Hōjō Masako
Hua Mulan
Hussite Wars, women in
Isabel de Conches
Isabel, countess of Buchan
Isabella de Lorraine
Isabella of France
Joanna of England
Johanna of Rožmitál
Julienne du Guesclin
Knyvet, Alice
Kong (first name unknown)
Liang Hongyü
Mabel of Bellême
Mahaut d'Artois
Margaret of Anjou
Margaret of Denmark
Maria of Pozzuoli
Marzia degli Ubaldini
Matilda Augusta of England
Matilda of Boulogne
Matilda of Tuscany
Meng, Consort
Nicolaa de le Haye
Pan (first name unknown)
Philippa of Hainault
Pingyang, Princess
Raziya, Sultan
Ren (first name unknown)
Rodiana, Onorata
Rodolfi, Camilla
Sforza, Caterina
Shalizhi
Shejie (first name unknown)
Sichelgaita of Salerno, duchess of Apulia
Stamira of Ancona
Tamar, queen of Georgia
Tomoe Gozen
Viking women warriors
Wulfbald, widow of
Xi (first name unknown)
Xian (first name unknown)
Xochitl
Yan (first name unknown)
Yang Miaozhen

List of Entries by Time Period/Conflict

Hundred Years' War (14th–15th Centuries)

Jeanne de Montfort
Jeanne de Penthièvre
Joan of Arc

Early Modern Period (16th–17th Centuries)

Albret, Jeanne d'
Arzamasskaia, Alena
Awashonks
Bankes, Lady Mary
Catherine of Aragon
Cavanaugh, Kit
Cezelly, Françoise de
Chevreuse, Marie, duchesse de
de Abreu e Lencastre, Dona Maria Ursula
De La Tour, Frances Mary Jacqueline
Delaye, Margot
Eger, women in siege of
Elizabeth I
Erauso, Catalina de
Forteguerri, Laudomia
Harley, Lady Brilliana
Henrietta Maria
La Guette, Catherine de Meurdrac
Lilliard of Ancrum
Liu (first name unknown)
Liu Shuying
Longueville, Anne-Geneviève de Bourbon-Condé, duchesse de
Mary I, queen of England
Mary of Hungary
Médici, Catherine de
Médici, Marie de
Montpensier, Anne Marie Louise d'Orléans, duchesse de
Mouron, Marie Magdelaine
Prémoy, Geneviève
Qin Liang-yü
Saint-Baslemont, madame de
Shen Yunying
Siena, women in siege of
Stanley, Charlotte, countess of Derby
Thirty Years' War, women in
Valois, Marguerite
Verchères, Marie-Madeleine de
Wa (first name unknown)
Zrinyi, Countess Ilona

18th Century

Begum Samru
Catherine II of Russia
Dahomey, women in army of
Drucour, Seigneuress Anne-Marie
Holkar, Ahilyabhai
Maria Theresa
Markina, Tat'iana
Snell, Hannah
Talbot, Mary Anne
Ward, Nancy
Wright, Prudence

American Revolutionary War (18th Century)

Bailey, Ann
Bailey, Anne Hennis Trotter
Brant, Mary
Corbin, Margaret Cochran
Darragh, Lydia
Hart, Nancy
Lane, Anna Maria
McCauley, Mary Ludwig Hays
"Molly Pitcher"
Motte, Rebecca
"Nancy Harts"
Samson, Deborah
St. Clair, Sally

List of Entries by Time Period/Conflict

French Revolution/Napoleonic Wars (1789–1815)

Aragón, Agustina
Barreau, Rose-Alexandrine
Bordereau, Renée
Bouillon, Rose
Chapuy, Reine
Charpentier, Marie
de Bruc (first name unknown)
de Fief (first name unknown)
Duchemin, Marie-Angélique Josephe
Durova, Nadezhda
Engle, Regula
Figueur, Thérèse
French Revolution and Napoleonic era, women soldiers in
Ghesquière, Virginie
Gouges, Olympe de
Julien, Sophie
Lacombe, Rose
Lanz, Katharina
Léon, Pauline
Malasaña, Manuela
Parent, Marie-Barbe
Petitjean, Madeleine
Prothais, Geneviève
Quatsault, Anne
Robin, Jeanne
Rochejaquelein, La Marquise de La
Rouget, Claudine
Schellinck, Marie-Jeanne
Théroigne de Méricourt
Tikhomirova, Aleksandra
Xaintrailles, Marie-Henriette Heiniken

19th Century

Allen, Eliza
Azurduy de Padilla, Juana
Bhai, Laxshmi
Bocanegra, Gertrudis
Bouboulina, Laskarina
Brewer, Lucy
Chief Earth Woman
Ehyophsta
Feitosa, Jovita Alves
García Morales de Carrera, Petrona
Lebstuck, Maria
Lozen
Masarico
Mavrogenous, Manto
Michel, Louise
Native American women
Nehanda
The Other Magpie
Paraguayan, women in the War of the Triple Alliance
Paris Commune, women's miliary roles in
Plater, Emilia
Quitéria de Jesus, Maria
Reachy, Ignacia
Secord, Laura Ingersoll
Soldaderas
Williams, Cathay
Woman Chief
Zaynab "Rustam-Ali"

American Civil War

American Civil War, women in
Bean, Mollie
Blalock, Sarah Malinda
Brownell, Kady
Budwin, Florena
Carroll, Anna Ella
Clark, Amy
Dix, Dorothea
Edmonds, Sarah Emma Evelyn
Etheridge, Anna
Galloway, Mary
Hart, Nancy
Hodgers, Jennie
Scaberry, Mary

List of Entries by Time Period/Conflict

Taylor, Susie King
Tubman, Harriet
Turchin, Nadine
Velazquez, Loreta
Vivandières
Wakeman, Sarah Rosetta
Walker, Mary
Walker, Nancy Slaughter

20th Century (General)

Algerian-French War, women in
Ani Pachen
Australian Defence Forces, women in
Barrera, Maria de la Luz Espinosa
Ben-Yehuda, Netiva
Ben-Zvi, Rachel Yanait
Best, Kathleen
Brasseur, Deanna
Buhrayd, Jamilah
CHEN
Clarke, Mary
Clay, Wendy
Duerk, Arlene B.
Erxleben, Heather
Etchebéhère, Mika
FANY Corps (First Aid Nursing Yeomanry Corps)
Foote, Evelyn (Patricia)
Foster, Jane
Grizodubova, Valentina
Gwynne-Vaughan, Helen
Hellstrom, Sheila
Herrera, Petra
Holm, Jeanne
Hopper, Grace Brewster Murray
Hungary, women in 1956 revolution
Ibárruri, Dolores
Israeli War of Independence, women's military roles
Jiménez, Angela
Kang Keqing
Libya, military women in
Mariner, Rosemary
Meir, Golda
Native American women
Nhongo, Joyce Mugari
Odena, Lina
OLK (Ochotnicza Legia Kobiet)
Order of Glory, women recipients
Order of St. George, women recipients
Order of the Red Banner, women recipients
Osoaviakhim
Phipps, Anita
Raskova, Marina
Ruiz, Petra
Sánchez Mora, Rosario
Shochat, Mania
Sosnowska-Karpik, Irena
Spanish Civil War, women in
Stratton, Dorothy Constance
Streeter, Ruth Cheney
Thatcher, Margaret
United Arab Emirates, military women in
WAAAF (Women's Auxiliary Australian Air Force)
Walker, Kath
Zagórska, Aleksandra
Zimbabwe, women in the national liberation war

First World War/Russian Civil War

Alekseeva, Ekaterina
Bashkirova, Kira Aleksandrovna
Bochkareva, Mariia Leont'evna
Bogat, A.P.
Dadeshkeliani, Kati
Dolgorukaia, Princess Sofiia Aleksandrovna
Ivanova, Rimma Mikhailovna
Krasil'nikova, Anna
Pal'shina, Antonina

xxix

List of Entries by Time Period/Conflict

Reisner, Larisa
Russian Civil War, women in
Samsonova, E.P.
Sandes, Flora
Shakhovskaia, Evgeniia
Smirnova, Zoia
WAAC (Women's Auxiliary Army Corps)
Women's Battalions of Death
Women's Legion
Women's Military Congress
Women's Military Unions
Zemliachka, Rozaliia Samoilovna

Second World War

122nd Aviation Group
125th Guards Bomber Aviation Regiment
46th Guards Bomber Aviation Regiment
586th Fighter Aviation Regiment
Armia Krajowa
ATA (Air Transport Auxiliary)
ATS (Auxiliary Territorial Service)
Bethune, Mary McLeod
Central Women's School for Sniper Training
Chaikina, Elizaveta Ivanova
Chechneva, Marina
Cochran, Jacqueline
Earley, Charity Adams
Egorova [Timofeeva], Anna Aleksandrovna
Emilia Plater Independent Women's Battalion
Fedutenko, Nadezhda (Natasha) Nikiforovna
Fomicheva-Levashova, Klavdiia
Gierczak, Emilia
Gower, Pauline
Hancock, Joy Bright
Hero of the Soviet Union, women recipients
Hobby, Oveta Culp
Khan, Noor Inayat
Khomiakova, Valeriia
Konstantinova, Tamara
Kosmodemianskaia, Zoia
Kovshova, Natal'ia/Polivanova, Mariia
Leska-Daab, Anna
Levchenko, Irina
Lewandowska, Janina
Litviak, Lidiia
Love, Nancy
Makarova, Tat'iana/Balik, Vera
Mazanik, Elena
McAfee (Horton), Mildred
Oktiabrskaia, Mariia Vasil'ievna
Onilova, Neonila Andreevna
Pavlichenko, Liudmila Mikhailovna
PLSK (Pomocnicza Lotnicza Służba Kobiet)
Reitsch, Hanna
Resistance movements, French women
Resistance movements, Italian women
Resistance movements, Soviet women
Resistance movements, Yugoslavian women
SPAR
Spiegel, Chaike Belchatowska
Stauffenberg, Melitta
WAC (Women's Army Corps)
WAC, African Americans in
WAFS (Women's Auxiliary Ferrying Squadron)
WASP (Women's Airforce Service Pilots)
WAVES (Women Accepted for Voluntary Emergency Service)
West, Harriet W.
Whittle Tobiason, Reba
Women marines
Women's Land Army
Zverova, Nina Ivanova

Vietnam War

Dho Minde
Le Tunn Ding
Nguyen Thi Dinh
Vietnamese Revolution, women in

Gulf War

Cornum, Rhonda
Rathburn-Nealy, Melissa
Rossi, Marie T.

List of Entries by Role/Branch of Service

Note: Some of the individuals listed may not have their own entry but appear in other entries (for example, within a group entry). Others may be mentioned in more than one entry. Check the index for precise locations.

Auxiliary Forces

ATA (Air Transport Auxiliary)
ATS (Auxiliary Territorial Service)
Cochran, Jacqueline
Gower, Pauline
Love, Nancy
OLK (Ochotnicza Legia Kobiet)
Osoaviakhim
WAAAF (Women's Auxiliary Australian Air Force)
WAAC (Women's Auxiliary Army Corps)
WAFS (Women's Auxiliary Ferrying Squadron)
WASP (Women's Airforce Service Pilots)
Women's Land Army
Women's Military Congress
Women's Military Unions
Zagórska, Aleksandra

Aviation Air Forces

122nd Aviation Group
125th Guards Bomber Aviation Regiment
46th Guards Bomber Aviation Regiment
586th Fighter Aviation Regiment
Brasseur, Deanna
Chechneva, Marina
Clay, Wendy
Dolgorukaia, Princess Sofiia Aleksandrovna
Egorova [Timofeeva], Anna Aleksandrovna
Fedutenko, Nadezhda
Fomicheva-Levashova, Klavdiia
Foster, Jane
Grizodubova, Valentina
Hellstrom, Sheila
Holm, Jeanne
Khomiakova, Valeriia

List of Entries by Role/Branch of Service

Konstantinova, Tamara
Leska-Daab, Anna
Lewandowska, Janina
Litviak, Lidiia
Makarova, Tat'iana
PLSK (Pomocnicza Lotnicza Służba Kobiet)
Raskova, Marina
Reitsch, Hanna
Shakhovskaia, Evgeniia
Sosnowska-Karpik, Irena
Stauffenberg, Melitta
Whittle Tobiason, Reba

Civilians
Bethune, Mary McLeod
Gwynne-Vaughan, Helen
Vivandières
Women's Legion

Defenders, Besiegers, ad Hoc Fighters
Adelaide of Turin
Agnes of Dunbar
Albret, Jeanne d'
Alruda Frangipani
Aragón, Agustina
Attendolo, Margherita
Badlesmere, Lady Margaret
Bankes, Lady Mary
Bianca di Rossi
Bianca Maria Visconti Sforza
Cezelly, Françoise de
Cloelia
Clotild
De La Tour, Françoise-Marie Jacquelin
Delaye, Margot
Drucour, Seigneuress Anne-Marie
Eger, women in siege of
Emma FitzOsbern
Forteguerri, Laudomia
Harley, Lady Brilliana

Hind bint'Utba
Isabel, countess of Buchan
Isabella de Lorraine
Jeanne de Montfort
Jeanne de Penthièvre
Joanna of England
Julienne du Guesclin
Knyvet, Alice
La Guette, Catherine de Meurdrac, madame de
Lanz, Katharina
Leubevére
Mabel of Bellême
Marzia degli Ubaldini
Matilda of Boulogne
Médici, Catherine de
Médici, Marie de
Nicolaa de le Haye
Saint-Baslemont, madame de
Secord, Laura Ingersoll
Sforza, Caterina
Siena, women in siege of
Stamira of Ancona
Stanley, Charlotte, countess of Derby
Telesilla of Argos
Verchères, Marie-Madeleine de
Zaynab "Rustam-ʿAli"
Zrinyi, Countess Ilona

Ground Forces/Armies
Alekseeva, Ekaterina
Algerian-French War, women in
Allen, Eliza
American Civil War, women in
Armia Krajowa
Australian Defence Forces, women in
Bailey, Ann
Bailey, Anne Hennis Trotter
Barreau, Rose-Alexandrine
Bashkirova, Kira Aleksandrovna
Bean, Mollie

List of Entries by Role/Branch of Service

Best, Kathleen
Blalock, Sarah Malinda
Bochkareva, Mariia Leont'evna
Bogat, A.P.
Bouillon, Rose
Brownell, Kady
Budwin, Florena
Buhrayd, Jamilah
Cavanaugh, Kit
Central Women's School
Chapuy, Reine
Charpentier, Marie
CHEN
Cimbrian women
Clark, Amy
Clarke, Mary
Claude des Armoises
Corbin, Margaret Cochran
Cornum, Rhonda
Crusades, Women in
Dadeshkeliani, Kati
Dahomey, women in army of
Duchemin, Marie-Angélique Josephe
Durova, Nadezhda
Earley, Charity Adams
Edmonds, Sarah Emma Evelyn
Emilia Plater Independent Women's Battalion
Erxleben, Heather
Etheridge, Anna
Feitosa, Jovita Alves
Figueur, Thérèse
Foote, Evelyn Patricia
Galloway, Mary
Ghesquière, Virginie
Gidinild
Gierczak, Emilia
Hart, Nancy (Revolutionary War)
Hart, Nancy (American Civil War)
Hobby, Oveta Culp
Hodgers, Jennie
Isabel de Conches
Joan of Arc
Julien, Sophie
Kovshova, Natal'ia/Polivanova, Mariia
Krasil'nikova, Anna
Lane, Anna Maria
Lathgertha
Lebstuck, Maria
Levchenko, Irina
Libya, military women in
Lilliard of Ancrum
Liu
Liu Shuying
Markina, Tat'iana
McCauley, Mary Ludwig Hays
"Molly Pitcher"
Mouron, Marie Magdelaine
"Nancy Harts"
Native American women
Oktiabrskaia, Mariia Vasil'ievna
Onilova, Neonila Andreevna
Parent, Marie-Barbe
Pavlichenko, Liudmila Mikhailovna
Petitjean, Madeleine
Phipps, Anita
Prémoy, Geneviève
Prothais, Geneviève
Qin Liang-yü
Quatsault, Anne
Quitéria de Jesus, Maria
Rathburn-Nealy, Melissa
Reisner, Larisa
Rodolfi, Camilla
Rossi, Marie T.
Rouget, Claudine
Russian Civil War, women in
Sandes, Flora
Scaberry, Mary
Schellinck, Marie-Jeanne
Shen Yunying
Smirnova, Zoia

List of Entries by Role/Branch of Service

St. Clair, Sally
Thirty Years' War, women in
Tikhomirova, Aleksandra
Tubman, Harriet
Turchin, Nadine
United Arab Emirates, military women in
Velazquez, Loreta
Wa
WAC (Women's Army Corps)
WAC, African Americans in
Wakeman, Sarah Rosetta
Walker, Kath
Walker, Nancy Slaughter
West, Harriet W.
Williams, Cathay
Women's Battalions of Death
Xaintrailles, Marie-Henriette Heiniken
Zemliachka, Rozaliia Samoilovna

Marines

Brewer, Lucy
Snell, Hannah
Streeter, Ruth Cheney
Women Marines

Medical

Dix, Dorothea
Duerk, Arlene B.
FANY Corps (First Aid Nursing Yeomanry Corps)
Ivanova, Rimma Mikhailovna
Samsonova, E.P.
Taylor, Susie King
Walker, Mary

Naval

Bouboulina, Laskarina
Hancock, Joy Bright
Hopper, Grace Brewster Murray
Mariner, Rosemary
Mavrogenous, Manto
McAfee (Horton), Mildred
Reisner, Larisa
SPAR
Stratton, Dorothy Constance
Talbot, Mary Anne
WAVES (Women Accepted for Volunteer Emergency Service)

Partisans, Resistance Fighters, Revolutionaries

Ani Pachen
Arzamasskaia, Alena
Azurduy de Padilla Juana
Barrera, Maria de la Luz Espinosa
Ben-Yehuda, Netiva
Ben-Zvi, Rachel Yanait
Bocanegra, Gertrudis
Bordereau, Renée
Bouboulina, Laskarina
Chaikina, Elizaveta Ivanova
Darragh, Lydia
Dho Minde
Engle, Regula
Etchebéhère, Mika
French Revolution and Napoleonic era, women soldiers
García Morales de Carrera, Petrona
Gouges, Olympe de
Herrera, Petra
Hungary, women in 1956 revolution
Israeli War of Independence, women's military roles
Jiménez, Angela
Kang Keqing
Khan, Noor Inayat
Kosmodemianskaia, Zoia
Lacombe, Rose
Le Tunn Ding
Léon, Pauline
Longueville, Anne-Geneviève de Bourbon-Condé, duchesse de
Malasaña, Manuela

Mazanik, Elena
Mavrogenous, Manto
Michel, Louise
Montpensier, Anne Marie Louise d'Orléans, duchesse de
Motte, Rebecca
Nechanda
Nguyen Thi Dinh
Nhongo, Joyce Mugari
Odena, Lina
Paraguayan, women in the War of the Triple Alliance
Paris Commune, women's military roles in
Phung Thi Chinh
Plater, Emilia
Reachy, Ignacia
Resistance movements, French women
Resistance movements, Italian women
Resistance movements, Soviet women
Resistance movements, Yugoslavian women
Robin, Jeanne
Rochejaquelein, La Marquise de La
Ruiz, Petra
Samson, Deborah
Sánchez Mora, Rosario
Shochat, Mania
Soldaderas
Spanish Civil War, women in
Spiegel, Chaike Belchatowska
Théroigne de Méricourt
Trung Trac and Trung Nhi
Vietnamese Revolution, women in
Wright, Prudence
Zimbabwe, women in the national liberation war
Zverova, Nina Ivanova

Political Leaders/Heads of State

Aethelburh
Aethelflaed of Mercia
Alice de Montmorency
Arsinoë III
Artemisia I of Halicarnassus
Artemisia II of Halicarnassus
Beatrice of Lorraine
Berenice II Euergetes
Berenice Syra
Bhai, Laxshmi
Boudicca
Brant, Mary
Brunehaut
Candace
Cartimandua
Catherine of Aragon
Catherine II of Russia
Chevreuse, Marie de Rohan-Montbazon, duchesse de
Cleopatra VII
Cynane
Deborah
Eleanor of Aquitaine
Eleanora of Arborea
Elizabeth I
Emma, queen of France
Ermengard, viscountess of Narbonne
Ermessend of Carcassonne
Eurydice II
Fredegund
Giralda de Laurac
Hatchepsut
Henrietta Maria
Holkar, Ahilyabhai
Hussite Wars, women in
Isabella of France
Johanna of Rožmitál
Kāhinah
Mahaut d'Artois
Margaret of Anjou
Margaret of Denmark
Maria Theresa
Mary I, queen of England

List of Entries by Role/Branch of Service

Mary of Hungary
Masarico
Matilda Augusta of England
Matilda of Tuscany
Mavia
Meir, Golda
Olympias
Philippa of Hainault
Raziya, Sultan
Semiramis
Tamar, queen of Georgia
Teuta
Thatcher, Margaret
Tomyris
Triaria, Empress
Valois, Marguerite de
Wulfbald, widow of
Xochitl
Zenobia, queen of Palmyra

Soldiers and Warriors
Aluzhen
Amazons
Awashonks
Begum Samru
Bona of Lombardy
Chi Zhaoping
Chief Earth Woman
Chiomara
de Abreu e Lencastre, Dona Maria Ursula
Ehyophsta
Elk-Hollering-in-the-Water
Erauso, Catalina de
Gao
Hōjō Masako
Hua Mulan
Judith
Khawlah, Bint al-Azwar al-Kindiyyah
Kong
Li Xiu
Liang Hongyü
Lozen
Lu
Mao, Empress
Maria of Pozzuoli
Meng, Consort
The Other Magpie
Pan
Pingyang, Princess
Ren
Rodiana, Onorata
Running Eagle
Shalizhi
Shao
Shejie
Sichelgaita of Salerno, duchess of Apulia
Tomoe Gozen
Ward, Nancy
Wetamoo
Woman Chief
Xi
Xian
Yan
Yang Miaozhen
Yarmūk, women in battle of

Strategists and Theorists
Carroll, Anna Ella
Christine de Pisan

List of Entries by Prisoner/POW Status

Note: Some of the individuals listed may not have their own entry but appear in other entries (for example, within a group entry). Others may be mentioned in more than one entry. Check the index for precise locations.

Ani Pachen
Armia Krajowa
Arzamasskaia, Alena
Begum Samru
Bianca di Rossi
Bocanegra, Gertrudis
Budwin, Florena
Casanova, Danielle
Cavanaugh, Kit
Chaikina, Elizaveta Ivanova
Chiomara
Cloelia
Cornum, Rhonda
Crusades, women in
Duchemin, Marie-Angélique Joséphe
Egorova [Timofeeva], Anna Aleksandrovna
Eleanor of Aquitaine
Figueur, Thérèse
Gérard, Georgette "Claude"
Hrozova, Erzsebet
Hungary, women in 1956 revolution
Isabel, countess of Buchan
Israeli War of Independence, women's military roles
Joan of Arc
Kashichkina, Liudmila
Khan, Noor Inayat
Kosmodemianskaia Zoia
Lebstuck, Maria
Lewandowska, Janina
Lozen
Magori, Maria
Mao, Empress
Marchiani, Irma
Margaret of Anjou
Marzia degli Ubaldini
Nehanda
OLK (Ochotnicza Legia Kobiet)

List of Entries by Prisoner/POW Status

Paraguayan Women in the War of the Triple Alliance
Petitjean, Madeleine
POW appendix
Rathbun-Nealy, Melissa
Raziya, Sultan
Reachy, Ignacia
Reitsch, Hanna
Resistance movements, French women
Resistance movements, Italian women
Resistance movements, Soviet women
Resistance movements, Yugoslavian women
Rudellat, Yvonne
Salabert, Erzsebet
Sánchez Mora, Rosario
Sebestyen, Maria
Sforza, Caterina
Shochat, Mania
Soldaderas
Spanish Civil War, women in
Sticker, Katalin
Szabo, Violette
Talbot, Mary Anne
Théroigne de Méricourt
Toth, Ilona
Valois, Marguerite de
Velazquez, Loreta Janeta
Walker, Mary Edwards
Whittle Tobiason, Reba
Wittner, Maria
Zenobia, queen of Palmyra

List of Entries on Groups and Organizations

Note: Some of the individuals listed may not have their own entry but appear in other entries (for example, within a group entry). Others may be mentioned in more than one entry. Check the index for precise locations.

122nd Aviation Group
125th Guards Bomber Aviation Regiment
46th Guards Bomber Aviation Regiment
586th Fighter Aviation Regiment
Algerian-French War, women in
American Civil War, women in
Armia Krajowa
ATA (Air Transport Auxiliary)
ATS (Auxiliary Territorial Service)
Australian Defence Forces, women in
Central Women's School for Sniper Training
CHEN
Cimbrian women
Crusades, women in
Dahomey, women in army of
Eger, women in siege of
Emilia Plater Independent Women's Battalion
FANY Corps (First Aid Nursing Yeomanry Corps)
French Revolution and Napoleonic era, women soldiers in
Hero of the Soviet Union, women recipients
Hungary, women in 1956 revolution
Hussite wars, women in
Israeli War of Independence, women's military roles
Libya, military women in
"Nancy Harts"
Native American women
OLK (Ochotnicza Legia Kobiet)
Order of Glory, women recipients
Order of St. George, women recipients
Order of the Red Banner, women recipients
Osoaviakhim
Paraguayan women in the War of the Triple Alliance

List of Entries on Groups and Organizations

Paris Commune, women's military roles in

PLSK (Pomocnicza Lotnicza Służba Kobiet)

Resistance movements, French women

Resistance movements, Italian women

Resistance movements, Soviet women

Resistance movements, Yugoslavian women

Russian Civil War, women in

Siena, women in siege of

Soldaderas

Spanish Civil War, women in

SPAR

Thirty Years' War, women in

United Arab Emirates, military women in

Vietnamese Revolution, women in

Viking women warriors

Vivandières

WAAAF (Women's Auxiliary Australian Air Force)

WAAC (Women's Auxiliary Army Corps)

WAC (Women's Army Corps)

WAC, African Americans in

WAFS (Women's Auxiliary Ferrying Squadron)

WASP (Women's Airforce Service Pilots)

WAVES (Women Accepted for Volunteer Emergency Service)

Women Marines

Women's Battalions of Death

Women's Land Army

Women's Legion

Women's Military Congress

Women's Military Unions

Yarmūk, women in battle of

Zimbabwe, women in the National Liberation War

Introduction: A Historical Overview of Women in the Military and in War

For more than thirty years in the late eleventh century, a woman commanded armies that defeated repeated attacks on the Italian papacy. Some say that **Matilda of Tuscany** ranks alongside William the Conqueror as a military figure, for her strategies protected Pope Gregory VII and his successors against no less an enemy than the holy Roman emperor, Henry IV.

During the period of the anti-British mutiny in India in the 1850s, **Laxshmi Bhai** became rani of the northern principality of Jhansi. A horsewoman and a good shot, the rani dressed in jodhpurs and wore a short sword and two pistols in her belt. When the British besieged her territory in 1858, the rani refused to surrender and rallied her subjects, including many women, to the defense of Jhansi. All observers report that the rani was personally in the thick of the fighting. When she was killed in June 1858, the regimental history of the British Eighth Hussars noted that "in her death the rebels lost their bravest and best military leader."

In September 1942, during the Battle of Stalingrad, a small, blonde female pilot, barely twenty years old, became the first woman in the world to shoot down an enemy aircraft. Fighter pilot **Lidiia Litviak** made two kills on her second combat mission and scored a total of fifteen kills during her brief career. Yet one searches in vain for her name in many of the books that have been written about military history.

When attention has been given to women in military roles, it has often focused on women like **Joan of Arc** who did not actually fight. Some women's exploits, like those of **Manuela Malasaña,** were wildly exaggerated for symbolic political purposes. Far more often, the accounts of military women like the hundreds in this book have been ignored or deliberately obscured. Are

Introduction

these women mere anomalies? The hundreds of stories told here (a small sample of the thousands that remain to be examined) reveal an amazing depth and breadth in the history of military women. Many people are unaware that throughout history women have served in the military as soldiers, sailors, physicians, and pilots. Their stories have truly been hidden from history. Many military women served disguised as men; their real identities were rarely acknowledged. Often military women were permitted to serve only briefly, during a national emergency or for the duration of a single campaign or even a single battle. Such women had no chance to achieve the high rank or to complete the sort of long and distinguished career that catches the attention of traditional historians. One fact is clear: these are not just exceptions that prove the rule; that much is demonstrated by the large-scale influx of women into military organizations when opportunity and incentive have coincided (the Soviet Union during the Second World War, for example). Military women constitute an important part of the diversity of military experience, and their history is inseparable from the history of war and the military. This introduction provides a brief survey of women's military roles throughout history.

ANTIQUITY

Among the earliest historical references to women in combat are stories of the **Amazons**. Writers such as Homer, Pindar, Herodotus, and Plato recorded tales of an independent community of women at war with the masculine world who lived in what is now the southern part of the Ukraine. A common modern interpretation is that Greek men created a myth of Amazons to keep Greek women in their place by devising a symbolic opposite of the ideal Athenian woman.[1] But does it really make sense for Greek men, in order to control their society, to create a legend of powerful military women, even if only to show how these women were defeated in the end? It seems far more likely that the Amazon legend has some basis in history; that it was the presence of real, rather than fictional, military women that needed to be explained.

Indeed, archaeologists have discovered ancient burial sites of women with weapons in Georgia, Ukraine, and Russia—areas of traditional Sarmatian culture. Dating primarily from the fourth and third centuries BCE, these women's graves contained swords, armor, lance heads, and archery gear.[2] Recent archaeological studies of women's skeletons in that region indicate that they spent much of their time on horseback, and many had sustained war wounds.[3] Herodotus believed that the Sarmatian women were the descendants of the Amazons; perhaps other, earlier societies existed that have left no trace or are as yet undiscovered. Amazons would be ever popular as a symbol of the power of queens. Anne of England (1665–1714), on a state visit to Bath, was surrounded by an honor guard of women archers attired as Amazons, and Prince Potemkin was reported to have costumed a regiment of "Amazons" to welcome **Catherine II** (the Great) when she visited the Crimea with the Austrian emperor Joseph in 1787.[4]

The Bible mentions a few women connected with war; entries on **Deborah** and **Judith** attempt to examine their historical basis. Other semilegendary figures are also considered, such as sisters **Trung Trac** and Trung Nhi, who led a first-century army in present-day Vietnam. Chinese records of the third–fourth centuries indicate that women such as **Li Xiu** and **Empress Mao** played a part in the endemic fighting and empire building of their day. But the women in ancient times whose accomplishments are best recorded are the queens. **Artemisia I** assisted the Persians in their wars against Greece; Macedonian queens like **Berenice II** and **Olympias** raised armies and vied for power. Recently the achievements of the female pharaoh **Hatchepsut** have come to light. Many centuries after her time, a series of Hellenistic queens who ruled in Egypt displayed military involvement, such as **Arsinoë III** and **Cleopatra VII**.

Women in classical Greek and Roman culture were generally restricted in their education and public roles and most often experienced war as observers or victims. Women like **Cloelia** are included in this collection who are not strictly military, but whose experience illustrates many issues that are connected with military women: the questions of becoming a prisoner of war, of rape, and of willingness to commit violence. There were more women who fought against Greece and Rome than for them. Tacitus remarked that women of the Germanic kingdoms were often present on the battlefield, and Ammianus Marcellinus claimed that these women were active combatants. He stated that the name of the Lombard people, which meant "longbearded," had originated in the women soldiers' practice of tying their long hair beneath their chins to simulate a beard, perhaps in order to deceive their opponents.[5] The **Cimbrian women** were said to have fought against Roman soldiers. Better known are the queens who opposed Roman domination, such as **Boudicca**, **Semiramis**, and **Zenobia**.

MIDDLE AGES

In Europe women were much more active participants in warfare during the Middle Ages. Women fought for loyalty, for honor, to defend their homes, and sometimes for power. Their involvement in war was diverse: both symbolic and physical, both strategic and tactical. Historian Megan McLaughlin notes, "So numerous ... are the references to women warriors in medieval sources, that the real surprise is how little attention has been paid to them by historians in the past."[6] McLaughlin observes that women's military activities were recorded almost as a matter of fact until the end of the eleventh century; it was only after that time that chroniclers seemed to be disconcerted by such activities and felt the need to explain them.[7] The opportunity for women to fill a military role was not simply offered up by chance—it was virtually a

Introduction

necessity in the Middle Ages. Husbands might be off on the quest for new property or away on crusade, and they often lived separately from their wives as they completed an education or pursued business. It was not uncommon throughout the period for men to spend lengthy periods as a prisoner or hostage.[8]

All these events required women—most frequently wives, but often mothers, sisters, or regents—to act as maintainer and defender of the family's holdings. If it was rare for women to be able to go to war, the shift away from mobile warfare and toward sieges instead brought war to women.[9] They could even be considered a sort of auxiliary military force. The existence of these "auxiliaries" was only tacitly acknowledged, but the willingness to draw upon them was evident. Philippe Contamine notes that "the participation of armed ladies . . . was considered, when everything is taken into account, as fairly normal . . . even ordinary women can be found here and there taking part in fighting."[10] It seems odd, if women could not avoid war, that they were never formally prepared for their responsibilities. They received no military education. On the contrary, the prevailing attitudes of the time portrayed women as inferior to men in any role involving power or authority.[11]

The "Truce of God" of the early eleventh century stated that women were supposed to be preserved from the horrors of war.[12] Knights are quoted by chroniclers as saying "we do not war on ladies."[13] This statement was patently untrue, for women defending their castles were frequently warred upon.[14] Despite the demands of chivalry, little allowance was made for the protection of women who defied a besieging force.[15] Women of the nobility could hope for somewhat better treatment than those of the lower classes. By the early modern period, however, there were few restrictions on the treatment of defenders. The seventeenth-century historian Hugo Grotius is quoted as saying, "the slaughter of infants and women is allowed to have impunity."[16]

The sources of the early period focused on the nobility, and few records exist for lesser-ranking individuals. High-ranking women both defended against and conducted sieges in the early centuries of the Middle Ages. Merovingian queens were regularly associated with the holding of towns.[17] There were bizarre armed feuds such as those between the Merovingian queens **Brunehaut** and Fredegund, and another between the nuns **Clotild** and Leubevére, all in the late sixth century.

Records in all parts of the world are problematic for the early periods of history. Some cases of women's military participation have been documented, however. For example, during the seventh century Arab women like **Hind bint'Utba** and **Kāhinah**—were drawn into the violent conflicts that accompanied the spread of Islam. In that same era, in China, a **Princess Pingyang** commanded an army of more than 70,000 and helped her father found the Tang dynasty.

In Europe between the ninth and eleventh centuries there is evidence that women warriors existed among the **Vikings**. The tenth century was in some ways a century of women; for a time in the 980s virtually all the important kingdoms in western Europe, including England, the Ottonian Empire, and France, were ruled by queens or queen-regents.[18] **Adelaide of Turin** and **Emma, queen of France,** were among the queens who were directly involved in military action. One of the greatest figures of this period was **Aethelflaed**. The daughter of Alfred the Great, Aethelflaed ruled Mercia independently in the early tenth century after the death of her husband. Her military activities centered around siege warfare; she built an entire series of fortresses. In the eleventh century the armies of Matilda of Tuscany were instrumental in preserving the papacy in Italy. Some historians see this period in European history as one in which there were "no really effective barriers" to women's power; "they appear as military leaders, judges, castellans, controllers of property."[19] McNamara and Wemple say that there are "numerous examples for the tenth and eleventh centuries of women leading military expeditions [and] going on crusade."[20]

In the Slavic countries, in addition to the legends of Amazons, there are tales of fighting women in the early Middle Ages: "Various early accounts describe Slavic women as warriors and hunters. Byzantine historians claimed that women among the Slavs dressed in masculine attire and fought alongside their men, and a smatter of archaeological findings adds some substance to these reports."[21] Russian folk literature sustained the woman-warrior tradition in the form of *polianitsy,* or warrior heroines, who were credited with powerful physiques and fighting prowess, conquering men in single combat.[22] But if these women warriors indeed existed, none were left by the ninth century. Patriarchy was solidly established by that time, and women's roles became increasingly restricted over the next millennium.[23] There were occasional exceptions such as **Tamar queen of Georgia,** who in the twelfth century directed her armies against the Turks.

From the eleventh to the twelfth centuries an increasing emphasis on primogeniture reduced the chances that women would inherit property. Female regency became increasingly rare in the later Middle Ages, further decreasing the opportunity for women to achieve the sort of power that had previously involved them in military activity.[24] But if the power of queens was reduced, the military responsibilities of noblewomen increased. **Emma FitzOsbern,** countess of Norfolk, held Norwich Castle in the absence of her husband during the rebellion against William the Conqueror in 1075. **Sichelgaita of Salerno,** duchess of Apulia and wife of Robert Guiscard, accompanied her husband on campaign in the late eleventh century; she is described as a large, fearsome woman who wore full armor and fought like a man. The **Crusades** of the eleventh and twelfth centuries also involved women. Queens like **Eleanor of Aquitaine** accompanied the crusading armies, and there were even reports of unnamed women fighting with the European armies. Fairly large numbers of women were involved in more traditional roles of support and logistics.

Introduction

The famous **Nicolaa de la Haye,** the hereditary castellan of Lincoln, held the castle during the rebellion against King John in 1216; he later appointed her sheriff. Despite a breach in the walls, Nicolaa's forces captured half the knights in the rebel army and won a virtually bloodless victory. Some of the bitterest fighting in the thirteenth century occurred in the crusade against the Cathars. When Simon de Montfort attacked the town of Lavaur in 1211 to cleanse it of heresy, it was defended by **Giralda de Laurac** in her husband's absence; the castle held out for two months. When Simon finally prevailed, Giralda was flung into a well and stoned to death. Yet there is still a pretense in the contemporary literature of this period that women were to be protected from the horrors of war. A fourteenth-century work specifically claimed that Simon de Montfort did not harm women.[25]

Fighting was endemic in Asia as well as Europe during this time. In twelfth-century Japan **Hōjō Masako** was the wife of the first shogun and was recognized as one of the principal architects of the Kamakura shogunate. Her contemporary **Tomoe Gozen** was a swordswoman and military leader. Chinese records also indicate military actions by many women such as **Liang Hongyü** and **Shalizhi**. In the thirteenth century a woman named **Shejie** led a tribal revolt against Chinese rulers, while **Yang Miaozhen** led her people in battles against the Mongols and others. On the other side of the world, in twelfth-century Mexico, the Toltec queen **Xochitl** called up women for military service and led them into a battle in which she died.

The fourteenth century seems to have occasioned more instances of women involved in siege warfare than any other period. The brutal wars between Robert Bruce of Scotland and the English in the early fourteenth century occasioned a number of sieges that placed women in military roles. **Isabel, countess of Buchan,** was a staunch nationalist, in direct opposition to her husband. She stood in for her brother, the earl of Fife (who was being held prisoner by the British), and crowned Robert I king of Scotland in 1306. She was besieged in the castle of Berwick by the English. When the castle was taken, she was imprisoned in a cage in the castle of Roxburgh. In September 1321, at the outbreak of civil war in England between the loyalists and the Lancastrians, King Edward II sent his queen, **Isabella of France,** to the castle of Lord Badlesmere at Leeds. **Lady Badlesmere** refused the queen entrance; fighting erupted and some of the queen's retinue were killed. King Edward opened siege on the castle in late October; by the thirty-first the castle fell. Retributions were harsh; many defenders were executed, and Lady Badlesmere and her children were sent to the Tower of London for a year.

One famous defender of a besieged castle was **Agnes of Dunbar** ("Black Agnes," the countess of March), who held Dunbar Castle in Scotland for nearly five months in 1338 against the English. When Agnes was told that her captured brother would be killed if she did not surrender, she is supposed to have replied that "in that case, she would inherit his earldom." Eventually she received provisions from the French, and the besiegers were forced to give up their siege.

The Hundred Years' War between the French and English from the mid-fourteenth to the mid-fifteenth centuries offers a number of examples of women involved in war. Two women were prominently involved in the battles for the succession to the duchy of Brittany in the mid-fourteenth century. The former duke's niece, **Jeanne de Penthièvre,** and his half brother, Jean de Montfort, both claimed the title. When Jeanne de Montfort was imprisoned, his wife **Jeanne de Montfort** conducted military operations and was known as a strategist. She defended the castle of Hennebon on the coast of Brittany. In addition to making symbolic appearances in armor and on horseback to rally the spirits of the defenders, Jeanne de Montfort took a direct and active role in military operations. Both Jeannes hired mercenaries, fought one another for towns and fortresses, and tried to gain the support of the people. The war that raged for the next twenty-two years, from 1342 to 1364, drew in the kings of England and France and desolated Brittany.

In the latter period of the Hundred Years' War the English besieged the city of Orléans on the bank of the Loire and unwittingly provided the opportunity for a woman to become the best-known figure in the history of siege warfare since Helen of Troy. Unlike Helen, Joan of Arc (Jeanne d'Arc) is well documented historically and was an active rather than a passive figure, and her actions ended rather than began a siege. By May 1429 Joan had persuaded a few people that she had a spiritual mission to help defeat the English in France. The king of France gave her armor and a horse, and she carried (but apparently did not wield) a sword. Though Joan did not actually fight, her courage was undisputed; she led a series of reckless attacks and freed Orléans. But chivalric practices, if they had ever influenced the treatment of women in war, had little effect by the late fifteenth century. Historian Pierce Butler noted that "Jeanne d'Arc herself . . . was no product of chivalry, and found no chivalry to shield her."[26] She was burned at the stake, not for her military actions, but for the crime of heresy and of wearing men's clothing.

Many examples of women who fought exist in this period, in addition to Joan of Arc. **Alice Knyvet,** the countess of Shrewsbury, found herself defending her husband's property against the king. The militant **Margaret of Anjou,** wife of Henry VI of England, is credited with promoting discord among the nobility that started the Wars of the Roses. The fifteenth century is an especially interesting time. **Christine de Pisan** wrote *The Book of Deeds of Arms and of Chivalry,* which became a standard work on the practice of warfare; she also wrote on the ethics of war.

Italian women, particularly of the Sforza family, were especially prominent in the fifteenth century. **Margherita Attendolo,** the sister of Muzio Sforza, was left in defense of Tricarico; her brother had been arrested in an attempt to force him to give up his castles. When the enemy demanded the surrender of

Introduction

Tricarico, she met the envoys dressed in armor and sword in hand and took them hostage, thus procuring her brother's release. **Bianca Maria Visconti Sforza** is credited with assisting her husband Francesco Sforza in besieging an enemy and in personally leading a force to the castle of Monza and achieving its surrender without a fight. By far the best known of the Sforza women was **Caterina Sforza,** who negotiated with Machiavelli and fought against Cesare Borgia. One historian claims "she was much feared by the men, whether mounted or on foot, because when she had a weapon in her hand, she was hard and cruel."[27]

But it was not only highly-born women who had the opportunity for military action. Francesco Sforza himself faced a "regular corps of women" who "fought as valiantly as the men" in defending Vigevano against his siege. After he captured the city, he reportedly liked to have the women parade in full armor with their commander, mercenary soldier **Camilla Rodolfi**. Women provided an important auxiliary force that supplemented, and sometimes substituted for, the trained military forces of the time. Often their role was one of directing and leading defensive actions; noblewomen were expected to take over the defense of the castle when the lord was absent or killed. But even ordinary women within a besieged city or castle did more than carry food and water to the soldiers; they often filled a breach in the defenses, pushed scaling ladders away from walls, and took over at the cannons and catapults.

THE EARLY MODERN PERIOD

Early modern history is a field in which historical study of military women is particularly wide open. As noted by contributor Joseph Patrouch, "There are a number of women who ruled during the Thirty Years' War, for example, about whom little has apparently been written, particularly in their capacities as military leaders." The wars of religion that ravaged Europe in this period drew in such high-ranking women as **Jeanne d' Albret, Françoise de Cezelly, Catherine de Médici** and **Marguerite de Valois** in France. Women were also involved in the **Thirty Years' War**.

Siege warfare continued to involve women as it was practiced on an increasingly brutal scale in the sixteenth and seventeenth centuries. As larger cities were fortified and attacked, and with the introduction of firearms, women appear to have been frequently active as defenders. For example, when the Turks breached the wall during the siege of Nice in 1543, Caterina Sigurana, a washerwoman, attacked their standard bearer and stole his flag.[28] Other instances in the sixteenth century include Kristina Gyllenstierna during the siege of Stockholm in 1520; Améliane du Puget in the siege of Marseilles in 1524; Marie Fourré de Poix in the siege of Saint-Riquier in 1535; townswomen in the siege of **Eger** in 1552; **Margot Delaye** in the siege of Montélimar in 1570; the Hasselaar sisters in the siege of Haarlem in 1573; and Madeleine de Saint-Nectaire, who defended Miremont from Henry IV's forces in 1575.

The Spanish women of San Mateo are reported to have been especially energetic in shooting at besiegers.[29]

The English civil wars of the seventeenth century were characterized by numerous "rough-and-ready sieges of small strongholds"[30] and provide a number of examples of women defenders of besieged castles: the countess of Portland at Carisbrooke Castle; Lady Mary Winter, who refused to surrender Lidney House to parliamentary forces; **Charlotte Stanley,** countess of Derby, who defended Lathom House against the Parliamentarians; and Blanche, Lady Arundel, who defended Wardour Castle for the royalists. Lady Brilliana Harley held Brampton Castle against the king for more than six months; when she died (probably of pneumonia), the garrison surrendered.[31]

One of the most famous defenders was **Lady Mary Bankes,** who held Corfe Castle while her husband stayed with the king in London. The first siege began in June 1643; Bankes successfully held out with only five soldiers and the women of the castle. She was besieged again in 1645 after her husband's death; this time a traitor apparently gave the enemy entry to the castle. Unidentified women continue to appear; for example, in 1645 it was reported that a woman corporal was taken prisoner among royalist troops captured by Major-General Poyntz of the New Model Army.[32] Women were also active participants in the civil wars in mid-seventeenth-century France, including the duchesses of **Longueville** and **Montpensier**.

Women are often described as being motivated to go to war in order to follow a lover. **Kit Cavanaugh** (seventeenth century) and **Hannah Snell** (eighteenth century) each allegedly put on male disguise and went to war to follow a husband who had abandoned her. However, as contributor Susanna Calkins notes, "neither one really seemed to pursue a path designed to catch him," and "it was not necessary for them to be disguised or to engage in battle just to be close to the soldiers," since women often accompanied men on campaigns during that era.

Elsewhere women were active in a variety of military events. In late-seventeenth-century Russia **Alena Arzamasskaia** joined the revolt of Stenka Razin; in male disguise she led several thousand troops. She was captured by tsarist forces, tortured, and burned at the stake. Chinese women such as **Liu, Shen Yunying,** and the general **Qin Liang-yü** also provide diverse examples of women's roles in war in the sixteenth and seventeenth centuries. In Africa the warrior Nzingha led the Angolans in their fight against the Portuguese. In prerevolutionary America both colonial and **Native American women** fought, primarily in defense of their homes or villages.[33]

THE REVOLUTIONARY AND MODERN PERIOD (EIGHTEENTH–NINETEENTH CENTURIES)

Many conflicts in the eighteenth century brought women into war, even in the age of "limited warfare." The empresses **Maria Theresa** of Austria and

Introduction

Catherine II (the Great) of Russia successfully directed wars of survival and expansion. Maria Theresa prevailed against the formidable Frederick the Great. Women served in disguise in European armies—even in the vaunted Prussian army. Frederick the Great, ruler of Prussia, learned of a woman disguised as a man who fought in his army during several campaigns of the Seven Years' War, reportedly with some distinction. He was neither amused nor appreciative (unlike the tsar of Russia when informed of **Nadezhda Durova**'s true identity). Frederick said, "It's contrary to nature. I don't want to have any women soldiers in my army. There might be some advantage from them in wartime, but the disorder would be all the greater in time of peace. The women have always had a hankering to wear men's pants, which would give rise to all sorts of confusion."[34] Interestingly, even Frederick did not criticize the woman soldier's performance, only the "disorder" he feared would result from the precedent she set. The woman of whom Frederick spoke may have been Anna Sophia Detzliffin, who at age nineteen reportedly fought in a regiment of cuirassiers, was wounded, later joined the grenadiers, was captured by the Austrians, escaped, eventually revealed her sex, and was discharged.[35] Her story remains unrecovered by historians. Similarly, in Russia the twenty-year-old **Tat'iana Markina** disguised herself as a man and served in the infantry of Catherine II, where she attained the rank of captain. An interesting case in the British army is that of Phoebe Hessel, who joined the army as a child. At first disguised as a drummer boy by her father, she later fought as a private soldier until she was wounded in the Battle of Fontenoy in 1745. She received a pension from George IV.[36]

A large number of women were attracted to life at sea during the Age of Sail, and many successfully disguised themselves as sailors. Hannah Snell served for five years as a British marine in the mid-eighteenth century; her travels took her to the East Indies and the Indian Ocean. When she was wounded at Pondicherry, she got a civilian to extract the bullet so that the surgeon would not discover her identity. After returning to England, she revealed her identity and obtained a pension. The more controversial **Mary Anne Talbot** also claimed to have served in male disguise for four or five years in the 1790s. And in late-eighteenth-century India several fascinating women emerged. **Ahilyabhai Holkar** was considered one of the most notable rulers of the Maratha Confederacy of central India in both political and military terms. In northern India **Begum Samru** led a mercenary army, eventually as an ally of the East India Company.

The revolutionary period of the late eighteenth century (extending into the early nineteenth in Europe) was a time when women had many opportunities to play a military role. The most extensive scholarly work has been done on the American Revolution (although far more women were actively involved in Europe). American women were most active on the battlefield in the eighteenth and nineteenth centuries. Women fought in the Revolutionary War and the Civil War; most adopted male identities and disguises. Women also served as military auxiliaries, especially in the Revolutionary War, and were commonly present on the battlefield in support roles.

The best work on military women in the Revolutionary War has been done by historians Linda Grant De Pauw and John Todd White.[37] De Pauw suggests that at least 20,000 women "were involved in active combat" in the War for Independence.[38] She defines three categories of women's participation. First were the "women of the army," a term she borrows from General George Washington. De Pauw believes that far from being mere camp followers, these women performed a military function; she notes that they were subject to military discipline and drew military rations. Camp followers, she contends, were civilians (including sutlers and prostitutes) who sold goods and services, but few of these attended the poorly paid Continental army. Women of the army, in contrast, were usually (but not always) soldiers' wives who competed for a limited number of female slots assigned to each company. De Pauw and White contend that the duties of these women have been trivialized as "battlefield domesticity." White notes that it is clear from personal documents that common soldiers did their own cooking and washing; women of the army worked primarily in the medical corps and the artillery. De Pauw points out that some women of the army who attempted to desert were subjected to court-martial and lashing; no civilian received such treatment.[39] De Pauw and White challenge the traditional view, such as that of Linda Kerber, that women of the army were "dependents who drifted after the troops . . . who had no means of support" and who "earned their subsistence by cooking and washing for troops."[40]

Some of the most famous women in the American Revolutionary War belong to the category of "women of the army." The name **Molly Pitcher** appears frequently, but this moniker was probably given to many female waterbearers. De Pauw notes that these women who served with artillery units have been erroneously portrayed as serving water to the soldiers when in fact they carried water to clean out artillery pieces between firings. She points out that it was well known in the army that drinking too much water during battle could induce the fatal "cold water disease" (a result of heat stroke), and that grog was the usual drink. On the other hand, a great deal of water was required to keep the artillery pieces firing properly, and by assigning women to carry water, men were freed for other duties. Women who performed this duty would undoubtedly become familiar with the procedure for loading and firing the guns and plausibly could have replaced fallen soldiers in actually firing the guns. De Pauw contends that there is "clear evidence that a handful of army women were engaged in loading or firing artillery pieces."[41] Many sources cite the story of **Mary Ludwig Hays McCauley** as one possible origin of the Molly Pitcher legend.

De Pauw's second category is women who fought as regular troops. She believes that **Margaret Corbin** has been mistakenly categorized as a "woman of the army" or a "Molly Pitcher" type when in fact she enlisted and served

Introduction

under her own name. Corbin accompanied her husband John and served in the artillery at the Battle of Fort Washington in November 1776. When her husband was killed, Corbin took over his position at the gun; she was wounded and disabled for life. De Pauw cites the fact that Corbin was captured (and later released), assigned to the Corps of Invalids that was posted to garrison duty at West Point, and later given a pension (including, apparently, male clothing) in 1780 as evidence that she served as a soldier and not an auxiliary.[42]

Deborah Samson was the most famous and best-documented woman soldier of the Revolutionary War.[43] Samson dressed in a man's suit, cut her hair, adopted the name of Robert Shurtliff, and enlisted in the Continental army. There was no physical examination; all she had to do was "pass muster," a cursory visual examination. Samson was tall, agile, and a good shot; when she was wounded by a musket ball in the thigh in 1782, she extracted the ball herself to prevent detection. Samson has attracted more attention than other women soldiers because her story is better documented. Newspaper accounts appeared soon after the war ended, letters concerning her have been preserved (including one from Paul Revere supporting her pension application), and her military records have been verified.[44]

Other women also fought during the revolution. Sally St. Clair passed as a male soldier until her death in the Battle of Savannah. Records in the National Archives indicate that **Ann Bailey,** also known as Samuel Gay, a corporal in the First Massachusetts Regiment, was discharged in 1777 for "being a woman, dressed in mens cloths" after three weeks' service. A different woman, "**Mad Anne Bailey,**" avenged her husband's death by serving as an army scout and courier in 1774. **Anna Maria Lane** received no recognition in her own time; she fought in disguise. She was granted a pension in 1808 because "in the revolutionary war, in the garb, and with the courage of a soldier, [she] performed extraordinary military services, and received a severe wound at the battle of Germantown."[45]

De Pauw identifies a third category of military women in the Revolutionary War: those who served in militia units. She offers some evidence to support this category, but documentation is even more problematic than for the other categories. One of the most interesting documents she mentions is a "classed return" (a militia list of people "capable of bearing arms between the ages of 15 and 50 years") drawn up in 1780; the names of nineteen women are interspersed with those of ninety-five men. De Pauw concludes that the women would have been expected to serve in an emergency.[46]

Few people today know that any women fought in the American Revolution. Some of their stories were briefly publicized in their own day, but were quickly forgotten. In 1840 Charles Francis Adams wrote that "the heroism of the females of the Revolution has gone from memory with the generation that witnessed it, and nothing, absolutely nothing remains upon the ear of the young of the present day."[47]

Introduction

The American Revolution was a fairly limited event in comparison with the French Revolution and the Napoleonic Wars that followed, and women's participation in revolutionary war in Europe was widespread. Women fought both for and against France in such diverse regions as Germany, Italy, and Russia. Contributor S.P. Conner recently uncovered information in French army archives on at least twenty-eight women who served in the French army as soldiers between 1791 and 1815. These include women like **Thérèse Figueur** of Lyon, who served at the siege of Toulon and later fought as a dragoon in Italy, Austria, Prussia, and Spain from 1793 to 1815. **Rose Barreau** fought alongside her husband and her brother; **Virginie Ghesquière** fought in the Peninsular War. **Marie Angélique Josèphe Duchemin** served from 1792 to 1799 and was noted for her bravery in the defense of Corsica; she was the first woman to receive the Legion of Honor. **Marie-Jeanne Schellinck,** a Belgian, fought for France in many campaigns and was wounded several times. Conner's entry on women in the **French Revolution and Napoleonic Era** provides an overview of the diverse military and political roles played by women.

A number of women fought against French forces during the same period. Some French women also rose in "Vendée" against the French Revolution, such as **Renée Bordereau,** who "fought like a lion," and **La Marquise de La Rochejaquelein,** wife of a counterrevolutionary general, who in her memoirs described some of the women who fought. In Spain, **Agustina Aragón** was one of many women who defended Zaragoza against French forces in 1808. Reportedly she rushed in to replace a fallen artillery sergeant, firing the 24-pound gun. The retreating citizens were shamed into reoccupying the battery, and they beat off the assault. She was given an artillery commission, and when she retired, she retained her rank, the right to dress as a soldier and bear arms, and military pay and pension. Also in 1808 Manuela Malasaña was martyred in Madrid, while in Tyrol **Katharina Lanz** fought against the French.

Many other women are reported to have fought against Napoleon, but their stories could not be investigated in time for inclusion in this volume. Mary Dixon reportedly fought in the British army for sixteen years and was in the Battle of Waterloo. Auguste Friedericke Krueger was one of at least seventeen women who fought in disguise during the German Liberation Wars against Napoleon of 1813–1815; she fought in the Ninth Prussian Regiment and received the Iron Cross and the Russian **Order of St. George**. Karoline Wilhelmine Gronten, Eleonora Prochaska, Anna Luehring, Johanna Stegen, and Louisa Scanagatti were just some of the other German women who fought in disguise.[48]

The most famous Russian woman soldier of the nineteenth century was Nadezhda Durova, who fought as an officer in the Russian cavalry during the early nineteenth century. In male disguise Durova participated in combat in 1807 and 1812–1814 in the wars against Napoleon. Her disguise was eventually discovered, but Tsar Alexander I permitted Durova to continue to serve

Introduction

due to her several decorations for heroism; she was the only woman to win the St. George Cross (Order of St. George) until the end of the First World War. Durova was unique in two ways. Once her true identity was discovered, she successfully appealed to the tsar to continue military service; and she later wrote about her experiences. The stories of women who kept their service secret are only beginning to come to light. For example, an Aleksandra Tikhomirova reportedly served for fifteen years in the Russian cavalry in the late eighteenth and early nineteenth centuries. Her identity was only discovered on her death in 1807.

Nationalistic revolutions continued throughout the nineteenth century and continued to involve women. Naval commanders Manto Mavrogenous and **Laskarina Bouboulina** became icons in the Greek War of Independence in the early nineteenth century; their statues stand in the War Memorial in Athens. **Emilia Plater** was one of several women to take an active part in the unsuccessful Polish insurrection of 1831 against Russia when at age twenty-five she organized military detachments and led a series of successful battles. The revolutions of 1848 featured a number of interesting women like the Croatian **Maria Lebstuck,** who fought against the Austrian Empire in male disguise. In Italy women like Louisa Battistati reportedly fought in revolutionary battles in 1848, Marietta Guiliani and Herminia Manelli served in Garibaldi's forces, and Sylvia Mariotti served as a private in the Eleventh Battalion of the Italian Bersaglieri, fighting the Austrians in the Battle of Custozza, but their stories could not be verified for this collection.[49]

In Africa women made up nearly one-half of the armed forces of the West African kingdom of **Dahomey** in the nineteenth century. Women were also prominent in wars of liberation. Mantatisi, queen of one of the clans displaced in the early 1800s by the wars of Shaka Zulu, engaged in military strategy and leadership. In 1896 Tayetu of Ethiopia led the very first defeat of a European power on African soil. **Masarico** was a warrior queen who led the Manes across much of the forest regions of West Africa, fighting the Portuguese and Dutch and subjugating indigenous populations in present day Sierra Leone and Liberia. **Nehanda,** a religious leader in what is now Zimbabwe, became a military leader when the British invaded; she was eventually captured and executed.

A number of women in India took up arms against the British during the 1857 Sepoy Rebellion, such as Laxshmi Bhai, the rani of Jhansi, mentioned earlier in this introduction, and Hazrat Mahal, the queen of Awadh during the Indian Mutiny, who, while not as actively involved in fighting as the rani of Jhansi, nevertheless led and commanded armies. In this period women in military roles appear even in Iranian history; **Zaynab** "Rustam-'Ali" fought at the siege of Zanjan in 1850.

In Latin America women continued to fight in colonial struggles. During the early part of the nineteenth century thousands of women participated in the Mexican War of Independence, including **Gertrudis Bocanegra**. Contributor Elizabeth Salas has chronicled hundreds of stories of Mexican *soldaderas*, including **Ignacia Reachy,** who started a women's battalion to defend against the French, fought in the Battle of Acultzingo in 1862, was captured by the French, spent a year in prison, escaped to fight again, and then commanded the Lancers of Jalsco until she was killed in battle in 1866. Further south, **Petrona García Morales de Carrera** was an active combatant alongside her husband in the revolution that brought him to power in Guatemala in 1837–1840, and **Juana Azurduy** fought as a guerrilla leader in Bolivia in the early 1800s. **Maria Quitéria,** a heroine of Brazilian independence, joined the army in disguise in 1822; although she was later discovered, she was commissioned as a second lieutenant. **Paraguayan women** fought in 1864–1870, when hundreds of women and children reportedly died at the last stand of Piribebuy in 1868.

In the United States during the nineteenth century the Civil War represented the historical peak of women's direct participation in combat. Contributor Elizabeth Leonard estimates that 500 to 1,000 women served in disguise. A group entry on women in the **American Civil War** provides a general context and entries on representative individuals. Documentation and verification have been difficult. Like other women who fought in disguise, these women took measures to keep their true identities secret, such as changing from one regiment to another repeatedly to avoid detection. Some, like Amy Clark (see American Civil War), were discovered only after being taken prisoner in battle. Others were discovered only after their deaths on the battlefield, and their true identities were never known. There is a report that when one member of the 14th Iowa Regiment was discovered to be a woman, she committed suicide.[50]

One of the most controversial female soldiers of the Civil War was **Loreta Janeta Velazquez**. The Cuban-born, New Orleans–raised Velazquez had sufficient wealth to outfit herself with supplies, equipment, male clothing, and a false beard and mustache. Velazquez adopted the name of "Lieutenant Harry T. Buford," picked up temporary assignments as courier, spy, and military rail conductor, and appears to have simply "tagged along" with regiments like the 8th Virginia Infantry. She became controversial after the publication of her memoirs in 1876. General Jubal Early believed Velazquez a complete fraud, and his refutation of her story tarnished her reputation with historians. One writer who recently reexamined the memoirs and other relevant documents believes there is sufficient evidence to show that Velazquez did in fact serve as Buford.[51]

The case of **Sarah Rosetta Wakeman** is easier to verify. Her letters to her family, written during the war, and a daguerreotype of herself in uniform (in male disguise) were recently uncovered, and her military and pension records in the National Archives (under her assumed name, Lyons Wakeman) confirm her term of service. Wakeman served with the 153rd Regiment of New York State Volunteers and fought in the Red River Campaign; she later died of illness, without fanfare.

Another fascinating woman whose biography has yet to be written is **Jennie Hodgers** ("Albert Cashier"), who has the longest documented length of service by a female soldier in the Civil War. Just over five feet tall, Hodgers was assigned primarily to foraging and skirmishing duty. She took part in forty battles and was never wounded; "Albert Cashier" is listed on the Vicksburg battlefield monument to Illinois soldiers. What is remarkable about Hodgers's life is that unlike most women soldiers who fought in disguise, she retained her male identity for fifty years after the war. We will never know the real identities of many other women whose gender was discovered only after their death. At least four women fought in the Battle of Antietam, for example, where two were wounded and one died.[52] Women also served openly in volunteer units with the Union army, often wearing male clothing but making no secret of their sex. Their presence was accepted because these volunteer units were often composed of neighbors and family groups. **Anna Etheridge,** who received the Kearny Cross for bravery under fire, later managed to win a pension for her service. At least two such women are reported to have actually led a charge during the Civil War. **Nadine Turchin,** wife of a colonel of the Nineteenth Illinois Infantry, led her husband's regiment when he was ill.

Dr. **Mary Walker** served with the Union army and was the first woman to be commissioned as a military doctor and lieutenant in the Medical Corps. Walker is noteworthy as the first woman to be awarded the Congressional Medal of Honor for her service and the four months she spent as a prisoner of war in Richmond. The award was withdrawn in 1917 when the army raised its criteria (some male recipients were also affected), but was restored in 1976. Walker was an unusual character; one writer claims that "her contemporaries counted her a freak . . . she has gone down in history, if at all, largely as an eccentric and a bore . . . due to her militant taste in clothes. She wore pants."[53] It seems surprising that Walker has not yet been the subject of a scholarly monograph.

One of the most interesting figures of the Civil War is **Anna Ella Carroll,** the daughter of a governor of Maryland and niece of General Winfield Scott. Her pamphlets supporting the Union cause impressed Abraham Lincoln, and Carroll was often asked to write materials for his government. Carroll later claimed that the Union victories at Forts Henry and Donelson were the result of her strategy. Carroll was not paid for her efforts and later petitioned Congress for reimbursement; it was a long and ultimately unproductive battle. Contributor Janet Coryell's study of Carroll, *Neither Heroine nor Fool,* is a scholarly examination of Carroll's real contribution and the subsequent historiographical controversy. The author began researching a doctoral dissertation in 1982, expecting to find that Carroll was a "much-maligned, forgotten heroine of American history." Instead, she uncovered a complicated woman, often admirable, sometimes unlikable, and a tangled story that was ultimately unsolvable. Coryell found little evidence to support Carroll's claims that she was the sole author of the Tennessee River strategy and believes that Carroll's true contribution lay in her work as a political publicist. Coryell's book is enlightening not only for the history it reveals, but for the methodology it makes explicit. Coryell found that Carroll has often had a symbolic value that led to the inflation of her role as a military strategist and deflation of her real achievement as a political writer. Coryell's book is one of the finest scholarly efforts to date on a "military" woman and should serve as a model for future studies.

African American women served on both sides in the Civil War, generally performing service functions but also as volunteer nurses. Only one left a written record: **Susie King Taylor,** a nurse with the Union army. **Harriet Tubman** played an interesting role in the Union army during the Civil War. She worked as a liaison between military authorities and former slaves in Federal-occupied areas of South Carolina in 1862. Later Tubman commanded a small unit that scouted for the army, and helped lead a raid in which 150 black troops destroyed bridges and rail lines along the Combahee River in June 1862. Tubman conducted similar missions in 1863–1864. It was only after the American Civil War that the first black woman can be verified as officially serving as a soldier: **Cathay Williams,** who, disguised as "William Cathey," served with the regular army from 1866 to 1868. A tall woman (5'9"), Williams passed a cursory medical examination and was enlisted in the 38th United States Infantry, a segregated all-black unit under white officers. Documented cases of American black women serving in military roles are rare until the twentieth century. At first their slave status limited their military participation; after emancipation there were few opportunities for any women to serve until the First World War. Only a few black nurses seem to have succeeded in military participation in this period. Partly because blacks were believed to be immune to typhoid fever, a handful of black nurses were recruited and served with the army during the Spanish-American War (thirty-two women) and the First World War (eighteen women).[54]

In the last half of the nineteenth century Native American women continued to fight in the last stages of the Indian Wars in the western United States. **Lozen,** a Warm Springs Apache woman and sister of Victorio, was renowned for her ferocity in combat and later in captivity. **The Other Magpie** was a Crow woman who fought with General Crook against the Sioux and Cheyenne at the Battle of the Rosebud in June 1876; she took a scalp and performed other heroic deeds. She may have been avenging her brother, who had been killed by the Sioux.

Women were also prominent once more in France in the later nineteenth century, where during the **Paris Commune** of 1871, for the first time in French history, armed women's units were formed. Women Communards like **Louise**

Introduction

Michel and Blanche Lefebvre, claiming full rights as citizens, built and defended barricades during the street fighting, and some died.

Throughout the eighteenth and nineteenth centuries women were active around the world in colonial liberation struggles. Women's traditional support roles in revolutionary conflict are well known. No revolution can function without support, supply, and logistics: someone to cook and launder, to provide shelter, to run errands and carry messages. Women were prominent in these areas and often ran risks as great as those of the combatants. A courier shot for violating curfew is just as dead as a soldier killed in the line of battle. Many women also took up arms in defense of revolutionary ideals. These women, like most military women, are often dismissed as anomalous and insignificant, but the scope of their participation, geographically and chronologically, is impressive.

TWENTIETH CENTURY

What makes women's military experience in the twentieth century unique is the emergence of total war. During the First World War, for the first time in history, whole societies made war their first priority. Militarism pervaded all facets of life, affecting women and children at home as well as soldiers at the front. The twentieth century marked a breakthrough in the creation of official military roles for women. But when women first entered formal military service in most countries, a separate sphere was created to demarcate and segregate military women from "real" soldiers. By formalizing and institutionalizing the means by which women could serve with (not in) the military, a path was opened up that permitted greater numbers of women to choose military roles in the twentieth century. At the same time, informal options were closed off. It was no longer possible for a woman to become a combatant by the simple expedient of putting on men's clothes, passing muster, and enlisting under a male name. Because of the difficulty of sustaining male disguise in the face of "modern" medical examinations, few women were able to participate in combat in regular armies of this period.

This was the path followed in the United States. In 1901 the Army Nurse Corps was established as an auxiliary organization; women had no military rank or benefits and were paid less than military men performing the same duties. The navy established its own Nurse Corps along similar lines in 1908. Except for nurses, women played a minimal military role during the first two years of the First World War. In 1916 Secretary of the Navy Josephus Daniels examined legal requirements for naval enlistments, discovered that women were not excluded, and decided to enroll women in the navy as "yeomanettes" to provide clerical assistance. When the United States entered the war in 1917, the navy enlisted more than ten thousand women. The Marine Corps admitted women only in 1918, two months before the war was over; the army did not permit women other than nurses to serve in the First World War. The navy's "yeomen (F)" enlisted personnel worked primarily in clerical fields. A total of 34,000 women served in the First World War, primarily in the nurse corps.[55] Although restricted to support duties, the women who served as yeomen (F) and marines (F) had full military status. Even these clerks were so threatening that in 1925 the wording of the Naval Reserve Act of 1916 was changed so that only "male citizens" could be enlisted, preventing the future enlistment of women.[56] There is no scholarly history on American military women in the First World War.

Most European countries restricted women's service to volunteer and auxiliary organizations. In 1907 Great Britain established **FANY Corps** (First Aid Nursing Yeomanry Corps), a volunteer corps of nurses who rode horseback between the battlefield and hospital. In 1917 the **WAAC/QMAAC** (Women's Auxiliary Army Corps/Queen Mary's Auxiliary Army Corps) was established to allow women to serve in traditional support roles such as clerks and telephone operators for the army. British women who wanted to fight, and even those who wished to serve in military support roles, were accused of wanting to cause all sorts of sexual confusion. It was not the risk of dying that was at issue; several WAAC members were killed in air raids. It was not even exposure to combat conditions that was an issue, for women volunteers who drove ambulances at the front were exposed to enemy fire and the sight and sound of soldiers with horrendous wounds. Instead, it was the demands of some women to be allowed to take up weapons that threatened many in conservative British society. The head of the Women's Volunteer Reserve in Britain during the First World War complained, "We had to contend with a Section of 'She-Men' who wished to be armed to the teeth."[57] The only British women who were able to serve as soldiers did so in the armies of other nations. **Flora Sandes** was an Englishwoman who served seven years in the Serbian army, first as a nurse but then in the infantry as a soldier. She was wounded several times and was decorated for bravery.

Russia (later the Soviet Union) stands out in world history in its willingness to use large numbers of women in combat. During both world wars and its civil war women fought on the front lines. In the period of the First World War Russia was the only country that used women in segregated military groups—some 6,000 women soldiers by 1917. Most were nurses, but other front-line, regular army roles included drivers, mechanics, and laborers, as well as combat roles. A number of young women appear to have joined the Russian army during the First World War; many joined fighting units, often in disguise, and some were decorated. These include many fascinating individuals such as **Ekaterina Alekseeva, Kira Bashkirova,** and **Princess Kati Dadeshkeliani** of Georgia. Rimma Ivanova and Antonina Pal'shina were among those who received the Order of St. George for bravery in combat. Several Russian women pilots also served, mainly in reconnaissance; they include **Sofiia Dolgorukaia** and **Princess Evgeniia Shakhovskaia**.

The most dramatic group was the **Women's Battalion of Death** commanded by **Mariia Bochkareva** in 1917. Bochkareva was a peasant woman who received permission from the tsar to enlist in the infantry as a line soldier early in the First World War. She served with some distinction, was wounded on at least three occasions, and received several medals. After the February Revolution Bochkareva conceived the idea of a Women's Battalion of Death: a unit of three hundred women that would "serve as an example to the army and lead the men into battle."[58] The battalion went to the front in July 1917. It participated in one abortive action, in which the women's battalion and a group of officers charged German lines, but were then left stranded when the troops refused to follow. They were eventually forced to retreat and sustained heavy casualties.

Women fought on both sides of the Russian Revolution and Civil War, though most prominently for the Communists. Women soldiers attempted to protect Aleksandr Kerensky from the Bolsheviks during the October Revolution in Petrograd. On the opposite side, many Bolshevik women carried weapons in the Red Guards. After the Bolsheviks came to power, their 1918 decree of universal military service permitted women to volunteer. During the **Russian Civil War** that ravaged the country from 1918 to 1921, more than 70,000 women served in the Red Army on a much broader scale and in a more organized fashion than during the First World War. Volunteers were allowed in from the start, and women were also conscripted for support duties by 1920. There were military nurses, but the Red Army, unlike its predecessors, instituted both political indoctrination and rifle training for its medical personnel. Women volunteers also served in combat roles on every front, as riflewomen, gunners, demolition troops, partisans, scouts, and spies. Unlike the female soldiers of the Provisional Government, women in the Red Army served in integrated units. As in many civil wars, prisoners were treated harshly by both sides, with women on both the receiving and giving end of torture. In 1919, three Red Army nurses captured by the Whites were hanged in their field hospital with their party pins stuck through their tongues. Instances were reported of women commissars in the Red Army who shot their prisoners joint by joint and used "the glove trick" of boiling the skin off their arms to make them talk.[59] Once the Bolsheviks consolidated their power, they instituted a universal paramilitary training program called **Osoaviakhim**. Tens of thousands of women and men were trained in the 1930s as drivers, snipers, parachutists, machine gunners, and pilots.

The interwar period was one during which women in most Western countries were mostly banned from military service. It was in regions where revolution was a factor that women were involved in war. In China women were part of the Long March (1934–1936) on propaganda teams, as telegraph decoders, and in political work, logistical support work, and transport work. There was a Women's Engineering Battalion, and some women fought on occasion.[60] Women were involved in the **Spanish Civil War** as well.

The Second World War provided the opportunity for women to be used in

Introduction

mass numbers. In Western nations these roles were invariably as noncombatants. The role of women in the twentieth-century American military was to free men for combat by filling support positions in administrative, clerical, and medical fields. As the demand for (male) soldiers increased, the value of women in the military rose; it was believed that there was more "difficulty getting enlisted men to perform tedious duties anywhere nearly as well as women will do it."[61] In the United States the WAAC (Women's Auxiliary Army Corps) was established in 1942, followed by the Navy Women's Reserve, or **WAVES,** and the Coast Guard women's volunteer reserve, **SPAR**. The auxiliary status of these groups was problematic from the beginning; after a year of debate and discussion the WAAC was replaced by the **WAC** (Women's Army Corps) and was granted full military status. Altogether, 140,000 women served in the WAC, 100,000 with the WAVES, 23,000 in the marines, 13,000 with SPAR, 60,000 in the Army Nurse Corps, and 14,000 in the Navy Nurse Corps. The nurses finally received full military status in 1944, a year after the WAC. When the WAAC was established in 1942, a goal of 10 percent black enlistments was set, but never met. By the end of the war the WAC included only 4,000, or 4 percent, black women. **WAC African Americans** served in segregated units under black female officers.[62] Perhaps the best-known American women of the Second World War were the **WASP** (Women's Airforce Service Pilots). Despite the excellent flying record of the WASPs, they were closed down before the war ended in 1944 and did not receive official military status until the 1970s.

In Europe women were similarly relegated to auxiliaries and noncombatant roles in most regular armies. In the United Kingdom, for example, women served in the **ATA** and **ATS**. Women were assigned to antiaircraft batteries, but they were not allowed to actually fire the guns. It was only in **resistance movements**—in France, Russia, Italy, and Yugoslavia, among others—that women were involved in fighting. In Poland women were a vital asset of the **Armia Krajowa** and fought during the Warsaw Ghetto Uprising (1943) and the Warsaw Uprising (1944). Women's roles as partisans and resistance fighters have been overlooked until recently, partly because those most active in such operations were Communists. Women were an essential part of the resistance support network, providing food and lodging and frequently acting as couriers. The British Special Operations Executive (SOE), formed in July 1940 to support resistance movements in occupied areas, trained women like **Noor Inayat Khan** to undertake hazardous work. Before, during, and after the war, women were active in Israel. Women like **Rachel Yanait Ben-Zvi** (whose husband became the second president of Israel) worked for Haganah. **Netiva Ben-Yehuda** worked with the Palmach during the **Israeli War of Independence**.

Only one regular army—the Red Army in the Soviet Union—used women in large numbers and also opened combat positions to them. Soviet women, in addition to their massive employment in combat support services, engaged in combat in all branches of service as well as in partisan units. By the end of 1943 more than 800,000 women were in military service; half of them were at the front.[63] They included thousands of snipers like **Liudmila Pavlichenko;** in fact, a **Central Women's School For Sniper Training** was established specifically to train women. There were machine gunners like **Neonila Onilova** and tank drivers like **Irina Levchenko**. The Soviet Union was also the first nation to allow women to fly in combat. Under the direction of **Marina Raskova,** the **122nd Aviation Group** trained the air and ground crews for three combat regiments—at first all-female, although two of the regiments later included some men. Fighter pilot Lidiia Litviak is just one example of the women who emerged from these units to prove their worth in combat. Soviet women were unique in being the only female soldiers to fight outside the borders of their own country.

Women were rapidly demobilized from every military organization after the Second World War and in general played a very minor role in regular forces until the 1970s. The countries that used women most extensively in combat, such as the Soviet Union and Israel, also had the greatest need to rebuild population, so women were once more relegated to a support role, at best, in the regular military. The Israeli creation of **CHEN** to oversee women's military roles in the postwar years is an example of the way women's roles "snapped back" to more traditional modes despite excellent wartime performance.

In the United States the Women's Armed Services Integration Act of 1948 abolished auxiliaries and gave women full military status. However, limitations were set on the number of women who could serve, the rank they could attain, and other benefits. Women in the air force and navy were also barred from serving in aircraft or ships considered to be "engaged in combat missions"; these stipulations came to be known as the combat exclusion rules.[64] Definitions of "combat" were ambiguous. For years the air force interpreted this provision as affecting all air force aircraft. Very little has been written on the postwar (1945–1978) service of American military women. Numbers were drastically reduced, and the range of assignments was generally curtailed as well.[65] Until the late 1960s the services were barely able to maintain a 1 percent proportion; nearly all these women were employed in traditional positions in health care and clerical work. In the 1970s some of the provisions of the 1948 act began to fall. The main issue of the 1980s and 1990s was the combat exclusion rule. There was never a statutory regulation prohibiting women from filling combat positions in the army, but in practice women have been barred from these roles.[66]

In the postwar world, while Western nations have debated whether women had the capability to fight—ignoring all historical examples—women continued to participate in combat in a variety of postcolonial and revolutionary situations. Women's roles as combatants in the **Vietnamese Revolution** are just beginning to be studied by scholars. Though little has been written by historians about Vietnamese women, their presence is noted in many memoirs and even in works of fiction. A Vietnam veteran commented in a recent short story, "Sometimes the women were the worst killers of all. Cut you down with a smile."[67]

In Africa women continued to participate in wars of liberation. Women were active in the **Algerian-French War** (1954–1962). Women have been trained for and used in combat by such revolutionary groups as the SWAPO in Namibia, the ANC in South Africa, FRELIMO in Mozambique, MPLA in Angola, and PAIGC in Guinea-Bissau. Women fought as part of the Mau Mau movement in Kenya. They fought in **Zimbabwe,** Eritrea, Ethiopia, and possibly many other countries.[68]

Major changes occurred in the 1990s when many Western nations began lifting the restrictions against women in combat. The European Union supports equal access, for example. Combat roles (with some exceptions such as special forces) are now open to women in many countries, including Belgium, Canada, Denmark, Eritrea, France, Israel, Luxembourg, the Netherlands, Norway, Sweden, the United Kingdom, the United States, Venezuela, and Zambia. Many other governments, such as that of Australia, are considering opening combat roles to women.

Many people view these developments as something entirely new, only made possible by technological developments that have made warfare more technical and less physical. However, numerous recent studies have shed new light on the demographics of war by showing that contrary to popular myth, wars have always involved women. Historians are beginning to realize that "the common vision of armies as all-male institutions strongly distorts past realities" and that "for thousands of years armies routinely comprised women as well as men."[69] In the end it is culture and opportunity—nurture and not nature—that determine whether women will fight.

NOTES

1. Abby Wettan Kleinbaum, *The War against the Amazons* (New York: New Press, 1983), 30.

2. Tadeusz Sulimirski, *The Sarmatians* (New York: Praeger, 1970), 33–34, 48, 105–106.

3. Jeannine Davis-Kimball, "Warrior Women of the Eurasian Steppes," *Archaeology* January/February 1997: 44–48.

4. Mary Zirin, introduction to Nadezhda Durova, *The Cavalry Maiden: Journals of a Female Russian Officer in the Napoleonic Wars,* trans. Mary Fleming Zirin (London: Angel, 1988), xxxv, citing G. Esipov, "A Regiment of Amazons during the Reign of Catherine II," *Istoricheskii vestnik* 1886, 1:71–75; Alfred G. Meyer, "The Impact of World War I on Russian Women's Lives," *Russia's Women: Accommodation, Resistance, Transformation.* Ed. Barbara Evans Clements et al. (Berkeley: University California Press, 1991), 219.

5. JoAnn McNamara and Suzanne F. Wemple, "Sanctity and Power: The Dual Pursuit of Medieval Women," *Becoming Visible: Women in European History,* ed. Renate Bridenthal and Claudia Koonz (Boston: Houghton Mifflin, 1977), 97.

Introduction

6. Megan McLaughlin, "The Woman Warrior: Gender, Warfare, and Society in Medieval Europe," *Women's Studies* 17 (1990): 196.

7. Ibid., 194.

8. Ann S. Haskell, "The Paston Women on Marriage in Fifteenth-Century England," *Viator* 4 (1973): 462.

9. Pauline Stafford, *Queens, Concubines, and Dowagers: The King's Wife in the Early Middle Ages* (Athens: University of Georgia Press, 1983), 118; Joseph Gies and Frances Gies, *Life in a Medieval Castle* (New York: Harper, 1979), 84.

10. Philippe Contamine, *War in the Middle Ages,* trans. Michael Jones (London: Blackwell, 1984), 241–242.

11. Mary Erler and Maryanne Kowaleski, eds., *Women and Power in the Middle Ages* (Athens: University of Georgia Press, 1988), 1.

12. Contamine, *War in the Middle Ages,* 272–274; Sidney Painter, *French Chivalry: Chivalric Ideas and Practices in Mediaeval France* (Ithaca: Great Seal, 1957), 105.

13. The duke of Bourbon to the captured duchess of Brittany, in the *Chronique de Loys de Bourbon,* cited by Painter, *French Chivalry,* 145.

14. Barbara W. Tuchman, *A Distant Mirror: The Calamitous 14th Century* (New York: Knopf, 1978), 62.

15. Christopher Duffy, *Siege Warfare: The Fortress in the Early Modern World, 1494–1660,* reprint ed. (London: Routledge, 1987), 251.

16. Cited by Duffy, *Siege Warfare,* 253.

17. Stafford, *Queens,* 118.

18. Ibid., 141.

19. McNamara and Wemple, "Sanctity and Power," 94.

20. JoAnn McNamara and Suzanne Wemple, "The Power of Women Through the Family in Medieval Europe, 500–1100," *Clio's Consciousness Raised: New Perspectives on the History of Women*. Ed. Mary S. Hartman and Lois Banner (New York: Harper, 1974), 100, n. 69, cite Aimée Ermolaef, *Die Sonderstellung der Frau im französischen Lehnrecht* (Bern: 1930), 56ff.

21. Dorothy Atkinson, "Society and Sexes in the Russian Past," *Women in Russia,* ed. Dorothy Atkinson, Alexander Dallin, and Gail Wanshofsky Lapidus (Stanford: Stanford University Press, 1977), 4.

22. Ibid., 10.

23. Ibid., 13–17.

24. Stafford, *Queens,* 194.

25. Painter, *French Chivalry,* 144–145, citing William of Tudela's *La Chanson de le croisade contre les Albigeois.*

26. Pierce Butler, *Women of Mediaeval France,* Woman in All Ages and in All Countries, vol. 5 (Philadelphia: Barrie, 1907), 422.

27. P.D. Pasolini. *Caterina Sforza,* cited in L. Collison-Morley, *The Story of the Sforzas* (New York: Dutton, 1934), 265.

28. Vezio Melegari, *The Great Military Sieges,* trans. Rizzoli Editore (New York: Crowell, 1972), 131–133.

29. Duffy, *Siege Warfare,* 145, 251.

30. Ibid.

31. Antonia Fraser, *The Weaker Vessel: Woman's Lot in Seventeenth-Century England* (London: Weidenfeld and Nicolson, 1984), 163–181.

32. Ibid., 197.

33. Jean Bethke Elshtain, *Women and War* (New York: Basic Books, 1987), 167–169. Elshtain relates an incident from 1677 that illustrates the capacity of colonial women for physical violence. Natalie Zemon Davis discusses similar episodes in "Men, Women, and Violence: Some Reflections on Equality," *The Role of Women in Conflict*

and Peace, ed. Dorothy G. McGuigan (Ann Arbor: University of Michigan, 1977), 19–30.

34. Christopher Duffy, *The Military Life of Frederick the Great* (New York: Atheneum, 1986), 295–296; John Laffin, *Women in Battle* (New York: Abelard-Schuman, 1967).

35. Minna Moshcherosch Schmidt, *400 Outstanding Women of the World and Costumology of Their Time* (Chicago: Schmidt, 1933).

36. Julie Wheelwright, *Amazons and Military Maids: Women Who Dressed as Men in the Pursuit of Life, Liberty, and Happiness* (London: Pandora, 1989), 42.

37. Linda Grant De Pauw, *Founding Mothers: Women in America in the Revolutionary Era* (Boston: Houghton Mifflin, 1975), and "Women in Combat: The Revolutionary War Experience," *Armed Forces and Society* 7.2 (1980): 209–226; John Todd White, "The Truth about Molly Pitcher," *The American Revolution: Whose Revolution?*, ed. James Kirby Martin and Karen R. Stubaus (Huntington, NY: R. E. Krieger, 1977), 99–105.

38. De Pauw, "Women in Combat," 209–210. See also the interview with De Pauw in Lavinia Edmunds, "Beyond Molly Pitcher," *Johns Hopkins Magazine* 39.1 (1987): 14–17; De Pauw explains how she arrived at this figure, which she believes is "a conservative estimate."

39. De Pauw, "Women in Combat," 212–213; White, "Molly Pitcher."

40. Linda K. Kerber, *Women of the Republic: Intellect and Ideology in Revolutionary America* (Chapel Hill: University of North Carolina Press, 1980), 55.

41. De Pauw, "Women in Combat," 215–216.

42. De Pauw, "Women in Combat," 219. See Robert H. Land, "Margaret Cochran Corbin," *Notable American Women, 1607–1950* (Cambridge: Harvard University Press, 1971) for the standard account. Land states that Corbin "probably" served as nurse, cook, mender, and washerwoman for the troops; he notes the pension and clothing allowance that were granted.

43. Various sources spell her name "Sampson/Samson" and her alias as "Shurtliff/Shirtliffe."

44. Curtis Carroll Davis, "A 'Gallantress' Gets Her Due: The Earliest Published Notice of Deborah Sampson," *Proceedings of the American Antiquarian Society* 91.2 (1981): 319–323; Julia Ward Stickley, "The Records of Deborah Sampson Gannett, Woman Soldier of the Revolution," *Prologue: The Journal of the National Archives* 4 (1972): 233–241.

45. Sandra Gioia Treadway, "Anna Maria Lane: An Uncommon Common Soldier of the American Revolution," *Virginia Cavalcade* 37.3 (1988), 134.

46. De Pauw, "Women in Combat," 220. A genealogist concludes, instead, that these women must have been recruitment agents because there are no records that they ever fought. See Edmunds, "Beyond Molly Pitcher," 17.

47. Linda Grant De Pauw and Conover Hunt, *Remember the Ladies: Women in America, 1750–1815* (New York: Viking, 1976), 9.

48. Marjorie P.K. Weiser and Jean S. Arbeiter, *Womanlist* (Saddle Brook, NJ: Stratford, 1981), 145–146.

49. Ibid., 145; Laffin, *Women in Battle.*

50. C. Kay Larson, "Bonny Yank and Ginny Reb," *Minerva: Quarterly Report on Women and the Military* 8.1 (1990): 40.

51. Richard Hall, *Patriots in Disguise: Women Warriors of the Civil War* (New York: Paragon House, 1993), 112.

52. Clara Barton reported treating a soldier whose real name was Mary Galloway; **Sarah Emma Edmonds** said that she cared for a female soldier in disguise who died of her wounds. Eugene L. Meyer, "The Soldier Left a Portrait and Her Eyewitness Account," *Smithsonian* 1995: 96; Lauren Cook Burgess, ed., *An Uncommon Soldier: The Civil War Letters of Sarah Rosetta Wakeman, alias Pvt. Lyons Wakeman, 153rd Regiment, New York State Volunteers, 1862–1864* (Pasadena, MD: Minerva, 1994), xii.

Introduction

53. Helen Beal Woodward, *The Bold Women* (Freeport, NY: Books for Libraries Press, 1958), 284.
54. Brenda L. Moore, "African-American Women in the U.S. Military," *Armed Forces & Society* 17.3 (1991): 363–384.
55. Jeanne Holm, *Women in the Military: An Unfinished Revolution*, rev. ed. (Novato, CA: Presidio Press, 1992), 10–12.
56. Ibid., 17.
57. Jenny Gould, "Women's Military Service in First World War Britain," *Behind the Lines: Gender and the Two World Wars*, ed. Margaret Higonnet et al. (New Haven: Yale University Press, 1987), 118.
58. Maria Botchkareva, *Yashka: My Life as Peasant, Officer, and Exile* (New York: Frederick A. Stokes Company, 1919), 157.
59. Richard Stites, *The Women's Liberation Movement in Russia: Feminism, Nihilism, and Bolshevism, 1860–1930* (Princeton: Princeton University Press, 1978), 318, 321.
60. Helen Praeger Young, "Women at Work: Chinese Soldiers on the Long March, 1934–1936," *A Soldier and a Woman: Sexual Integration in the Military*, ed. Gerard J. DeGroot and Corinna Peniston-Bird Longman, 2000), 83–99.
61. Susan M. Hartmann, *The Home Front and Beyond: American Women in the 1940s* (Boston: Twayne Publishers, 1982), 35, quoting General John H. Hildring.
62. Debra L. Newman, "The Propaganda and the Truth: Black Women and World War II," *Minerva: Quarterly Report on Women and the Military* 4.4 (1986): 72–92; Martha S. Putney, *When the Nation Was in Need: Blacks in the Women's Army Corps during World War II* (Metuchen, NJ: Scarecrow, 1992).
63. Vera Semenova Murmantseva, *Zhenshchiny v soldatskikh shineliakh* (Moscow: Voenizdat, 1971), 9. See also Anne Eliot Griesse and Richard Stites, "Russia: Revolution and War," *Female Soldiers—Combatants or Noncombatants? Historical and Contemporary Perspectives*, ed. Nancy Loring Goldman (Westport, CT: Greenwood Press, 1982), 73. The Soviets consistently have given these figures for the past fifty years; however, precise statistical breakdowns are not available. Some corroboration is possible through examination of Komsomol histories (the Komsomol claims to have mobilized 500,000 women in five major mobilization drives) and through extensive reading of memoir literature. The opening of Soviet archives on this period should finally permit confirmation of the numbers. Even if the actual number of women who fought was far lower than that claimed (and there is no evidence to suggest that the Soviets have exaggerated in any significant way), it would still be an impressive figure, far outweighing the participation of women in any other country.
64. Charles Moskos, "Army Women," *Women in the Military*, ed. E.A. Blacksmith (New York: Wilson, 1992), 42.
65. Two useful studies are Ann Allen's Ph.D. dissertation on the WAC and Mary Stremlow's official history of women marines. See E. Ann Allen, "The WAC Mission: The Testing Time from Korea to Vietnam," Ph.D. dissertation, University of South Carolina, 1986; Colonel Mary V. Stremlow, *A History of the Women Marines, 1946–1977* (Washington, DC: History and Museums Division, HQ USMC, 1986).
66. Beverly Ann Bendekgey, "Should Women Be Kept out of Combat?" *Women in the Military*, ed. E.A. Blacksmith (New York: Wilson, 1992), 18.
67. Dean Koontz, "Chase," *Strange Highways* (New York: Warner, 1995), 430–551.
68. Contributor Tanya Lyons of the University of Adelaide provided information on women's participation in African revolutionary movements.
69. Barton Hacker, "Where Have All the Women Gone? The Pre-Twentieth Century Sexual Division of Labor in Armies," *Minerva: Quarterly Report on Women and the Military* 3.1 (1985): 107, 132.

ALPHABETICAL ENTRIES

A

ADELAIDE OF TURIN (fl. 1036; died 19 December 1091, Canavese, Italy). Countess of part of the March of Ivrea and of Turin in northwestern Italy. Reputedly donned armor to defend her inheritance.

Adelaide of Turin has often been compared to her close contemporary **Matilda of Tuscany** as an example of a powerful medieval woman. She was the oldest daughter and heir of Olderic Manfredi (of the Ardouin family) and Berta of Turin. She is often incorrectly associated with Susa. Her family, an antipapal dynasty, was always closely tied to the fortunes of the German emperors. Adelaide outlived her three husbands: Hermann, duke of Swabia (died 1038), Enrico degli Aleramici of Monferrato, and Oddone of Savoy, son of Humbert I the White-Handed (died 1060). She also survived her five children, who were all dead by 1080. From the time she inherited her paternal estates in 1035, she governed a vast territory, at times through regencies. Her death in 1091 left a power vacuum; various cities rebelled, and a war of succession began.

Contemporaries describe her virile strength and military aptitude. Her confessor Peter Damian eulogized her "masculine strength," writing that "like Debora, without male help, you supported the whole weight of the state" (Cognasso, 77). The chronicler Arnulf noted "the wisdom of the countess Adelaide, very much a soldierly mistress." In 1070 she used military force to impose a bishop of her selection upon Asti. Yet modern scholars seem to have overlooked her military capacity, focusing instead on her power and influence in the political and ecclesiastical spheres. Adelaide's military endeavors remain understudied.

Bibliography: Arnulf, "Gesta," 1848; Cognasso, *Storia di Torino*, 1959; Gherner, "La contessa Adelaide," 1992; Valeri, *Storia d'Italia*, vol. 1, *Il Medioevo*, 1965.

—Gloria Allaire

AETHELBURH (fl. early eighth century), also known as Aethelburg. Queen of Wessex, head of military forces, England.

Nothing is known of Aethelburh's origins or early life. She is usually identified as the wife of Ine, king of Wessex (688–726). A brief entry in *The Anglo-Saxon Chronicle* states that in 722 Aethelburh razed the fortress of Taunton in Somerset (built by her husband on the northern frontier of his kingdom). This event, which is the sole military action ascribed to Aethelburh, is not given a historical context by the *Chronicle*, nor is it mentioned in other sources. The reasons for Aethelburh's action, as well as her position with regard to Ine and the political situation in Wessex at that time, are therefore unknown. These unanswered questions have puzzled modern students of the early history of Wessex. Consequently, although various theories have been advanced in order to place the sack of Taunton in some kind of credible context, no strong consensus has yet emerged, largely due to the absence of a coherent historical record of Ine's reign.

In 722, according to *The Anglo-Saxon Chronicle*, a man named Ealdberht "the Exile" seems to have risen in revolt against Ine. Aethelburh's destruction of Taunton in the same year may have been an initial military response to the situation on Ine's behalf, perhaps because Ealdberht had managed to seize Taunton for himself. One scenario is that Ine may have been embroiled at that time in a war with the Britons on his southwestern border, leaving responsibility for defending the realm in Aethelburh's hands. Aethelburh perhaps ousted Ealdberht from Taunton but may have had insufficient troops with which to garrison such an outlying fort, deciding instead to raze it lest it fall again into the hands of an enemy. The *Chronicle* reports that Ealdberht fled to Sussex, where in 725 he was slain in battle with Ine.

An alternative scenario is that Aethelburh's destruction of Taunton was unconnected with the actions of Ealdberht but was rather a response to a Welsh or British attack from the north or west in which Taunton was seized. This would fit well with the tradition that during Ine's reign Wessex was engaged in fighting on all fronts. Yet another possibility is that Aethelburh herself, like Ealdberht, rose in revolt against Ine, an explanation that at least provides a simple military motive for her destruction of the king's recently built border fortress at Taunton. Such a revolt could have arisen after a major dynastic or internal political crisis in which Ine and his queen stood on opposing sides. It has been additionally suggested that Aethelburh might not in fact have been Ine's wife but should instead be envisaged as the head of a disaffected or rival branch of the royal kindred, proclaimed as queen and military leader by a rebellious faction. The latter scenario contradicts the strong tradition, dating back to at least the time of William of Malmesbury in the twelfth century, that Aethelburh and the king were always on good terms.

Despite the paucity of information relating to the sack of Taunton, the very fact that it received mention in *The Anglo-Saxon Chronicle* may indicate that it was perceived by the chroniclers as an event of some significance in the early history of Wessex. Whether or not its significance was due in part to the active military role played by a woman is unknown, but it is worthwhile to note that Aethelburh's action appears exceptional, given that contemporary reports from Merovingian Gaul seem to imply that queens involved in siege warfare were usually fulfilling a defensive rather than an aggressive function.

Bibliography: *The Anglo-Saxon Chronicle,* 1954; Cunliffe, *Wessex to AD 1000,* 1993; Major, *Early Wars of Wessex,* 1913; Reed, *The Rise of Wessex,* 1947; William of Malmesbury, *De gestis regum Anglorum,* 1887.

—Tim Clarkson

AETHELFLAED OF MERCIA (born ?; died 12 June 918, Tamworth, England), also known as Ethelfleda, Athelflad, and so on, Edelfrida in Irish sources. "Lady of the Mercians." Military leader of Mercia during the Scandinavian invasions of England, from 911 or earlier until her death.

Aethelflaed of Mercia fortified the Mercian frontiers, erecting defensive earthworks and garrisoning ten towns. In the offensive against the Midlands Danes, she gained control of two of their five major strongholds. She launched a successful expedition against the Welsh king of Byrcheiniog. According to Irish annals of doubtful authenticity, her strategic direction was responsible for the defeat of Dublin **Vikings** at Chester and at Corbridge (in the north). The latter victory won her widespread fame.

The brief outline of Aethelflaed's activities in the *Mercian Register* is the only contemporary account. Anglo-Norman chroniclers from William of Malmesbury onwards trivialized her achievements. William recognized, however, her vital contribution to the eventual conquest of the Danelaw by her brother, Edward of Wessex. The near absence of reference to her in the Wessex (Parker) chronicle is attributable to Edward's suppression of Mercian autonomy after her death. Frequently described by later writers as a "formidable" leader, she is, simultaneously, celebrated for her loyal subordination of Mercian interests to Edward's advancement of Wessex supremacy.

Aethelflaed was the eldest child of Alfred the Great of Wessex. Through her mother, Ealhswith, she was descended from the Mercian royal house. Presumably her childhood was shaped by Alfred's campaigns against the Danes, which resulted in his recovery of Wessex, but left much of Mercia in Danish control. Some time after 886 Alfred gave Aethelflaed in marriage to his fellow campaigner, Ealdorman Aethelred, ruler of Mercia.

When Aethelred died in 911, Aethelflaed was accepted by the Mercians as sole ruler and acknowledged military leader; however, she was effectively in command some time before 911. The *Mercian Register* dates her construction of forts from 910. The Irish annals suggest that Aethelred was incapacitated by 905 and perhaps as early as 902. Charters reveal her involvement in Mercia's defense in the 890s. The fortification of Worcester was jointly ordered by Aethelflaed and her husband (BCS579), and both were party to discussions with Alfred on the refurbishment of London (BCS577). Some regard her as continuing Aethelred's policies. Others stress their joint adherence to Alfred's vision of a Wessex-dominated united front and his recognition of the strategic value of the fort as an offensive and defensive unit.

Aethelflaed's fortresses, erected 910–916, joined up with her brother Edward's to form a line running diagonally across England from coast to coast, which prepared the ground for the 917–918 offensive against the Midlands Danes. The 916 expedition into Wales suggests that Bridgnorth (912) and Chirbury (915) were directed against the Welsh. Eddisbury (914) and Runcorn (915) were probably a response to the Dublin Viking threat.

In 917, while Edward was heavily engaged in fighting to the south, Aethelflaed captured the Danish stronghold at Derby. It has been suggested that she took advantage of the garrison's engagement in the south, but her success involved heavy fighting, and the reputation of her army may account for the fact that when Edward attacked Stamford in 918, Leicester surrendered to her without a fight. This facilitated the capture of the remaining Danish strongholds. Aethelflaed's sudden death deprived her of the chance to share in Edward's victory six months later. Subsequently Edward annexed Mercia and deposed Aethelflaed's daughter (his niece), whom the Mercians had accepted as her successor.

The *Mercian Register* leaves no doubt that the fortification of Mercia took place under Aethelflaed's supervision (e.g., she "went to Tamworth with the whole Mercian army and built the fortress there"). Her role in framing the strategy employed against the Midlands Danes is an open question. Wainwright concludes that she may merely have been a loyal agent of her brother's policy, but Stafford intimates that her willing acceptance of Mercia's subordination to Edward's quest for control of all England has been too readily assumed. Edward's emergence as the ultimate beneficiary of Aethelflaed's success was not the result of a strategic master plan on his part, but depended on two fortuitous circumstances: her untimely death and the absence of a male Mercian heir. Further, the defense of the frontiers was initiated by Aethelflaed's fortification of Bremesbyrig in 910.

Wainwright, however, does regard Aethelflaed as the architect of victory in the north. His case rests on his handling of the authenticity of the Irish annals. They relate that Aethelflaed repaired the fortifications at Chester and garrisoned her armies there, in anticipation of an attack by Irish-Norwegian settlers. In the account of the battle (ca. 905) Aethelflaed and her bedridden husband are presented as joint authors of the tactics employed by the garrison. In the account of the 918 victory at Corbridge against Dublin Vikings led by

Ragnall, however, she appears as sole strategic commander. Wainwright dismisses the details of both accounts as legendary, but regards the annals as evidence that she was the acknowledged leader of an alliance concluded between Ragnall's opponents, including the Britons and Scots. His claim is supported by the *Mercian Register*'s statement that Danish-occupied York submitted voluntarily to Aethelflaed shortly before her death.

Aethelflaed's active presence on the battlefield is undocumented. Anglo-Saxon sources affirm that the overlord's presence on the battlefield was vital to morale. But evidence is similarly lacking for the military involvement of many of Aethelflaed's male counterparts, and most historians assume that she participated in expeditions, whether as commander or combatant. A few suggest that Aethelflaed's nephew Aethelstan, Edward's adolescent son who was entrusted to her upbringing, must have been the real power behind her military achievements. More plausibly, like her male counterparts, Aethelflaed depended on the experience of her veterans and the caliber of her troops, not on the advice of a teenager. Her achievements rested upon her ability to inspire the cooperation that characterized her own collaboration with her brother, whether from loyalty or from intelligent self-interest.

Bibliography: Charters (BCS): Birch, *Cartularium Saxonicum*, 1885–1893; Irish Annals: Radner, *Fragmentary Annals of Ireland*, 1978; Mercian Register: *The Anglo-Saxon Chronicle*, 1954; Stafford, *Unification and Conquest*, 1989; Wainwright, "Aethelflaed, Lady of the Mercians," 1959.

—Stephanie Hollis

AGNES OF DUNBAR (born ca. 1300; died ca. 1369 in Scotland), also known as Agnes Randolph, "Black Agnes" because of her dark complexion. Countess of Dunbar and March; later countess of Moray. Defended Dunbar Castle, Scotland, during the Second Scottish War of Independence, 1332–1341.

Agnes of Dunbar led the defense of Dunbar Castle against a vastly superior force of English soldiers in 1338. Through her efforts the castle held out for five months until the siege was raised. Such actions were not untypical for aristocratic women of the period; the previous countess of Dunbar had also defended the castle against the English in 1297. However, the bravery and panache with which Agnes carried out the defense caught the attention of contemporary chroniclers and of Scottish historians for centuries afterwards.

Agnes's family was active in Scottish resistance to English attempts to conquer the country in the fourteenth century. Her father, Thomas Randolph, earl of Moray, was regent from 1329 to 1332. Her brother John became joint regent in 1335, but was captured by the English shortly afterwards. By 1324 Agnes had married Patrick, ninth earl of Dunbar and March, governor of Berwick. When Berwick was occupied by English forces in 1333, Patrick went over to the English side, and Edward III granted the couple English lands. Patrick gained permission from Edward to refortify his castle of Dunbar near Edinburgh and probably used grants of money from the king for this purpose. The following year he switched back to Scottish allegiance. From Dunbar Castle, now strongly fortified, men harried Edward's supporters throughout southeast Scotland, provoking an English invasion. Already angry at her husband's changing loyalties, Edward was also suspicious of Agnes. When she wrote to her captive brother in the Tower of London in 1337, he ordered her letter examined for anything "sinister." Edward was determined to take Dunbar and end the troublesome activities of the earl and countess; in 1338 he sent William de Montagu, earl of Salisbury, and a force of 20,000 men to besiege the castle. In her husband's absence, the defense of the castle was conducted by Agnes.

The besiegers arrived on 13 January 1338 and soon cut off the castle both on land and with two galleys at sea. Agnes was not intimidated and throughout the siege taunted the attackers. When stones were hurled against the castle walls, she and her ladies appeared on the battlements with white handkerchiefs to wipe off the dust from their impact. A machine called a "sow" was built to act as a shield to bring the English soldiers up to the walls of the castle. The *Liber Pluscardensis* reports that Agnes taunted Montagu, saying, "For all the power you may have, I shall shortly cause your sow to farrow against her will." Her soldiers dropped a huge rock on the machine, crushing many of those below and sending the rest out squealing like pigs.

Montagu tried to persuade Agnes to surrender by bringing her captive brother to the siege and threatening to kill him. Unmoved, she replied that the castle was her husband's and she could not surrender it without his permission. Moreover, if her brother was killed, she would inherit the earldom of Moray. Montagu next tried to win the castle by stealth, bribing a guard to open a gate at night. The guard took the money, then informed his mistress, who set an ambush. When Montagu's men crept into the castle, they were surrounded by soldiers, and the portcullis dropped behind them. Montagu escaped just before the portcullis closed, and fled with Agnes's derisive words "Adieu, adieu, Monsire Montagu" ringing in his ears.

With the help of local allies, Agnes was able to arrange to smuggle goods into the castle at night and keep her forces well supplied. Finally, on 16 June the besiegers used the excuse of joining Edward III's expedition to France at the opening of the Hundred Years' War to retire. A truce was arranged, but

---

An unknown female Flemish mercenary fought in the English army during its invasion of Edinburgh 1335 (Battle on the Borough Moor, Second Scottish War of Independence):

For this campaign the English had hired numerous bands of foreign mercenaries, and in particular a large body of Flemish soldiers of fortune, steel-clad cavalry, commanded by Guy, Count of Namur. . . . while the conflict was at its height, a Scottish esquire, named Richard Shaw, and a Flemish warrior, both of whom were conspicuous for valour, challenged each other to combat in the full view of the field. They set their spears in rest and clapped spurs to their steeds . . . at the first crash of encounter they transfixed one another with their lances, and both falling from their saddles, gasped out their lives on the bloody turf. . . . when the victors came to scan the scene of contest on the Borough Moor, they discovered to their astonishment that the Flemish soldier who fought and died with Richard Shaw was a woman. Of her name and history nothing was ascertained.

—Robert Scott Fittis, *Heroines of Scotland* (London: Alexander Gardner, 1889), 41–42.

the English never returned. Agnes Randolph had won. The siege of Dunbar had cost the English crown almost £6,000 and gained it nothing. On the earl of Moray's death in 1346, the earldom passed to the Crown, but Agnes and her husband were allowed to take the titles of earl and countess of Moray, thus fulfilling Agnes's earlier words to Montagu. She died about 1369.

Bibliography: Androw of Wyntoun, *Orygynale Cronykil of Scotland*, 1872–1879; Bain, *Calendar of Documents Relating to Scotland*, 1881–1888; *The Chronicle of Lanercost*, 1913; Graham, *A Group of Scottish Women*, 1908; *Liber Pluscardensis*, 1876.

—Elizabeth Ewan

ALBRET, JEANNE d' (born 16 November 1528, Saint-Germain-en-Laye, France; died 9 June 1572, Paris, France). Queen of Navarre during the Wars of Religion, sixteenth-century France.

Jeanne d'Albret, queen of Navarre from 1555, became involved in the Wars of Religion on the Protestant side upon the death of her husband Antoine de Bourbon; she fought without respite to defend her kingdom and her faith and to protect the interests of her son, who would one day become Henry IV, king of France.

Jeanne's father, Henry II d'Albret, never lost hope of regaining Upper Navarre, confiscated by Spain in 1512. Her mother, Marguerite d'Angoulême, a sister of Francis I of France, was a patron of the Reformers. These two areas of conflict were the driving forces of Jeanne's political life. Agrippa d'Aubigne said that Jeanne was "a woman only in body, her entire soul is manly, her heart invincible to adversity." At age twelve she defied Francis I in refusing to marry the duke of Cleves. At age twenty she was married to Antoine de Bourbon, first prince of the blood, whose lineage placed him in the front ranks of those eligible for the French crown if the Valois line failed. She became queen of Navarre two years after the birth of her son Henry. Her immediate challenge was to maintain the independence of her small realm (Béarn and Lower Navarre), nestled between France and Spain, at a time of increasing religious troubles. She supported the Reformers in Béarn and converted to Calvinism, but could not prevent her husband from returning to Catholicism.

Portrait of Jeanne d'Albret. (Perry-Castañeda Library)

Antoine de Bourbon was killed in combat the same year that the Wars of Religion (1562–1598) began. With his death, Jeanne became, on behalf of her nine-year-old son, ruler of the last great French feudal state. From that time on she involved herself more and more openly in battle, aligning herself against **Catherine de Médici,** who was to a certain extent her counterpart on the Catholic side.

From 1562 she had as an adversary Monluc, who was charged by Catherine de Médici with maintaining order in Guyenne. Monluc was eager to invade Béarn and also publicly declared his desire to find out "if it was as good to go to bed with queens as with other women." But Jeanne escaped from him and after 1563 slowly imposed Calvinism in her kingdom, despite having to subdue a number of rebellions. Confronted with increased threats from France against herself and her son, she took refuge in 1568 at La Rochelle, where with Condé she directed the war against the army of King Charles IX. When Condé was killed in combat at Jarnac in 1569, she became the true leader of the Protestants, even if officially her young son Henry of Navarre and Henry of Condé (under the authority of Coligny) were chiefs. That same year she repulsed an attempted invasion of her kingdom by the Catholics, and despite the Peace of Saint-Germain, the Protestants remained on a war footing.

At the beginning of 1572 Jeanne agreed to the marriage of her son to **Marguerite de Valois,** daughter of Catherine de Médici and sister of Charles IX. Jeanne traveled to Paris for the ceremony, but died there of tuberculosis two months before the celebration of the marriage. One rumor, revived and augmented by Alexandre Dumas in *The Queen Margot* (1845), made Catherine de Médici responsible for her death by the agency of poisoned gloves. As for Henry, he escaped the St. Bartholomew's Day Massacre (24 August 1572) by agreeing to convert to Catholicism. He remained essentially a prisoner at court until 1576 but later prevailed against his competitors and reigned as King Henry IV.

Jeanne d'Albret was considered, along with Catherine de Médici, one of the most remarkable political personalities of her time. A heroine to Protestants and a dangerous heretic to Catholics, Jeanne d'Albret played a key role in the Wars of Religion.

Bibliography: Albret, Jeanne d', *Lettres d'Antoine de Bourbon et de Jehanne d'Albret,* édition par le marquis de Rochambeau (Paris: 1877); Roelker, Nancy Lyman, *Jeanne d'Albret, reine de Navarre (1528–1572)* (Paris: 1979); Roman d'Amat, Charles, *Dictionnaire de biographie française,* tome 18 (Paris: 1991); Ruble, Alphonse de, *Antoine de Bourbon et Jeanne d'Albret,* 4 vols. (Paris: 1881–1886), and *Jeanne d'Albret et la guerre civile* (Paris: 1897).

—Marie-Thérèse Lalaguë-Guilhemsans (trans. by Annette Parks)

ALEKSEEVA, EKATERINA (born 1895; died ?, Russia), also known as Aleksei Sokolov. Infantry, Russia, First World War.

Ekaterina Alekseeva, a domestic servant of peasant origin, disguised herself and volunteered for active duty as soon as mobilization commenced in August 1914. Her strong and stocky physique allowed her to enter a reserve battalion as Aleksei Sokolov with relative ease. Motivated to join the fighting by strong patriotic and religious fervor and imbued with a highly romanticized conception of war, Alekseeva went into combat seeking adventure and excitement. She participated in a number of battles until she was wounded at the end of 1914 and medical attention exposed her true sex. Alekseeva's male comrades had been impressed by her courageous and capable performance in combat, especially after learning of "Aleksei's" true identity, and they treated her with respect and care. Alekseeva's further activities after convalescence remain unknown.

Bibliography: "Odna iz mnogikh," *Zhenskoe Delo* 15 January 1915: 19–20.

—Laurie Stoff

ALGERIAN-FRENCH WAR, WOMEN IN. Algeria, 1954–1962.

Algerian women engaged in a wide variety of combat and combat-related activities during their war of independence against the French. Although only one woman is positively known to have served formally as a regular, full-time combatant, female guerrillas numbered in the hundreds. These women served as armed fighters in the countryside, urban bomb placers, guides, and arms and supply-cache supervisors. Other women served as nurses as well as in such support roles as cooks and secretaries for the independence movement. One of the most famous was Djamila Bouhired (Jamilah Buhrayd), who was captured and tortured by the French in the late 1950s.

The role of women in the Algerian-French War has acquired an almost mythic status both in Algeria and internationally. Archival work by National Liberation Front veteran Djamila Amrane has begun the process of separating fact from myth on the subject of the actual extent of women's participation in the war.

Bibliography: Amrane, *La femme algerienne,* 1980; Coughlin, "Women, War, and the Veil," 2000; Helie-Lucas, "Women, Nationalism, and Religion in the Algerian Liberation Struggle," 1990; Ladewig, "Between Worlds: Algerian Women in Conflict," 2000.

—William Henry Foster

ALICE DE MONTMORENCY (died ca. 1221). Military organizer and recruiter, France, Albigensian Crusade (1209–1218).

Though overshadowed by her famous husband, Simon de Montfort (leader of the Albigensian Crusade), Alice de Montmorency deserves to be called a lieutenant of the crusade in southern France. Montmorency assisted her husband and hence the crusade in several key ways. Because only forty days' service was required in the south to gain the papal indulgence that was the main motivation for crusade volunteers, the crusade faced constant personnel shortages as individuals and groups served the required time and then returned home. Montmorency served as a recruiter in the north of France and acted as supply master for the crusade, ferrying troops and supplies from the north to various sieges and other locations in Languedoc. She regularly attended baronial councils, dispensing advice to her husband, and more than one source notes that she personally pleaded with French nobles on several occasions to remain in the south with their contingents longer than forty days.

Montmorency was present on several important campaigns during the endemic siege warfare of the crusade, among them Termes in 1210 and the second siege of Toulouse, 1217–1218. She was close by when her husband won the battles of St. Martin-la-Lande in 1211 and Muret in 1213. Like her contemporary **Giralda de Laurac,** there is no record that Montmorency participated directly in combat. Montmorency represented an exemplar of female conduct of the early thirteenth century and went beyond it by providing consistent, direct military assistance. She was the most important of Simon de Montfort's lieutenants on the crusade, his best and most constant helper. Montfort would have lost in the south long before had Alice de Montmorency not aided him so effectively. Ironically, Simon de Montfort was reportedly killed during the siege of Toulouse (1218) by a lucky shot from a female artillery crew. The anonymous narrator of *Song of the Cathar Wars* wrote that "a stone was launched. It fell right on the steel helmet of Simon de Montfort. His forehead burst open, his jaw shattered, his brains and eyes spurted from his head. The count, blood-soaked, fell to earth. He was dead."

Bibliography: Sumption, *The Albigensian Crusade*, 1978; Vaux-de-Cernay, *Historie albigeoise*, 1951; William of Tudela and Anonymous, *The Song of the Cathar Wars*, 1996; Zerner, "L'épouse de Simon de Monfort et la croisade albigeoise," 1992.

—Laurence W. Marvin

ALLEN, ELIZA (born 27 January 1826, died?), also known as Eliza Billings, George Mead. Volunteer soldier, United States Army, Mexican War.

Known only through an autobiography published in 1851, Eliza Allen served as a private soldier in the Mexican War. She participated in the Battles of Monterrey and Cerro Gordo and was among the American soldiers who occupied Mexico City.

According to her autobiography, Allen was born to wealthy parents in 1826 in Eastport, Maine. At the age of twenty she fell in love with William Billings, a respectable but poor young man. Allen's parents opposed their marriage, and Billings left, joining a volunteer company to fight in the war against Mexico.

Allen had read about **Deborah Samson,** who had disguised herself as a man and fought in the American Revolution, and **Lucy Brewer,** who had similarly fought in disguise as a marine on the frigate *Constitution* during the War of 1812. Emulating these models, Allen cut off her hair, dressed in men's clothing, and sailed to Portland to enlist as a volunteer under the name George Mead. She met an officer who needed one last man to fill his company and therefore waived the usual physical examination. The company sailed to the Rio Grande, where it joined General Zachary Taylor's army. Allen participated in the American victory at Monterrey in September 1846. Afterwards her company was transferred to the army of General Winfield Scott at Vera Cruz. During the advance on Mexico City, Allen participated in actions against Mexican guerrillas. Her left arm was slashed in the Battle of Cerro Gordo. While being treated, she found William Billings. Allen was not able to rejoin her company until the fall of Mexico City. Her last military action was a small street skirmish. Allen then participated in the occupation of Mexico City until the peace in 1848. Her unit sailed, via New Orleans, to New York, where it was paid off and discharged. However, Allen continued her disguise, working as a man to help Billings until they achieved financial security in 1849, when they married.

Allen's autobiography resembles other nineteenth-century memoirs about women succeeding in male society while disguised as men; such books were popular among women at the time. While there is no independent confirmation that Allen served in combat, other women are documented as having served on both sides during the Mexican War, and their exploits echo those listed by Allen.

Bibliography: Billings, *The Female Volunteer,* 1851; Johannsen, *To the Halls of the Montezumas,* 1985; Salas, *Soldaderas,* 1990.

—Tim J. Watts

ALRUDA FRANGIPANI (died 1177). Countess of Bertinoro. Military leader, Italy, Byzantine-Venetian conflict (1170–1177).

On horseback Alruda Frangipani led an army that helped liberate Ancona from imperial siege in 1172. Her heroic action and warlike spirit were recorded by Muratori, Levati, and Givhiano Saraceni. A member of the powerful Frangipani family of Rome and wife of Bertinorese count Rainier Cavalconte, Alruda was widowed in 1144 and afterwards ruled the city of Bertinoro for many years as regent for her children. Her court was elegant and refined. Alruda also had close ties with a female relative of the Eastern emperor Manuel Comnenus.

The port city of Ancona had come under the control of Comnenus and was a political threat to the German emperor Frederick Barbarossa. In 1172 or 1173 Ancona was subjected to a lengthy siege by Archbishop Christian of Mainz to prepare the way for an invasion by the Germans. The normally anti-imperial Venice sided with the archbishop against Ancona and blockaded the port. Bertinoro and Ferrara were both in the anit-imperial Lombard/Romagnolo confederation that rallied to help its ally Ancona. Alruda led troops from Bertinoro, while Guglielmo di Bulgaro Marchesella degli Adelardi, of an important Guelf family, led forces from Ferrara. Alruda apparently harangued the troops before battle. She then bluffed the besieging army into withdrawing by a clever military strategy: her troops attached two or three torches to each lance and marched in a wide formation, giving the appearance of a much larger army. Although combat was avoided, this encounter took on the aura of legend and has been compared to the Battle of Legnano (1176). It was an important victory for the papal supporters and for the doomed Eastern Roman Empire and testified to the growing strength of small localities versus the powerful Western emperor.

Bibliography: Gatti, *Bertinoro: Notizie storiche*, 1968–1971; Polverari, *Ancona e Bisanzio*, 1992; Salvatorelli, *L'Italia comunale*, 1940; Saracini, *Notitie historiche della città d'Ancona*, 1968.

—Gloria Allaire

ALUZHEN (fl. 1217, China). Mistress Commander Duke. Commander of a local Jurchen troop during the Jin dynasty (1115–1234), founded by the Jurchen (the Tungus tribes of the modern Heilongjiang Province). Leader, military commander, China.

Aluzhen was the daughter of Chencong, a remote relative of the royal family. In 1217 her father, marshal of Shangjing, was captured during the rebellion of Puxian Wannu (died 1233). In preparation for a siege Aluzhen, a widow who commanded more than 1,000 soldiers, strengthened the ramparts, repaired armories, and stored supplies for self-defense. Puxian Wannu asked a mediator to call for her surrender, but she refused. Thereupon he shot a letter written by her father into the ramparts, but Aluzhen tore it into pieces and said that it was a fake meant to trick her. She put on men's clothes and, with her son, led her troops in fierce fighting against Puxian Wannu, killing several hundred and capturing more than a dozen of the enemy before Wannu's troops retreated and dispersed. Later she sent other officers to assault Puxian Wannu and captured one of his commanders. For these military actions the emperor conferred upon her the title of Mistress Commander Duke.

Bibliography: Chen Menglei, *Gujin tushu jicheng (Qin ding)*, 1993; Tuotuo et al., *Jin shi*, 1975.

—Sherry J. Mou

AMAZONS. Legendary nation of women warriors who reportedly lived in the Black Sea region during the Greek Bronze Age (ca. 1600 BCE–ca. 1100 BCE) and on the fringes of the known world during the historic period.

Most ancient writers localized the original homeland of the Amazons along the northern Black Sea, between the Sea of Azov and the Caucasus Mountains. The Amazons later migrated to the Thermodon River region on the southern littoral of the Black Sea (modern northeastern Turkey), where they founded a kingdom ruled by queens. From their capital city of Themiscyra, they staged attacks on various areas of the eastern Mediterranean and were said to have founded many cities along the Aegean coast of modern-day Turkey. Other accounts place the original homeland of the Amazons along the Thermodon River.

Their chief deities were the war god Ares (from whom they claimed descent), Taurean Artemis, and the mother earth goddess Cybele. Their primary pursuits were hunting and war. Fierce warriors who gave no quarter, the Amazons were especially famous for their skill with the bow. According to many ancient writers, Amazon girls had their right breasts seared in childhood so as not to interfere with their drawing a bow, although this seems unlikely. The idea probably came from the etymology of the term "Amazon," which many historians believe is derived from the Greek *a* (without) *mazos* (breast).

Scene of an Amazon from a frieze at the Mausoleum at Halicarnasus. (Library of Congress)

The society was supposedly ruled by a dual monarchy, one queen to command the armies and one to rule at the capital. The Amazons fought both from horseback (according to legend, the first peoples to do this) and on foot, were skilled in the use of the double-headed battle axe (*labrys* or *sagaris*), bow and arrow, and short sword, and carried a crescent-shaped shield; they apparently adopted the lance after combat against the Greeks.

The Amazons appear frequently in the exploits of various legendary Greek Bronze Age heroes. When the Amazons invaded Lycia, Bellerophon helped King Iobates drive them out of his kingdom. Jason and the Argonauts on their voyage to Colchis in search of the golden fleece passed along their territory, and Heracles stole the golden sword-belt of Hippolyte, the Amazon queen, during one of his twelve labors. Theseus, who had accompanied Heracles to Themiscyra, later returned and abducted Antiope, the sister of Hippolyte. In retaliation, an Amazon army invaded Greece and laid siege to Athens. Almost successful, they were finally defeated and driven from Greece. The Trojans under the leadership of the young Prince Priam, future king of Troy, also fought against the Amazons when they attacked the Phrygians. During the Trojan War, however, the Amazons came to the aid of King Priam against the Greeks, and Penthesilea, the Amazon queen, died in combat against Achilles after she had killed several Greek heroes.

The position of men in Amazon society is not clear. According to some ancient Greek writers (Diodorus Siculus, 2.45ff.), Amazon society was a matriarchy: the women were the rulers and warriors, the men took care of domestic needs. Society and politics as the Greeks knew it were reversed. Other writers, such as Strabo (*Geography*, 11.5ff.), gave a radically different model of Amazon society: in this account, which was the popular version in antiquity, men played no role in Amazon society. Only for the purpose of procreation did Amazon women associate with men. They passed two months each year in the Caucasus Mountains near the territory of the Gargareans, and by night the Amazon women would mate indiscriminately with Gargarean males. Female children born of these unions were raised by their mothers; male children were either sent back to the Gargareans to be raised, were maimed, or were killed.

With the end of the Age of Heroes and the dawn of written history, the Amazons were no longer located in the Thermodon region. According to Herodotus (*Histories*, 4.110ff.), the Amazons were almost wiped out by Greek raiders, who took the surviving Amazons on board their ships. The captives rebelled and killed their captors, beaching the ships near the Sea of Azov in the Crimea region of the northern Black Sea. Initially fighting the Scythian tribesmen who lived there, the Amazons eventually took husbands from among the Scythians and moved beyond the Don River to the steppes of southern Russia and Kazakhstan, where they established an independent kingdom. In this society women shared equal rights with the men, who fought alongside the Amazon women. Herodotus writes that the offspring of these Scythian men and Amazon women were the ancestors of the Sauromatians, who in the historic period ranged from the northern Black Sea region of the Ukraine to northwestern Uzbekistan and whose women were reported to be skilled in the arts of war and horsemanship. In other accounts the Amazons, under military pressure from their neighbors, migrated north or northeast of the Black Sea, but retained their traditional female society.

Throughout the Greco-Roman era many continued to believe in the existence of Amazon warrior women in Asia, and stories of encounters with Amazons in various parts of Asia appeared during the historic period. For example, in the late fourth century BCE Alexander the Great was said to have met the Amazon queen on the frontiers of Afghanistan, and the Romans believed that the Amazons had fought against them on the side of Mithridates during their campaigns in the Black Sea area during the first century BCE. On the maps of Roman imperial cartographers, such as Pomponius Mela, one could locate the Amazons in the lands to the north and northeast of the Black Sea.

The Amazons and especially the Amazonomachy (Amazons battling Greeks) were also favorite subjects of representation by sculptors and painters in antiquity. For example, the Amazonomachy appeared on the shield of Athena Parthenos's statue as well as on the metopes of the Parthenon on the Athenian acropolis. Other commonly appearing themes were Heracles fighting the Amazon women and Achilles battling Penthesilea. The ancient Greeks built numerous monuments to the Amazons throughout Greece.

What gave rise to the stories of the Amazons has been the subject of much discussion. In the nineteenth century many scholars accepted as fact that matriarchal societies had actually existed in the distant past and that the stories about warrior women who controlled their own societies, such as the Amazons, likely had a historical basis. Today very few specialists believe in the existence of ancient matriarchies, and many scholars now think that the stories about the Amazons are wholly mythical. Yet recent archaeological excavations in the steppes of Kazakhstan and southern Russia now suggest that there may in fact be some kernel of truth behind at least part of the stories concerning the Amazons: namely, that women were trained as warriors in the Black Sea area. Russian and American archaeologists, while excavating Sauromatian-era graves (ca. 600 BCE–ca. 400 BCE), found that some women were buried with swords, daggers, bronze arrowheads, quivers, and whetstones for sharpening their weapons. Bowed leg bones indicate that these women spent long hours on horseback; some female remains show evidence of wounds and perhaps death inflicted in combat.

While graves of warrior women do not demonstrate the existence of a matriarchal society—there were far more male warrior graves than female warrior graves—they do give credence to the stories that Sauromatian women were trained in the arts of war and horsemanship and fought and rode with their husbands in war gear on the steppes of modern Ukraine, southern Russia, Kazakhstan, and Uzbekistan. Although these graves are later by several hundred years than the earliest literary references to Amazons in Homer's *Iliad*, it is possible that during the Bronze Age women of the steppes had also been trained to ride and fight alongside their husbands. Stories of warrior women brought back to Greece by traders and travelers to the Black Sea area could have been the building blocks for future Greek storytellers who created the tales of the Amazons handed down to us. Interestingly, one Russian writer believes that the term "Amazon" is a Greek adaptation of the Slavic name *omuzhony* or "masculine women." Given the archaeological evidence for the historical existence of Slavic "Amazons," this might be a more logical origin for the term than the standard "breastless" Greek derivation.

Bibliography: Blok, *The Early Amazons*, 1995; Bothmer, *Amazons in Greek Art*, 1957; Davis-Kimball, "Warrior Women of the Eurasian Steppes," 1997, and *Warrior Women: An Archaeologist's Search for History's Hidden Heroines*, 2002; Sulimirski, *The Sarmations*, 1970; Tyrrell, *Amazons*, 1984.

—Ralph F. Gallucci

AMERICAN CIVIL WAR, WOMEN IN. United States, 1861–1865.

Women served in a wide variety of roles in the American Civil War. Thousands filled traditional roles as nurses and **vivandières,** while others engaged in espionage. A unique female role was that of "daughter of the regiment." These young women were revered by their regiment as guardian

Fredericksburg, Virginia. Nurses and officers of the U.S. Sanitary Commission. Photograph from the main eastern theater of war, Grant's Wilderness Campaign, May–June 1864. Photo by James Gardner. (Library of Congress)

angels or symbols of good luck; they served as color bearers and nurses and carried water to the wounded and dying on the battlefield. Although their role was that of a mascot, they often marched and ate with their units, enduring all the same privations as the soldiers, and many went onto the battlefield to help motivate their comrades. In the Union army, in particular, there were large numbers of women who marched with their husbands or lovers or served alone. Many of the volunteer units that served in the Union army insisted on permitting some women to accompany them to war, especially their designated "daughters of the regiment."

Other women fought as soldiers, most often in male disguise. Many of the women who served as Civil War soldiers have been lost to history, in part because they fought in disguise and were unable to prove their service in later years. Many women refrained from mentioning their wartime experiences in order to avoid possible scandal that might have tarnished society's perceptions of their femininity and honor. Aside from a few published memoirs, most of the information on these women appears in scattered newspaper and eyewitness accounts. One of the greatest risks for a woman serving in disguise was to end up in a hospital. A Mary Scaberry, who had been serving under the name "Charles Freeman," was expelled from an Ohio regiment when she was hospitalized for fever and her true identity was discovered; the reason given for her discharge was "sexual incompatibility." Nurses like Clara Barton and Mary Livermore both reported discovering wounded soldiers who turned out to be women in men's uniforms; for example, Barton said that she nursed a soldier at Antietam who was at first reluctant to be treated, and "he" turned out to be a Mary Galloway.

Soldiers on both sides note having captured enemies who turned out to be women. Many passing references are found in letters and newspaper stories. For example, in a letter dated 29 May 1864, A. Jackson Crossley wrote from the Army of the Potomac headquarters to his friend Samuel Bradbury that a woman dressed in male clothes had been captured among the Confederate troops who "was mounted just like a man and belonged to cavalry though she was taken as a spy; she wore her hair long and did not like to have our men looking at her." Similarly, a Florence, South Carolina, newspaper reported that a "Florena Budwin was brought to Florence from Georgia along with other Federal prisoners captured by Confederates in an encounter. She wore the uniform of the Union soldier and it was not for some time that it was found she was a woman." According to the *Richmond Whig* of 20 February 1865, a Mollie Bean was picked up in men's uniform not far from Richmond on the night of February 17; she claimed that she had served with

Portrait of Mary Walker. (Library of Congress)

the 47th North Carolina for two years and had been wounded twice. Estimates of the number of women who actually fought in the Civil War vary from 400 to 1,000.

There were some differences between the Union and Confederate armies in their use of women, due at least in part to cultural differences between North and South. The best-documented cases of women who served in male disguise were nearly all in the Union army (see **Sarah Emma Evelyn Edmonds, Jennie Hodgers,** and **Sarah Rosetta Wakeman**)—probably due to the fact that there was less material destruction in the North, and more family and other records were preserved. Other women associated with the Union army include **Anna Ella Carroll, Dorothea Dix, Susie King Taylor, Harriet Tubman,** and **Mary Walker.**

In the Confederate army it has been more difficult to verify cases of women who actually fought. The best known is **Loreta Janeta Velazquez,** although there is dispute about her activities. There are many references to other women in letters and newspapers that are now difficult or impossible to verify. For example, Henry Besancon, a musician with the 104th New York Infantry, made a diary entry on 26 May 1864 that "a female dressed in Rebel uniform was taken this morning" (manuscript department, William Perkins Library, Duke University). Women who supported the Confederate cause had opportunities to act as guerrilla fighters (see under "Representative Participants").

For example, Nancy Slaughter Walker was reported to have fought as a raider with William Clarke Quantrill, staying with his group after the war and fighting in its raids in Indiana and Texas.

A great deal of research has recently been undertaken to investigate women's roles with Civil War armies. The best single scholarly source on women who served in the Civil War is Elizabeth Leonard's 1999 book *All the Daring of the Soldier*.

Representative Participants

Blalock, Sarah Malinda Pritchard (born 1839; died 1903), sometimes spelled "Blaylock," who enlisted as Sam Blalock in the 26th North Carolina Infantry and served with her husband. She served only a few weeks but appears to have performed well in drill and, according to one source, in battle. When her husband was medically discharged, she revealed her true identity to the unit commander and was dismissed from service. She and her husband later assisted the Union resistance in North Carolina.

Clark, Amy (also known as Anna Clark). The thirty-year-old Clark disguised herself as a man to enlist with her husband (one source says a lover) in a Confederate cavalry regiment. Clark's husband was killed at Shiloh, but she remained in Confederate service, reenlisting in the 11th Tennessee Infantry. She was wounded and captured in late 1862, whereupon her sex was discovered and she was sent to Federal prison and required to dress in feminine attire. She was apparently later released and may have enlisted a third time.

Hart, Nancy, fought as a guerrilla in West Virginia. Hart may have been romantically connected with Perry Connally, the commander of the infamous Moccasin Rangers of Wood County, West Virginia. In an engagement along the West Fork River, Hart was captured, but charmed her captors into releasing her as harmless. She returned to her group, and on 2 January 1862, when the Moccasins were surprised at Welch Glade, West Virginia, Connally was severely wounded and brutally beaten to death by the Yankees; Hart escaped. She was again captured by the Federals in July 1862 and imprisoned in Sum-

---

A soldier in the Texas Rangers describes an encounter with a female soldier in 1863 in a letter dated 7 August 1863 from Tyner's Station, Tennessee (original spelling):

Pa—among all the curiosities I have seen since I left home one I must mention, a female Lieutenant. I had heard of her deeds of bravery in several battles and a few evenings ago, I went to the Station about a quarter of a mile distant from camp. I discovered quite a crowd, approaching the crowd I inquired what was up. One of the soldiers directed my attention to a youth apparently about seventeen years of age, well dressed with a Liutinants badge on his collar. I remarked that I saw nothing verry strange, he then told me that the young man was not a man, but a female. It is said that she volunteered with her husband as a private fought through the battles of Shiloh where her husband was killed, she performing the rites of burial with her own hands. She then continued with Braggs army in Ky fighting in ranks as a common soldier, until she was twice wounded—once in the ankle and then in the breast when she fell a prisoner into the hands of the Yankees, her sex was discovered by the federals and she was regularly paroled as a prisoner of war, but they did not permit her to returne until she had donned female apparel. She has since her return I suppose been promoted to the office of Lt.

—Letters of Robert Hodges, from Fort Bend County, who served in Terry's Texas Rangers. Galveston and Texas History Center, Rosenberg Library, Galveston, Texas.

mersville, West Virginia. Hart escaped by tricking a Union guard into letting her hold his loaded musket; she then shot the hapless guard, raced from the house, and stole a horse. Later she rejoined the Moccasin Rangers. Hart survived the war and settled in Greenbrier County, West Virginia.

Bibliography: Clinton and Silber, *Divided Houses,* 1992; Larson, "Bonny Yank and Ginny Reb," 1990, and "Bonny Yank and Ginny Reb Revisited," 1992; Leonard, *All the Daring of the Soldier,* 1999; Massey, *Bonnet Brigades,* 1966, and *Women in the Civil War;* 1994; Meyer, "The Soldier Left a Portrait and Her Eyewitness Account," 1995.

—James S. Baugess and Susannah U. Bruce

ANI PACHEN (born 1933, Gonjo, Tibet; died 2 February 2002, Dharmsala, India), also known as Pachen Dolma, Nun Big Courage. Guerrilla leader, Struggle to Free Tibet, Tibet.

When the Chinese invaded Tibet in the 1950s, Ani Pachen, a Buddhist nun, sat at the side of her father, a clan chieftain, in the war councils as her clan decided to resist. After her father died in 1958, Ani Pachen led about 600 resistance fighters who fought on horseback in mountainous terrain, attacking Chinese convoys and encampments. The Chinese overran Gonjo, and Ani Pachen was taken prisoner as she and her family fled over the Himalayas into India. She spent the next twenty-one years in prison, where she endured shackles, torture, and long periods of solitary confinement. After her release in 1981, she remained in Lhasa for seven years and took part in a number of large demonstrations. In 1988 she fled into India to avoid being arrested again. She remained at Dharmsala, where she joined the supporters of the Dalai Lama in exile, until her death.

Bibliography: Ani Pachen and Donnelley, *Sorrow Mountain,* 2000.

—Valerie Eads

ARAGÓN, AGUSTINA (born 1788; died 1857, Spain). Born Agustina Zaragoza y Domenech; also known as "la Artillera" and "Maid of Saragossa." Defender during French assault on Zaragoza, Spain, 1808.

On 4 July 1808 French troops assaulted the Portillo battery guarding an entrance to the besieged city of Zaragoza, the strategic key to northern Spain. The French succeeded in killing the men serving the battery and had begun to storm into the city when Agustina Aragón saved the day. She stepped into the breach, wrenched a lighted match from the dead hand of one of the Spanish artillerymen, and fired the twenty-four-pound cannon into the oncoming French. The blast of grapeshot reputedly saved the city, which held out until February 1809, and served to rally a movement of popular resistance that eventually helped drive Napoleon out of Iberia.

Born in Catalonia, Agustina was present in Zaragoza because her husband, an artillery sergeant, had been stationed in the city in 1808. After the death of her husband in the siege, Agustina dropped her patronymic and took the name Aragón. The Spanish commander in Zaragoza, José Palafox y Melci, who had witnessed her bravery, rewarded Agustina with a decoration and the rank of officer in the Spanish artillery. In this capacity she served with her unit in Seville in 1809 and later participated in the defense of Tortosa in Catalonia.

Agustina's international fame brought English and Spanish officers on pilgrimages to meet "la Artillera" during the war. In the fall of 1808 Agustina sat for the artist Juan Galvéz, who sketched her as a beautiful young woman in feminine attire. Later portraits showed her in military uniform striking Byronic poses or as an allegorical figure, as in Goya's portrait titled *What Courage!* (number seven in his series *The Disasters of War*). These portraits made Agustina the most visible icon of Spanish resistance to Napoleon.

In fact, Agustina was just one of many Zaragozan women who distinguished themselves in the siege. In sieges and wars of occupation, when front lines run through traditionally feminized spaces like homes, churches, and marketplaces, women are more likely to take up arms. This was certainly the case in Zaragoza. For example, María Consolación de Azlor y Villavicencio, the Condesa de Bureta, organized a battery to stop a French assault on 4 August 1808 when her home came under French fire. María Agustín was wounded supplying ammunition to the troops, and Casta Alvarez fought with the men on the city walls; both received pensions and decorations for bravery.

The belligerency of the women of Zaragoza amazed contemporaries, like Lord Byron, who met Agustina and wrote of her in *Childe Harold*. Spanish romantic nationalists employed Agustina and the other women of Zaragoza as symbols of national unity at the dawn of a long period of disunity and constant civil war. Indeed, in the long run, the military contributions of the women of Zaragoza to the war against Napoleon, valuable as they were, may have been less important than their iconic function as embodiments of patriotic unity in defense of a Spanish nation that did not exist for most Spaniards in 1808 and that remained contested terrain into the late twentieth century.

Bibliography: Tone, "Women in the Resistance to Napoleon," 1998; Toreno, *Historia del levantamiento, guerra, y revolución de España*, 1851.

—John Lawrence Tone

ARMIA KRAJOWA (AK), WOMEN IN. Couriers, medics, sappers. Poland, Second World War.

Women performed many important roles in the European resistance movements during the Second World War. In Poland they comprised approximately one-seventh of the overall force of the Polish Home Army, the Armia Krajowa (AK). By far the largest number of these women served as *łazniczki* (liaison couriers) and as *sanitaruszki* (medics). There were also two units comprised completely of women: the *minerki,* a unit of demolition experts, and the *kanalarki,* a specialty unit created during the Warsaw Uprising for duty in the city's sewer system.

Owing to the underground nature of the resistance and the absolute necessity for security, most AK officers only knew their immediate counterparts, superiors, or subordinates. Therefore, reliable, secure communications between the various elements of the resistance became a crucial need for the AK. The women soldiers who filled this role were known as the *łazniczki*. They were the foundation for a nationwide communication system for the effective dissemination of standard correspondence on a daily and weekly basis. Their network was also used as an early warning system for the other members of the underground because of the intelligence they were able to gather during the completion of their rounds. In addition to the military use of the liaison network, these women also provided postal services and the delivery of the underground newspaper (*Biuletyn Informacyjny*), helping to provide reliable information to the Polish nation as well.

In preparing for the Warsaw Uprising, which began on 1 August 1944, the *łazniczki* were the primary means of notifying all commanders and soldiers of the date, times, places, and code words for the initiation of open hostilities. In addition to acting as couriers, the *łazniczki* also transported arms and ammunition to distribution points throughout the city. During the uprising these women were instrumental in effective coordination between units engaged in combat, often completing their rounds under fire in order to deliver essential information and supplies to the infantry, frequently without the protection of personal weapons. Many of the *łazniczki* were additionally cross-trained as *sanitaruszki*.

The *sanitaruszki* were assigned to every unit throughout the AK and performed a variety of roles. In addition to providing first aid to soldiers, they conducted training sessions in medical assistance and made, collected, and maintained medical supplies. In the nationwide uprising during the spring and summer of 1944 known as Akcja Burza (Operation Tempest) and in the Warsaw Uprising these women braved enemy fire on countless occasions to care for and retrieve their fallen comrades.

The *minerki* were a small group of women who, during most of the occu-

---

From an interview with Ida Kasprzak, who served with the Polish Home Army, the Armia Krajowa (AK), during the Second World War:

Ida was an officer in charge of a group of women medics. She and her staff were responsible for carrying the wounded to safety and applying first aid. "They didn't give us guns, because they would not have helped us. Only those behind the barricades were given weapons. Because there weren't enough to go around, they rarely offered the girls any. If it happened that an AK soldier was killed, the girl next to him would grab his gun and start shooting. Out in the streets we were open targets, dodging bullets, with no time to shoot anyone. . . . People were screaming and moaning. My girlfriend and I spotted a man severely wounded in the head. She grabbed his legs and I grabbed his arms and we ran through the streets. He was a huge man, he must have weighed two hundred pounds. We were running, and I saw a piece of shrapnel fly into her leg. . . . Just at that moment I saw a piece of metal coming straight towards my stomach. There was a gap of about six inches between his head and my stomach, and it was coming straight towards us. I thought, my God, what can I do? If I drop him he will die as soon as his head hits the pavement. Then, just as the shrapnel came within an inch of my stomach it fell down, between me and the wounded man's head. It was as though someone had punched it from above. I screamed, "Dunca, stop running, put him down gently.' I ran back and picked up the shrapnel. It was still hot. I carried it with me through the uprising. . . . I still have it today."

—Shelley Saywell, *Women in War* (Markham, Ontario: Viking, 1985), 116–117.

pation, conducted demolition raids against the German rail and communication network. Upon the initiation of the Warsaw Uprising the *minerki* opened passageways between the cellars of the housing blocks in the area of the city held by the AK. After completing this task within the first several days, upon their own initiative they began offensive operations, using their skill with mines to attack German tanks and other vehicles. One of their most effective tactics was to attach a rope to either end of a mine and then maneuver it beneath an advancing vehicle. As the Germans became wise to this tactic, the *minerki* adopted another method, employing hollow objects that could hold explosives and be rolled at oncoming vehicles.

During the Warsaw Uprising protected passage for the infantry, couriers, and medics was a primary concern. One of the most effective solutions to this problem was the city's sewer system. The *kanalarki,* a specialty unit comprised completely of women volunteers, reconnoitered, cleared, and marked routes and also provided guides for those who had to pass through these underground passages. When the Germans realized that the AK was making effective use of the sewers, they began to place booby-trapped obstacles at openings to the sewers in their area of control. The Germans then would pump in toxic fumes from vehicle exhausts as well as randomly toss in grenades to combat the AK's use of the tunnels.

Resistance leader Jan Karski believed that women were sometimes better than men at undercover work; he felt that they were more careful and deliberate and less conspicuous. He praised in particular the women couriers, noting that one of his couriers made more than 200 points of contact within a week, thus exposing herself to great risk. He felt that the couriers had one of the most difficult and dangerous jobs in the resistance.

Upon the surrender of the AK forces on 2 October 1944 after the failure of the Warsaw Uprising, all AK soldiers in the city were recognized as full-fledged combatants by the Germans and treated as prisoners of war as per the Geneva Convention. However, the women prisoners were taken to the Ravensbrück camp and, upon arrival, were reclassified as civilian auxiliaries and shipped off for forced labor duties throughout the remnants of the German Reich. Even after the end of German occupation many of the soldiers who managed to evade capture in Warsaw or who had served elsewhere in Poland voluntarily kept their wartime service secret because the Soviets and the Polish Communist Party were openly hostile to the AK, due to the possibility that resistance might continue against the Soviet occupation of Poland. Those soldiers who were identified as AK members were, in many cases, subjected to repeated interrogations by the Communist Office for State Security (Urząd Bezpieczeństwo) as to their activities in the resistance, who and where their commanders were, and so on. In 1956 a general amnesty was proclaimed by the Polish Communists, but many AK veterans, male and female, kept their past secret until the situation in Poland began to liberalize in the mid-1980s. Because of this long tradition of secrecy and the impossibility of publishing

AK memoirs under the former Communist regime, information on women's roles in the AK has only recently come to light.

Bibliography: Karski, *Story of a Secret State*, 1944; Laska, *Women in the Resistance and in the Holocaust*, 1983; interviews by author.

—W. James Dixon

ARSINOË III PHILOPATOR (died 204 BCE). Queen of Egypt and commander at the Battle of Raphia.

Arsinoë III, with her brother/husband Ptolemy IV Philopator, led the Egyptian forces in the Battle of Raphia in Palestine against Antiochus the Great in 217 BCE. She may have commanded a section of the infantry phalanx. Regarded as one of the turning points in the history of the Mediterranean world after the death of Alexander, the battle involved 55,000 troops under Arsinoë and Ptolemy and 68,000 under Antiochus. Both sides employed cavalry, elephants, and specialized troops such as archers, as well as the traditional Macedonian phalanx, which played a particularly important role in this battle.

For the first time since the conquest of Egypt by Alexander, the Macedonian rulers included native Egyptian troops in their army (in addition to Greek and Macedonian colonist-soldiers). Arsinoë and Ptolemy reviewed the army, sometimes speaking through interpreters, offering rich rewards in the event of victory. They were stationed on the elect wing when Antiochus's Indian elephants caused a rout there. Arsinoë again entreated her soldiers not only with the inducements of courage and family but also with the promise of two minas of gold each if they won. The army regrouped, counterattacked, and defeated Antiochus. No army again threatened Ptolemaic Egypt until Octavian fought Antony and Cleopatra at Actium nearly two centuries later.

Bibliography: III Maccabees 1:1–4; *Hellenistic Queens*, 1975; Macurdy, Polybius, *Histories*, 5.79–87, 15.25a, 15.25b.

—Alexander Ingle

ARTEMISIA I OF HALICARNASSUS (fl. fifth century BCE). Queen of Halicarnassus, commander of naval force, Battle of Salamis, Persian Wars.

Artemisia was the daughter of a Halicarnassian father, Lygdamis, and a Cretan mother; she came to the throne of Halicarnassus (modern Bodrum in Turkey) upon the death of her unnamed husband. As Halicarnassus lay in Ionia, along the western coast of Turkey, it was under the suzerainty of the Persian Empire. Thus when in 480 BCE Xerxes organized his expedition to invade Greece, he called on all his subjects to provide military aid. As the Ionians were descendants of Greek colonists, it is difficult to know how motivated Artemisia may have been, but she provided five ships that she commanded. She had a son of sufficient age to rule and command, but the Greek historian Herodotus says that she led this force because of her "spirit of adventure and manly courage."

Artemisia I of Halicarnassus

Artemisia I holding an urn containing the ashes of her husband and overseeing building of his marble tomb. Folio 35V of 1505 manuscript *La Vie des Femmes Celebres* (Life of Famous Women), by Antoine du Four. (The Art Archive/Musée Thomas Dobrée Nantes/Dagli Orti)

Artemisia distinguished herself in the campaign's first major naval action, off the coast of Euboea. As Xerxes' army was forcing the pass at Thermopylae, the Persian fleet was skirmishing with the Greek fleet based at Artemisium. After losing some ships to bad weather and to raids, the Persian fleet closed with the Greeks, and both sides lost heavily. News came of the Spartan defeat at Thermopylae, convincing the Greeks to withdraw their fleet. No details are given of Artemisia's skill during this first encounter, but no one contradicted her when she alluded to it in conference with Xerxes.

It was at that conference that Artemisia began to make herself remarkable. Contradicting the advice given Xerxes by his other advisors, Artemisia told him not to attack the Greek fleet. They were too skilled, she argued, and there was no need either for hurry or a naval battle. The Greek fleet would disperse, she argued, as many of the provincial leaders had little desire to fight to save Athens; also, food was becoming scarce and the Greek defense could not last much longer. She advised that if Xerxes threatened targets on land, the Greek forces would disintegrate. Xerxes admired Artemisia's courageous stand, but conceded to the majority and ordered his fleet to pursue the Greek fleet into the narrow channel between the island of Salamis and Attica.

The division among the Greeks almost proved their undoing, as the naval commanders could not agree on a strategy. One of the Athenian commanders, Themistocles, sent a secret message to Xerxes claiming a desire to abandon the squabbling Greeks and join the Persians. Hearing this, Xerxes ordered his ships to attack, and the battle was joined on 20 September 480 BCE. The offer of Athenian assistance was a ruse to force the battle, and the fight soon justified Artemisia's caution. The narrow waters off Salamis negated the superior numbers of the Persian fleet, and the small and more maneuverable Greek ships soon gained the upper hand.

In the midst of the ensuing Persian confusion, Artemisia stood out. Chased by an Athenian ship, she took the desperate measure of ramming an allied ship. This convinced the pursuers that she was either a Greek commander or had changed sides, so they turned away in search of other targets. This saved her from possible capture or death and at the same time was noticed on shore. Xerxes sat on a golden throne on high ground overlooking the battle; Artemisia's action was pointed out to him, and he assumed that the ship she rammed was Greek. Seeing her success in the midst of his fleet's defeat, he made the famous remark "My men have turned into women, my women into men." Her escape cost Ameinias, the Greek who had been chasing her, 10,000 drachmae, for the Athenians had put a price on her head for daring to "fight like a man."

The Persian ships that could escape returned to their base at Phaleron. Artemisia made one final appearance before Xerxes. His general Mardonius had suggested that the emperor leave for Persia and leave behind 300,000 men with which he would deliver the Greeks in chains. Should he take this course? Artemisia advised that he go home, for Mardonius could be blamed if the Greeks were not defeated, but Xerxes could take credit if they were. He found this counsel agreeable and sent Artemisia to Ephesus with his sons while he made the final dispositions for his return to Persia. What became of Artemisia after the great naval battle, like her life previous to it, is unrecorded. Apparently she returned home to Halicarnassus and turned the reins of power over to her son. Her name was bestowed on many later queens in the area.

Bibliography: Herodotus, *The Histories*, 8.87–88.

—Paul K. Davis

ARTEMISIA II OF HALICARNASSUS (died 351 BCE). Queen of Halicarnassus and other cities in Caria. Naval strategist and commander.

Queen Artemisia II of Halicarnassus (a city-state on the southwest coast of Turkey) commanded a fleet and played a role in the military-political affairs of the Aegean after the waning of Athenian naval supremacy in the mid-fourth century BCE. More than one hundred years after her namesake **Artemisia I,** Artemisia II ruled in Halicarnassus with her brother and husband, Mausolus (who is best remembered for the tomb Artemisia completed for him, the Mausoleum). When Mausolus died (ca. 353–352 BCE), Artemisia inherited the throne, prompting a revolt in some of the island and coastal cities under her sway. According to Vitruvius, the island republic of Rhodes objected to the fact that "a woman should rule all the cities of Caria." Rhodes sent a fleet against Artemisia without realizing that Mausolus had built a secret harbor. Here Artemisia hid her ships, rowers, and marines and allowed the Rhodians to sail into the main harbor. She had her citizens greet the enemy fleet from the city walls and invite the Rhodians into the city. When the ships' crews had disembarked, Artemisia sailed her fleet through an outlet to the sea and into the main harbor. She captured the empty Rhodian ships, and the Rhodian crews were killed in the marketplace of Halicarnassus. Artemisia then put her own troops on the Rhodian ships and sailed back to Rhodes, where they were welcomed into the Rhodian harbor. Artemisia's forces took the island. After putting to death the leading citizens, she erected two statues, one of the city of Rhodes and one of herself burning the brand of slavery on Rhodes.

Although the seizure of Rhodes under Artemisia is a historical fact, some scholars question the account given by Vitruvius, primarily because they find it difficult to accept Artemisia's role as military strategist and leader. It is not, however, implausible. In smaller ancient polities such as Halicarnassus, military command and absolute rule were often the same thing. In this case, sculpture (including the famous statue of Artemisia and Mausolus in the British Museum) and inscriptions all attest that Artemisia shared in the rule of Caria even while Mausolus was alive. For her to play an active military role after his death would not be unique for a ruling queen of Halicarnassus, especially in light of the legacy of Artemisia I.

Bibliography: Diodorus Siculus, *Historical Library*, 14.2.16; Hornblower, *Mausolus*, 1982; Vitruvius, *On Architecture*, 2.8.13–15.

—Alexander Ingle

ARZAMASSKAIA, ALENA (born?, Volga region, Russia; died 1670), also known as "Temnikovskaia." Soldier and leader, Russia, peasant revolts, 1670–1671.

Two centuries after the death of **Joan of Arc,** in Russia rather than in France, another peasant girl put on men's clothing and led soldiers in battle. Like Joan, she was executed at the stake, partly for her heretical defiance of church rules regarding proper clothing for women. Unlike Joan, Alena Arzamasskaia did not fight for a king, but in rebellion against one.

Alena Arzamasskaia was born to a peasant family in the Volga region of Russia in the mid-seventeenth century. After an early marriage she entered the Nikolaevskii Monastery as a nun, but was unhappy with the strictly regulated life of the convent. In 1669 she ran away, cut her hair, donned a male disguise, and joined the peasant revolt led by Cossack Stepan ("Stenka") Razin. Razin's revolt was initially successful in southern Russia. Arzamasskaia claimed to be a (male) Cossack commander and began rallying forces for Razin; eventually she led as many as 6,000 soldiers. Her troops subjugated the fortress city of Temnikov and held it for two months. Arzamasskaia's leadership was affirmed by local rebels, and Arzamasskaia was acknowledged as an intelligent leader and excellent archer. In 1670 she was captured by tsarist forces and tortured, but refused to name other peasant leaders. She was legally condemned not only for her rebellion against the tsar but also because she committed the crime of dressing like a man. Arzamasskaia was burned at the stake; an eyewitness reported that she showed no fear and made no sound when she died.

Bibliography: Pushkareva, *Women in Russian History,* 1997, 82–83; Smirnov and Chistiakova, *Alena Arzamasskaia-Temnikovskaia,* 1986.

—Reina Pennington

ATA (AIR TRANSPORT AUXILIARY). Founded in Great Britain, September 1939; disbanded November 1945.

The Air Transport Auxiliary (ATA) served as a ferry service for new and repaired aircraft to active squadrons during the Second World War. This was a civilian organization that employed both men and women as pilots and other personnel. Gerard d'Erlanger, director of British Airways, founded the ATA as a taxi and freight service to employ licensed male pilots who were ineligible for active service in the Royal Air Force (RAF) because of age or unfitness. Almost immediately it evolved into a delivery service for new planes to active airfields. In January 1940, four months after ATA started, pilot Pauline Gower, "a person of great force, clear thinking and by all standards most efficient and knowledgeable" (Curtis, 17), was given permission to recruit eight women. The first eight women pilots (Rosemary Rees, Winnie Crossley, Margie Fairweather, Joan Hughes, Mona Friedlander, Marion Wilberforce, Margaret Cunnison, and Gabby Patterson) received a fair amount of media attention. Although these female pilots were among the most experienced in Britain, at first they were only allowed to fly Tiger Moth trainers to RAF flying schools.

U.S. first lady Eleanor Roosevelt (*right*) talks with American women ferry pilots serving with the Air Transport Auxiliary at an airfield in England in October 1942 during World War II. Standing next to the first lady is Pauline Gower, commandant of the ATA service. (AP Photo)

By July 1941 there were about 30 women in the ATA, and the first combat aircraft, a Hurricane, was officially flown by a woman ferry pilot. Over the course of the war a total of 166 women pilots flew with the ATA (and 1,152 men), or a little more than 12 percent of total pilots. Lettice Curtis, who ferried nearly 1,500 aircraft for the ATA, provides a list of all female ATA pilots in an appendix to her book, *The Forgotten Pilots.*

As word spread about the ATA, and as air forces like the RAF, United States

Air Force (USAF), and Royal Canadian Air Force (RCAF) rejected applications by women pilots, women applied by the hundreds to join the ATA. The ATA was truly multinational; there were women pilots from the British Commonwealth (including Canada, New Zealand, and South Africa), as well as women from Chile, Poland, Holland, and the United States. Three Polish women who had been military pilots before the war joined the ATA (**Anna Leska-Daab,** Stefania Wojtulanis-Karpinska, and Jadwiga Pilsudska-Jaraczewska); they were members of the **PLSK,** a Polish unit within the British Women's Auxiliary Air Force (WAAF). Helen Harrison Bristol was the first Canadian woman to join the ATA; she served from 1942 until March 1944 and flew about 500 hours. Canadian pilot Violet Milstead Warren flew with the ATA from May 1943 to August 1945, accumulating 700 hours on 29 different single-engine aircraft and 17 different twin-engine aircraft.

A number of American women also flew with the ATA. **Jacqueline Cochran** decided that if she sponsored a group of female American volunteer pilots in the ATA, they could prove the capability of American women for ferrying duty. According to Cochran, the British Air Commission requested that she personally recruit American women pilots to serve with the ATA. In late 1941 Cochran arrived in England with twenty-five volunteer women pilots from the United States. Cochran was officially a member of the ATA only from January through July of 1942. Her presence was resented by many women in the ATA, some of whom believed that Cochran did not pull her weight in terms of flying actual missions. Lettice Curtis, a British ATA pilot, wrote that Cochran "had little time for flying. Our main memories of her are of someone who lived at the Savoy Hotel, wore a lush fur coat, and arrived in a Rolls Royce" (Curtis, 142). Cochran clearly viewed American participation in the ATA as a stepping-stone to creating a women's aviation unit back home in the United States, and after returning home she got permission to form the **WASP.** The other American women were better accepted, and most served faithfully with the ATA for the duration of the war.

In the first months women pilots ferried the light aircraft to destinations that were nearly always in Scotland, a distance of at least 300 miles. Many of these small aircraft had open cockpits, and temperatures at flight altitude were often subzero. The limited range of these aircraft and weather and security considerations often required diversions, stopovers, and long waits en route. It could take several days or even as much as two weeks to deliver the airplane and return to base, including long hours on trains during the return (eventually taxi aircraft would round up pilots after deliveries). When female pilots occasionally were forced to stop at RAF airfields, they found that there were no provisions for them. Women were not allowed in the mess; trays would be brought to their rooms. There were often no bathrooms for women. Anna Leska-Daab recalled that once she had to go behind an isolated hangar; she said, "If the King of England himself had walked by I couldn't have done nothing about it!" (Curtis, 20).

Although women had been accepted by the ATA to fly only noncombat aircraft, they were soon assigned to fly all types, including loaded bombers and fighters. Until January 1945 female pilots were restricted to transferring planes between factories, repair depots, and operational fields on British soil only. The first operational overseas flights actually occurred in September 1944, and by the end of October continental ferrying was well established, but formal recognition and orders only came several months after the fact.

Lettice Curtis described the working conditions of the ATA: "It meant so training [the pilots] that they could fly without wireless and finding their way by map, through worse weather than they had ever hoped to see from the air. It meant 'converting' them on to new types in easy stages so that they could eventually fly any type any time. It meant that at the peak a pilot could and did fly any one of well over 100-odd types in any one day—perhaps four in one day, from a four-engined bomber to a new type of fast jet, and then a pre-war light aircraft with a third the approach speed of the jet flown but an hour before."

One famous pilot in the unit was Amy Johnson, who was nearly as well known internationally as Amelia Earhart. Johnson had completed many record-breaking flights before the war. Her achievements included a solo flight from England to Australia in 1930 and a 1931 London-to-Tokyo flight. She was killed in January 1941 when she ran out of fuel attempting to deliver an aircraft in bad weather; icing and cloud cover kept her from locating a safe landing strip. Johnson was observed parachuting into the ocean, but rough weather prevented rescuers from reaching her.

When the ATA was disbanded at the end of the war, none of the civilian personnel received military honors for their services. Statistics about the organization, however, reveal that its pilots consistently supplied the RAF with operational equipment throughout the entire war, regardless of adverse weather, dangerous ground conditions, and enemy action; doubtless this service was critical to the successful Allied war effort. Service in the ATA was not without risk; 173 pilots died, including 15 women.

Bibliography: Cheesman, *Brief Glory*, 1946; Cochran, "Final Report on Women Pilot Program," 1945; Curtis, *The Forgotten Pilots*, 1985; Gower, "A.T.A. Girls," 1943; King, *Golden Wings*, 1956; Render, *No Place for a Lady*, 1992; Williamson, "The Air Transport Auxiliary," 1943; Wills, *Free as a Bird*, 1974.

—Patricia Bowley and Reina Pennington

ATS (AUXILIARY TERRITORIAL SERVICE) (9 September 1938–1 February 1949). British women's army corps, Second World War.

The ATS was created to release men for combat. It was designed to administer women who trained and served operationally in other army branches. The ATS originated in both the War Office's early 1938 realization that it would need women in the event of war and the sustained pressure of women

ATS (Auxiliary Territorial Service)

> *Restrictions were imposed on British women serving in the Auxiliary Territorial Service in antiaircraft batteries during the Second World War; in order to preserve the noncombatant status of the women, they were not allowed to load or fire the guns:*
>
> The most important sexual distinction involved the issue of who fired the guns. A 1938 Royal Warrant prohibited women from using any weapons in combat. According to [General Sir Frederick A.] Pile, this "was a political issue: we were quite prepared to let them fire light anti-aircraft guns, but there was a good deal of muddled thinking which was prepared to allow women to do anything to kill the enemy except actually pull the trigger." As Pile argued, "There is not much essential different between manning a G.L. set or a predictor and firing a gun: both are means of destroying an enemy aircraft." As the war progressed, more advanced guns were fired by remote control when on target. Since women on the command post aimed the gun it was absurd to claim that they were not firing it. But despite this blurring of responsibility, the sophistry which separated women from men was maintained. Even late in the war, when mixed batteries were deployed against pilotless V-1s, women were not allowed to "kill" inanimate missiles.
>
> —Gerard DeGroot, "I Love the Scent of Cordite in Your Hair: Gender Dynamics in Mixed Anti-Aircraft Batteries during the Second World War," *History* 265 (1997): 83.

veterans of the First World War. ATS members were considered noncombatants, but held jobs ranging from traditional support roles (clerks, drivers, and the like) to positions in the Anti-Aircraft Command, where women served in mixed batteries.

The premature announcement of the ATS's formation during the Munich crisis in 1938 meant that it started life without proper administrative or command structures, a fact the press noted. Not until 3 July 1939 was Dame **Helen Gwynne-Vaughan** appointed its first director. Nonetheless, the ATS was at its establishment strength of 17,000 in September 1939. Continuing administrative and image problems led to the appointment of a stylish young director, Jean Knox, in 1941. Knox had first joined ATS in September 1938 and was a company commander at the outbreak of war, then served in the War Office before becoming inspector and then director of the ATS in 1941. She served as director for more than two years (July 1941–October 1943). Leslie Whateley, who had been in the ATS since 1938, was appointed deputy director of ATS in 1941 and became director in October 1943, serving until the summer of 1946.

At the outbreak of the war ATS members served only as domestics, clerks, cooks, and drivers. Following the defeat of France, the number of trades open to ATS members increased to more than eighty. In the summer of 1941 the ATS was fully incorporated into the army. Its members served in every theater of war. For the first time, women in Britain were enlisted in the ranks, earned commissions in the officer corps, and were subject to a modified version of military law, and a few served operationally under trained female officers. The introduction of conscription for women in March 1942 drastically increased the size of the ATS, which peaked in late 1943 at about 207,000. The ATS continued in existence until the creation of the peacetime Women's Royal Army Corps on 1 February 1949. Princess Elizabeth (later Queen Elizabeth II) was commissioned into the ATS.

The most important military contribution of the ATS was to the Anti-Aircraft Command, which by 1943 employed 57,000 women as spotters, predictors, and radar operators; some served in all-female searchlight companies, while others served in mixed batteries. A typical mixed battery had 189 men and 299 women, with the actual gun crew consisting of 6 women and 2 men;

women were forbidden to actually fire the guns. Commanders like General Sir Frederick Pile were enthusiastic about their capabilities and preferred mixed units to all-male batteries. Pile later expressed his opinion regarding the prohibition against firing, saying, "There was a good deal of muddled thinking which was prepared to allow women to do anything to kill the enemy except actually pull the trigger."

As was common with administratively segregated women's services, there were conflicts between the administrative chain of command (through the ATS) and the operational chain. Similar power struggles occurred in the United States over the jurisdiction of the **WASP** and **WAC**. There was resentment among male officers about the military's lack of disciplinary power over ATS members (although ATS rules were quite strict), which reinforced the idea that women were not real soldiers. However, disputes flared up when full absorption into the military was suggested. For example, Pile believed that the women commanders of the ATS saw him as a threat when he suggested that the women should be absorbed into the Royal Artillery.

Despite its contribution to the war effort, the ATS remained noncombatant because of lack of public support for female soldiers. The ATS suffered from persistent rumors that its members were licentious, a universal cultural problem affecting all women's services in the Second World War. A special commission, the Markham Committee, found that sexual activity of ATS members was similar to that of civilian women in the same age group, and illegitimate births were actually lower. Winston Churchill wrongly predicted that the ATS would permanently change women's roles. However, it did help prepare the way for a permanent female presence in the British army.

Bibliography: Bigland, *Britain's Other Army*, 1946; DeGroot, "I Love the Scent of Cordite in Your Hair," 1997; Pile, *Ack-Ack*, 1949; Terry, *Women in Khaki*, 1988; Thomas, "Women in the Military," 1978; Whateley, *As Thoughts Survive*, 1949.

—C.H.N. Hull

ATTENDOLO (SFORZA), MARGHERITA (fl. 1415–1416, Italy). Military commander, Italy, fifteenth century.

Margherita Attendolo belonged to a family of well-to-do farmers from Cotignola, near Ravenna, in the Romagna region of Italy. The surname Attendolo would later yield to the nickname "Sforza." The Attendoli/Sforza are famous for their participation in Italian history as professional warriors. The earlier generations served the Angevin kings of Naples, while later ones became rulers in their own right.

Margherita grew up in a spartan domestic setting: the homestead was sparsely furnished and functional, with shields and cuirasses ornamenting the walls. Meals were irregular, and mattresses without coverlets served for sleep. Her parents were both from combative families, and her twenty siblings matched each other in ferocity, whether male or female.

Margherita is remembered for effecting the release of her brother Muzio from the Castel dell'Ovo prison in Naples. When Queen Joanna II married the Bourbon count James of La Marche, Muzio, who had formerly served the queen, was imprisoned and tortured as a possible threat to the new ruler. At the time of his arrest Margherita, with her husband Michelino Catti of Ravenna and certain relatives, was holding the Attendolo fortress at Tricarico in Basilicata. In response to this and to a revolt in Naples, loyal Attendolo followers assembled an army with Margherita in command.

James of La Marche sent envoys to demand the surrender of the castle, and they began negotiations with the captains and populace. The fully armed Margherita burst into the proceedings, announced herself as the sister of Muzio and the true commander of Tricarico, rejected their terms, and took them hostage despite their safe-conduct pass. She threatened to hang them all if any harm came to her brother. Alarmed by this action, the families of the representatives interceded for their safety, persuading James to release Muzio without reprisals. Margherita's militant behavior foreshadowed that of her more famous relative, **Caterina Sforza.**

Bibliography: Durant, *The Renaissance*, 1953; Faraglia, *Storia della Regina Giovanna II d'Angiò*, 1904; Trease, *The Condottieri*, 1971.

—Gloria Allaire

AUSTRALIAN DEFENCE FORCES (ADF), WOMEN IN.

By 1994, 7,620 women were serving in the Australian Defence Forces (ADF), or 12.3 percent of the total. This percentage was comparable with that in the United States, where American women represented about 12 percent of military forces in 1993–1994. In the same year, women held fewer than 7 percent of positions in the British armed forces.

Before 1941 Australian women were not permitted to serve in the ADF even in a noncombatant capacity. Historically, however, Australian women were no strangers to fighting or the use of firearms. In Australian frontier legend pioneer "bushwomen" are depicted as strong and uncomplaining and able to use a rifle to hunt or to protect their homes and children against dangerous animals or unwelcome humans. Australian "bushmen" came to be regarded as the apotheosis of the soldier—for example, the members of the Australian and New Zealand Army Corps (Anzacs) who fought on the Gallipoli Peninsula in 1915. However, there was no combatant role for bushwomen. During the South African War (1898–1901) and the First World War Australian women were confined to support roles undertaken either by trained professional nurses or by enthusiastic amateurs.

Before the Second World War the only Australian women's service in existence was the Australian Army Nursing Service (AANS). The AANS had a distinguished record of service in the First World War and again in the Second World War. In the 1939–1945 war, however, the way was cleared for women

Australian Defence Forces (ADF), Women in

to enlist in nonmedical, noncombatant services that were regarded as "auxiliaries" of the army, navy, and air force. By the end of 1945, out of 993,000 service personnel and an Australian population of around 7 million, 64,833 women had enlisted in the women's services, including the AANS and the Australian Army Medical Women's Service (AAMWS). Of these, 26,591 enlisted in the Women's Auxiliary Australian Air Force (**WAAAF**) and 24,160 in the Australian Women's Army Service (AWAS). While the vast majority were Anglo-Australians, some women from ethnic minorities enlisted, including Aboriginal women like the well-known poet and Aboriginal rights activist Kath Walker (born Oodgeroo Noonuccal), who served in the AWAS.

Woman workers stacking casings for 25-pound artillery shells at the munitions plant in Salisbury, Australia, May 12, 1942. (©Bettmann/Corbis)

Upon the establishment of the WAAAF in 1941, the politically conservative United Australia Party government announced the formation of a "women's home army." Women would be appointed as cooks, clerks, orderlies, telephonists, canteen attendants, driver mechanics, etc. in order to free more men for fighting units. Enlistment began in January 1942 and was remarkably popular, to the surprise of the authorities. Numbers peaked at 24,000 in July 1944—4 percent of the army's strength. The Women's Royal Australian Naval Service (WRANS) was also established in 1941; its membership never exceeded 3,000. WRANS served in intelligence and as cooks, stewardesses, drivers, clerks, and coders. Most were employed in wireless telegraphy and by the end of 1944 had replaced Royal Australian Navy (RAN) personnel at most Naval Signal Stations.

Despite the importance of their work and the fact that they often performed identical duties to those of men, women always played a subservient role in the ADF during the Second World War and for many years afterward. Servicewomen's maximum pay was two-thirds the male rate. Many single women who enlisted were responsible for the support of aged parents, yet there was no provision for dependents' allowances. Women were grudgingly allowed the same rank as men, but they were kept entirely separate from male troops and did not perform combat duties.

As the war continued, servicewomen were allocated tasks that more directly affected the war effort. In early 1943 the first AWAS antiaircraft searchlight

Australian Defence Forces (ADF), Women in

RAAF's first female pilot, Joanne Mein in the cockpit of an A97 Lockheed Hercules C130, June 1994. (©Commonwealth of Australia 2000)

unit entered service, but AWAS members did not operate antiaircraft batteries. Originally, women could serve only in restricted geographical areas. On 15 November 1944 the War Cabinet approved the posting of AWAS volunteers to New Guinea, where they arrived in May 1945. AWAS members also served in Malaya, Singapore, and Borneo.

All of the women's services underwent considerable postwar changes. Initially they were discharged after the failure of discussions to bring women into the permanent ADF. In 1950 Women's Services were reintroduced but were kept separate from the mainstream male services, and women's employment was restricted primarily to clerical work. The Women's Royal Australian Army Corps (WRAAC) was founded with **Kathleen Best** as its commanding officer. In five years, Best raised a corps of more than 1,000 enlisted members. WRAAC officers received about two-thirds the pay given men, and marriage was regarded as a barrier to career; it was assumed that most women would not serve until retirement age. The WRAAC existed until women were permitted to join regular units of the Australian army in 1979. Integrated training of male and female recruits commenced in new facilities at Bonegilla, Victoria, in 1981, but integrated officer cadet education began only after the opening of the Australian Defence Force Academy at Canberra in 1986.

The Women's Royal Australian Air Force (WRAAF) also was reformed in 1950. WRAAF was not regarded as an "auxiliary," as had been its wartime predecessor. Still, there were many restrictions. Originally, WRAAF members were employed in traditional female occupations—cooks, stewards, clerks, drivers, orderlies, and telephone or teleprinter operators. Officers were not offered permanent commissions until 1965, and they were given different titles from Royal Australian Air Force (RAAF) officers: squadron officer rather than squadron leader, and group officer rather than group captain (following RAF precedent). Promotions were depressingly few until WRAAF career prospects were widened in the early 1960s, again, because the RAAF could not attract sufficient numbers of qualified men. WRAAF members were not permitted to serve overseas until 1967. Pay and conditions remained less than those of airmen, and until 1969 WRAAF members who married were immediately discharged. After 1969 married women were permitted to remain in the WRAAF but were not recruited. The WRAAF and the RAAF were integrated in 1977.

The WRANS—the only branch of the women's services to retain its wartime title—was reconstituted in 1950. As in the other branches of the Women's Services, WRANS members' pay and conditions were poor compared with those of RAN. Unlike the WRAAF, it took several years to achieve the initial recruitment total of 300. Although the number of occupations and the range of postings open to women gradually broadened, WRANS members were banned from serving at sea, except for brief familiarization periods. The WRANS was abolished in 1985, and its members were integrated into the navy.

The 1984 Sexual Discrimination Act was a significant step toward increasing the number of women in the ADF and broadening the career choices available to them. Even so, Section 43 of the act exempted the ADF from admitting women to combat-related duties. Yet the proportion of females in the ADF rose from 4.4 percent in 1974 to 12.3 percent in 1994. In all branches of the services, the greatest influx of women occurred in the late 1980s and early 1990s. By the beginning of 1994, 83 percent of all army positions were open to women, although infantry, armor, artillery, and combat engineer units were still denied to females. In December 1992 the RAAF opened 99 percent of its positions to women, and the navy followed suit in mid-1993. The percentage of women in the ADF has remained fairly constant since 1991, with the air force employing the greatest proportion of female members (15.4 percent in 1994).

The role of women in the ADF has undergone immense changes in the past fifty years. If Australian forces were involved in another war, there seems little doubt that some women would be employed as combat troops.

Bibliography: Dennis, *The Oxford Companion to Australian Military History*, 1995; McKernan, *All In!*, 1983; Shephard, *A Compendium of Australian Defence Statistics*, 1995; Urry and Wilcox, *Women and the Australian Armed Forces*, 1994.

—Bobbie Oliver

AZURDUY DE PADILLA, JUANA (born 8 March 1781, Chuquisaca, Bolivia; died 25 May 1862, same). Lieutenant colonel, anti-Royalist guerrilla fighter, Latin American independence forces in Upper Peru (Bolivia), 1811–1816, and Argentina, 1816–1825.

Juana Azurduy de Padilla led a group of anti-Royalist guerrillas in the east central region of Upper Peru in the early nineteenth century. The region was divided into small pockets of resistance called "republiquetas" held by patriot guerrilla forces. With her husband, Manuel Ascencio Padilla, Azurduy organized attacks against the Spanish from the republiqueta of La Laguna. The two kept the road between Buenos Aires and Chuquisaca open for several years, providing a vital transportation corridor for independence forces.

Azurduy de Padilla, Juana

Born to affluent parents, Azurduy spent her formative years accompanying her father on his rounds of their hacienda. Due to the early death of her only brother and the untimely passing of her mother, she became her father's close companion, learning to ride, shoot, and speak various indigenous languages with skill. Upon her father's death in 1797, she and her younger sister became the charges of relatives, who sent the unruly Juana to a convent. Chafing under the restrictive rules and regulations, Azurduy left the convent early. In 1805 she married her childhood friend, Manuel Padilla. They formed a partnership that went beyond the domestic sphere, sharing experiences on and off the battlefield. While her life as a guerrilla leader was a series of successes, her personal life was a series of losses: her parents at a young age, her three sons and a daughter to malnutrition and disease in the guerrilla camps, and ultimately her husband, captured and beheaded in the Battle of "Villar" on 14 September 1816.

In her first major battle at Pintatora in 1815, she left the battlefield to give birth but quickly returned to capture the standard of the Spanish troops and rally the patriot forces. During a subsequent battle she came to the aid of the general commander of the patriot forces, General Belgrano. He showed his gratitude by awarding Azurduy the full command of the legion "Macha." Azurduy's ability to "fight like a man" impressed the troops, and her valor, courage, and levelheadedness ensured their loyalty. She had organized the "Leales" battalion in 1813, commanded the "Húsares" cavalry unit in 1815, and was often accompanied by a personal guard of twenty-five women, referred to as "Amazonas."

Juana Azurduy's bravery on the field was rewarded on 13 August 1816 when the independent government of Buenos Aires awarded her the title of "teniente coronel" in recognition of her distinguished service. Her greatest military success occurred earlier that year outside the village of Tarabuco during the Battle of Jumbati. Azurduy led outnumbered and poorly armed Indian troops in a victory against the infamous Spanish battalion, the "Verdes."

Juana Azurduy fought in more than sixteen major actions against the Spanish, including the Battles of "Pocoma," "Tarbita," "Aiquile," "Carreta," "Laguna," "Poopó," and "Presto." The fall of the republiquetas and the resurgence of Spanish forces in Upper Peru forced her to flee to the Republic of Argentina in 1818, where she continued to fight for independence in Salta under the orders of the caudillo leader Martin Güemes. With the independence of Upper Peru in 1825, she returned to Chuquisaca, newly named Sucre. Azurduy was granted a pension, but it was later revoked. At eighty-one years of age Juana Azurduy de Padilla died alone and penniless in Sucre. She was buried without military honors. While national recognition of her great contribution to Bolivia's independence came posthumously, the Indians of Tarabuco have remembered the Battle of Jumbati and their teniente coronela in a celebration called Pujjllay every 12 March since 1816.

Bibliography: Gantier, *Doña Juana Azurduy de Padilla,* 1946; Mitre, *Historia de Belgrano,* 1902; O'Donnell, *Juana Azurduy,* 1994; Valencia Vega, *Manuel Asconcio Padilla y Juana Azurduy,* 1981.

—Heather Thiessen-Reily

B

BADLESMERE, LADY MARGARET (born ca. 1286, died 1333), also known as Baddlesmere. Defender, England.

The daughter of Thomas de Clare and widow of Gilbert de Umfreville, Lady Margaret subsequently married Bartholomew Badlesmere, an important landowner in Kent. He belonged to the so-called Middle Party, which sought compromise in the power struggle between King Edward II and the earl of Lancaster; it aimed to enforce the Ordinances of 1311 restricting the authority of the king. After the Treaty of Leake (1318) Badlesmere was appointed steward of the royal household, but incurred the anger of both the king and Lancaster and became politically isolated after the breakup of the Middle Party in 1321. Edward took revenge by taking advantage of a superficially minor incident. In September 1321, while traveling to Canterbury, Edward's wife, **Isabella of France,** sought hospitality at the royal castle of Leeds, Kent, of which Badlesmere was constable. Lady Badlesmere, in her husband's absence, refused to receive the queen on the grounds that she could admit no one without his permission. A struggle broke out between their respective servants in which six of Isabella's retainers were killed. It has been suggested that the queen was sent with the intention of gaining control of the castle without the need for a campaign, but this has not been proven. Events escalated; on 17 October a siege force commanded by the earls of Pembroke, Norfolk, and Richmond arrived and was soon joined by the king with reinforcements. Lady Badlesmere held the castle for two weeks, but it fell on 31 October. There followed violent reprisals noted as being the "first application of martial law to internal discords." The garrison was executed, and Lady Badlesmere and her children were sent to the Tower of London. A year later she was released from custody into the care of a convent; on 1 July 1324 she was allowed to go free.

The incident is significant because, as noted by Natalie Fryde, "The execution of the Leeds garrison opens a new episode in English history when opponents of the king could seriously expect to lose their heads if they were defeated." Later Isabella and her lover rebelled against Edward II, who would be seen by history as a tyrant.

Bibliography: *Calendar of Close Rolls, 1318–1322;* Cockayn, *The Complete Peerage,* 1910–1959; *Dictionary of National Biography;* Fryde, *The Tyranny and Fall of Edward II, 1321–1326,* 1979; Maddicott, *Thomas of Lancaster,* 1970; Page, ed., *Victoria County History, Kent,* 1908–1932.

—David S. Green

BAILEY, ANN (fl. 1777), also known as Samuel Gay or Gray, Nancy Bailey, Bayley. Corporal in the Continental army. Soldier, United States, Revolutionary War.

Ann Bailey disguised herself as a man to enlist in the First Massachusetts Regiment of the Continental army during the Revolutionary War. She was promoted to the rank of corporal. When she was discovered to be a woman, she was dismissed from the army and eventually brought before the Suffolk County court to face charges of fraud. Bailey's enlistment is an early example of the lengths to which American women sometimes went in order to enlist in regular military forces and of the consequences that awaited them if they were found out.

Little is known about who Ann Bailey was or why she decided to disguise herself as a man and join the army during the Revolutionary War. Her age at the time of her enlistment on 17 February 1777 is uncertain, but she was probably over the age of twenty. Court records describe Ann Bailey as a "spinster" and a resident of Boston. It is certain that Bailey was so motivated to join the army, probably both by patriotism and a desire to earn money, that she took on "the outward appearance of a young man" and presented herself to Nathaniel Barber, the muster master of Suffolk County, Massachusetts. She joined thirty-one other men who signed up at the same time to serve "for the duration of the war" and was assigned to Captain Abraham Hunt's Company commanded by Colonel John Patterson. By 1777 the states had trouble meeting their enlistment quotas, so Bailey and the other recruits received a bounty payment of fifteen pounds, ten shillings, upon their entrance into the army. "Samuel Gay" quickly showed himself to be a trustworthy and accomplished soldier, since records show that "he" had already been promoted to the rank of corporal by 3 March 1777, when Captain Hunt first discovered that "Samuel" was actually Ann Bailey in disguise.

Once she was discovered, Ann disappeared, and the captain swore out a warrant for her arrest on 10 March 1777. The General Court of Sessions for

Suffolk County issued a summons for Bailey on 13 March, demanding that she pay sureties of fifty pounds "to answer for the Offence of having appeared in Man's Apparel . . . [and] more especially for defrauding this State of the Sum of Fifteen pounds ten Shillings." Ann may have spent time in jail until 26 August 1777, when the court found her guilty of defrauding the public and ordered her to pay sixteen pounds and serve an additional two-month jail term. It is significant that the court was offended by Bailey's wearing of men's clothing almost as much as by her taking state money, for her actions would have been viewed as upsetting the "natural" order of society established along the lines of gender norms.

No more can be known about Bailey than the few scraps of information that survive from enlistment and court records. As an unmarried "spinster" in eighteenth-century Boston, Bailey possessed the power to control her own fate in a way denied to most married women, but she also had to earn her own living. Both factors probably influenced her attempt to serve in the military, and both factors disturbed government authorities once her disguise was discovered. Whatever the case, Ann Bailey wanted to be a soldier, and she was willing to risk reputation, money, and even imprisonment for the chance to fight.

Bibliography: De Pauw, "Women in Combat," 1980; Leonard, "Ann Bailey," 1993; *Massachusetts Soldiers and Sailors in the War of the Revolution,* 1900, 6:340.

—Sarah J. Purcell

BAILEY, ANNE HENNIS TROTTER (born 1742, Liverpool, England; died 1825, Gallia County, Ohio), also known as "Mad Anne," "The Pioneer Heroine of the Great Kanawha Valley," and "The White Squaw of the Kanawha." Frontier militia scout, Revolutionary War; scout during campaigns against Indian tribes on the western Virginia frontier in the 1790s; United States.

Anne Bailey joined the militia after her husband, Richard Trotter, was killed during Lord Dunmore's War against the Indians on the Virginia frontier in 1774. Bailey served as a scout and courier for American forces during the

---

Some interesting historical facts about American women and the military:

The first woman awarded a disability pension by Congress for wounds incurred during military service was **Margaret Corbin**.

During the War of 1812, two women served as nurses aboard the *United States,* Stephen Decatur's flagship.

During the **American Civil War** women disguised as men served on both sides. Women also served as spies and nurses—including onboard at least one hospital ship—and one, Dr. **Mary Walker,** received the Congressional Medal of Honor.

Women have served in the American armed forces for more than 100 years—since 1901, when the Army Nurse Corps was established. The Navy Nurse Corps soon followed in 1908.

Women who were not nurses were first enlisted in the navy and Marine Corps during the First World War. Only nurses served in the army during this war, but the army did hire about 200 civilian women who were fluent in both English and French to serve as telephone operators. These women, often referred to as the "Hello Girls," were later given veterans' status.

Four hundred and thirty-two American military women were killed during the Second World War. Eighty-eight were prisoners of war, all but one of these in the Pacific theater.

Seven women died in the line of duty while serving in theater during the Vietnam War. Their names can be found inscribed on the Vietnam Veterans Memorial.

Almost 41,000 women served in theater during the Persian Gulf War. Thirteen women were killed, and two were taken as prisoners of war.

—Women's Research and Education Institute, 1750 New York Avenue, NW, Suite 350, Washington, DC 20006.

Revolutionary War, and she continued her military services in the late 1780s with her second husband, John Bailey, a prominent soldier in southwest Virginia. Anne Bailey patrolled the frontier against Indians in the service of the U.S. military.

She had emigrated to Virginia from England after both her parents died in 1760 and settled in the frontier area of the Shenandoah Valley, where she married Richard Trotter, a British soldier. After Trotter was killed in a campaign against the Indians in 1774, Anne decided to join the militia herself, this time in opposition to the British. She dressed in the traditional rustic garb of male frontier soldiers and wielded her rifle as she harangued her neighbors to fight the Indians and the British. She soon began to serve as a scout on missions between remote western areas and eastern Virginia forts; scouts were often more effective than regular soldiers in remote areas because of their survival skills. During and after the Revolutionary War Bailey took part in campaigns against the Shawnee, who admired her fighting prowess so much that they dubbed her "Mad Anne" as a symbol of her ferocity.

In 1788, at the age of forty-six, Anne married John Bailey, and the couple was posted to Fort Lee. Anne Bailey frequently served as a scout and messenger between Fort Lee and frontier posts such as Point Pleasant. In 1791 Bailey was credited with preventing an Indian attack on Fort Lee and with saving lives by delivering ammunition to the fort.

After being widowed again, Bailey moved further into the frontier to Gallia, Ohio, in 1818. Despite being more than seventy years old, she worked as an express rider and hunted extensively. Bailey died in Ohio in 1825 at age eighty-three; in 1901 the Point Pleasant Battle Monument Commission reinterred her remains in the Point Pleasant, West Virginia, State Park.

Bibliography: Cole, "Anne Bailey," 1980; De Pauw, "Women in Combat," 1980; Everett, *The Dixie Frontier*, 1948.

—Sarah J. Purcell

BANKES, LADY MARY (died April 1661). Defender, England, English Civil War.

Lady Mary Bankes, wife of Sir John Bankes, defended Corfe Castle against parliamentary forces in 1643 and 1645. By 1643 Corfe Castle was the last Royalist garrison on the Dorsetshire coast. Her husband was away serving the king in London and Oxford; Lady Bankes remained at Corfe with her children, servants, and a small troop of soldiers. A first attempt by parliamentary forces to take Corfe occurred in May 1643 when forty seamen demanded the surrender of the castle's four pieces of ordnance. With her small group of soldiers and maidservants, Lady Bankes turned the cannon on the enemy and drove them away. She later gave up the ordnance so she could gain time to resupply the castle. On 23 June 1643, 500–600 parliamentary troops began the first siege of Corfe. The besiegers lost 100 men in an assault on the castle in August;

Lady Bankes herself, with her daughters, maidservants, and only five remaining soldiers, dropped stones and hot embers on the attackers and drove them back. After this defeat, and hearing reports of an approaching royalist force, the besiegers withdrew so quickly that they left horses, ordnance, and their dinner behind. The castle was safe for nearly two years, and Lady Bankes stayed either there or in London (her husband died in December 1644). Parliamentary forces tried repeatedly but unsuccessfully to take the castle in the summer of 1645; in December the castle was again besieged, Lady Bankes again leading the defense. This second siege of Corfe ended in February 1646 when a traitor allowed 50–100 enemy soldiers into the castle, pretending that they were a royalist relief force. Having taken by stealth what they could not take by force, the Parliamentarians allowed Lady Bankes and her children to leave safely. Known as the "Heroine of Corfe Castle," she lived on until 1661.

Bibliography: Bankes, *The Story of Corfe Castle*, 1853; *Dictionary of National Biography*, 1:1043–1044; Fraser, *The Weaker Vessel*, 1984; Hutchins, *The History and Antiquities of the County of Dorset*, 1973, 504–507.

—Robert W. Gee

BARREAU, ROSE-ALEXANDRINE (born May 1773, France; died 24 January 1843, Avignon, France), also known as "Liberté" Barreau. Grenadier, soldier, infantry, France, French Revolution.

Rose Barreau was one of the most widely known women soldiers who enlisted in the French revolutionary armies during the initial year of the War of the First Coalition in 1792. Barreau, who was born in the region of the Tarn in France, enlisted with her husband François Leyrac and with her brother. Her original enlistment, the day after her marriage on 5 March 1792, was in the 2nd Battalion of the Tarn (Army of the Pyrenees), which was later amalgamated into the 63rd Demi-brigade. She was inscribed in the military register as the "son of Jacques and Jeanne Barreau," bearing the name "Liberté" in honor of the French Revolution.

She distinguished herself at Bréaton in July 1793 in combat against Spanish incursions on the French border while under the command of Latour d'Auvergne. When her brother was killed at her side and her husband was grievously injured, she moved forward with the troops. According to contemporary stories of the incident, she rushed through cross-fire to avenge herself against Spanish soldiers, firing her husband's weapon until she had no more ammunition. Then she wielded her saber, and finally she relied on hand-to-hand combat. Only after the fighting ended did she return to bandage her husband's wounds and to mourn the death of her brother. "Showing herself to be more than a man," she had behaved as a soldier first, the report of her actions noted. Then she had turned back to care for others.

After serving one year and two months, Barreau was discharged. Experiencing severe headaches, phlebitis, and impaired sight and hearing, she was also six months pregnant. Although she received an award of 300 livres from the National Convention of France for her valor, she was not granted a retirement pension until 1806. In the intervening years she had followed her husband as part of the army train rather than as a fellow soldier; and according to reports in 1809, she was the mother of five children. Later treated as the subject of prescriptive literature for young French women, Barreau was heralded as a mother "who had not relinquished the virtues of her sex. So it was that after having made war, she set about remedying the depopulation it caused."

Invalided later in life, Barreau petitioned tirelessly to be admitted to veterans' care. Although her service injuries were documented, not until 1832 was she admitted together with her husband to the veterans' hospital in Avignon, where she died at the age of seventy. During her service, in addition to the monetary award by the National Convention of France, Barreau was featured in a period engraving by Duplessis-Berteaux (August 1793) and was treated as a national hero in the government-sponsored *Annales du civisme et de la vertu* (1793).

Bibliography: Brice, *La femme et les armées,* n.d., Hennet, "Rose Barreau"; Dossier XR48, "Barreau," Service historique de l'Armée, Vincennes, France.

—S.P. Conner

BARRERA, MARIA DE LA LUZ ESPINOSA (born Yautepec, Morelos, Mexico; fl. 1910–1920, Mexico), also known as "La Coronela." Lieutenant colonel, Zapata's army. Soldier, Mexico, 1910 Revolution.

As a young woman, Espinosa Barrera had been sentenced to five years in jail for killing a woman who was having an affair with her husband. After serving her time, she joined her father to fight in the Mexican Revolution. Skilled with horses, Barrera soldiered under Emiliano Zapata, 1910–1920. Disguised as a male soldier, she recalled her first battle "as the chattering of her teeth and nervous jingling in her spurs," but proved her mettle during a decade of fighting. Zapata acknowledged her bravery by signing papers promoting her to lieutenant colonel. Espinosa Barrera later qualified for a veteran's pension and became a traveling vendor of clothing.

Bibliography: Jaiven and Escandon, *Mujeres y Revolucion,* 1993; Macías, *Against All Odds,* 1982; Salas, *Soldaderas,* 1990, 69, 75.

—Elizabeth Salas

BASHKIROVA, KIRA ALEKSANDROVNA (fl. 1914), also known as Nikolai Popov. Infantry and cavalry, Russia, First World War.

A student at the Mariinskii Women's Gymnasium in Vilna, Bashkirova disguised herself as a scout in October 1914 and was accepted into the active army. For the next month she participated in a number of dangerous and important reconnaissance missions. Maintaining her male disguise under the pseudonym Nikolai Popov, in December she volunteered for service in a Siberian infantry regiment where she was assigned to cavalry reconnaissance. For courageous performance in a mission on 20 December, she was awarded the St. George's Cross (the **Order of St. George**). Soon after, her commander discovered her true identity and sent her back to Vilna. Bashkirova refused to accept this as the end of her military career; rather than return home, she joined another regiment in a different area of the front. After fighting several battles with this new unit, she was wounded and sent to a field hospital, where her sex was once again revealed. Further information on Bashkirova's fate is not known.

Bibliography: "Zhenschiny i voina," *Zhenskii Vestnik* March 1915: 73; April 1915: 93–94.

—Laurie Stoff

BEATRICE OF LORRAINE (born ca. 1020, northeastern France; died 18 April 1076, Pisa, Italy), also known as Beatrice of Tuscany, Beatrice of Canossa, Stateswoman, Investiture Controversy, Italy.

After the death of her father, Duke Frederick II of Upper Lorraine, in 1033, Beatrice and her sister Sophia were taken to live at the imperial court by their maternal aunt, the empress Gisella. In her teens Beatrice became the second wife of Marquis Boniface of Tuscany sometime between mid-1036 and mid-1039. Little else is known about her life before the murder of Boniface in 1052. During the minority of Boniface's only surviving child, **Matilda of Tuscany,** there is frequent mention of Beatrice in the records of the turbulent papal and imperial politics of the time, although these sources do not always distinguish between the actions of Beatrice and those of her second husband, Godfrey the Bearded, the dispossessed duke of Upper Lorraine. Much of his authority in Italy derived from his marriage to Beatrice.

In 1059 Beatrice escorted Pope Nicholas II into Rome. The hostile commentator Benzo of Alba writes that she set up an idol as pope. From 1061 to 1065 she took action against the antipope, Bishop Cadalus of Parma (Honorius II), preventing him from reaching Rome by means of traps and ambushes, according to Benzo, or ousting him after he did so. In 1064 she protected Pope Alexander II at the Council of Mantua. As Cadalus's troops approached, the city was in a state of panic, and the imperial representative, Archbishop Anno of Cologne, fainted. Beatrice undertook a successful attack on Cadalus's camp. In 1067 Beatrice and Matilda were with the army before the battle against the Normans at Aquino, but the sources do not describe a role in the battle.

Beatrice's husband Godfrey was not present during most of her military actions; he was in Germany in late 1061 and early 1062, and it is known that he was absent during her defense of the Council of Mantua in 1064. He died in 1069. Five years later, in June 1074, Beatrice and Matilda brought a large contingent to Pope Gregory VII at Cimmino, but the intended campaign against the Normans did not take place. Gregory VII also planned a campaign to the East against the Muslims during which Beatrice was to remain in Italy to safeguard the interests of her daughter and of the pope; however, that campaign also never took place.

In 1074–1076 Beatrice was a key negotiator in the dispute between the pope and her kinsman, King Henry IV of Germany, over certain rights in episcopal appointments. An escalation to the open warfare of the Investiture Controversy was thus delayed for some years. Beatrice's daughter Matilda of Tuscany would find herself embroiled in an even more militant life than her mother's, once again in defense of the papacy.

Bibliography: Bertolini, "Beatrice di Lorena," 1960; Cowdrey, *Pope Gregory VII, 1073–1085,* 1998; Duff, *Matilda of Tuscany, la Gran Donna d'Italia,* 1909; Glaesener, "Un mariage fertile," 1947; Goez, *Beatrix von Canossa und Tuszien,* 1995.

—Valerie Eads

BEGUM SAMRU (born 1753?; died 1836). Mercenary commander and political figure, eighteenth-century India. Fought both for and against the British East India Company.

The exploits of Begum Samru are closely associated with those of the European mercenaries who joined the struggles over the remains of the Mughal Empire in northern India at the end of the eighteenth century. A considerable amount is thus known about her life, although it is scattered among numerous sources, and no major work has yet been written that focuses primarily upon her.

Begum Samru was probably born in 1753 in Kutana, a small town in Meerut District near Delhi. Her father was an Arab merchant, and her mother may have been one of his Kashmiri concubines. When her father died, both she and her mother were persecuted by her half brother, and they fled to Delhi in 1760; Begum Samru's mother probably became a courtesan and trained her daughter for a similar career. At the age of twenty-two Begum Samru became a camp follower of a mercenary army commanded by Walter Reinhard, an Austrian soldier who had deserted from the military forces of the Compagnie des Indes Orientales to pursue a tumultuous career training and leading the armies of various north Indian princes. Reinhard, like most European mercenaries in India, had adopted a false name and was known to his fellow soldiers as Sombre, or Samru.

Reinhard took command of a brigade of mutineers from the East India Company's army during the war between the Mughal emperor and the company (1763–1765). After surviving the disastrous Battle of Buxar by a timely retreat, he entered the service of the Jat rajah, Jawahir Singh. In 1775, during the wars in which the Maratha-led imperialist coalition attempted to reinstate the authority of the Mughal emperor in northern India, Samru defected to the imperial camp of Shah Alam along with his highly trained infantry and artillery. Shortly afterward he was given a large body of irregular cavalry and a *jagir* (revenue-collection rights to be used for the support of troops) over the estate of Sardhana to the northeast of Delhi. At some time during this period Begum Samru became first his concubine and then his wife. When Reinhard died in 1778, the Sardhana estate and command of the army passed into her hands rather than those of his son by a former mistress, Zafariyab Khan. At the time, the Sardhana forces consisted of five battalions of regular infantry, forty cannon, and some 400 Rohilla cavalry under the direction of 300 European and Eurasian officers and gunners. The training of the army was carried out under an Italian soldier named Paoli, but he was assassinated during an attempted palace coup at Delhi in 1783.

The first test of Begum Samru's personal military and political skills came a few years later. Ghulam Qadir, a rebel general, advanced toward Delhi with a strong contingent of Sikh irregulars and attempted to remove the obstacle of Begum Samru's trained battalions through bribery. Samru refused the offer and marched to protect the emperor while his Maratha overlords stood idle, thus winning the close affection of Shah Alam. The next year, 1788, her troops formed the core of an imperial army sent to Gokalgarh to crush the rebel general Najif Quli Khan. Shah Alam deployed his least steady troops to advance the trenches against the enemy fortifications, and Najif Quli Khan was able to overwhelm them through a surprise night attack by his cavalry. At the same time, additional rebel forces struck and routed the imperial army's rear guard. During this crisis Begum Samru drew up her forces and called the emperor and his household to the relative safety of her headquarters. Together with an Irish mercenary captain, George Thomas, she personally led an assault on the fort that threw the enemy into confusion and purchased sufficient time for Shah Alam's officers to rally the disorganized Mughal troops. For her role in the imperialists' victory, she was granted a khelat, or robe of honor, together with the titles "daughter of the emperor" and *Zeb-ul-Nissa,* or "ornament of her sex," as well as additional lands for the support of her troops.

George Thomas, who had served Begum Samru since 1787, resigned his post in 1792 upon discovering that she had secretly married his rival Le Vaisseau, one of her French artillerymen. Thomas joined the Marathas, received a *jagir* near Sardhana, and raised and trained his own small army. In 1795 his troops collided with those of the Begum, and she immediately marched to destroy him. Some of her officers who were friends of Thomas led a revolt en route, and Begum Samru and Le Vaisseau fled to Sardhana. Command of her troops was given to Zafariyab Khan, Walter Reinhard's son, who marched to arrest the Begum and her French husband. As the army approached Sardhana, Samru persuaded Le Vaisseau to seize the treasury and flee with her to British territory, and the two apparently agreed to commit suicide if they were captured.

Numerous accounts offer conflicting versions of what happened next. Most agree that upon being intercepted, Le Vaisseau killed himself and the Begum tried to fake her own suicide. However, she was captured and tortured by her enraged troops before being placed under house arrest by a sympathetic French officer named Saleur. She managed to communicate secretly with George Thomas, forming an alliance with her former officer and enemy, and the Irish mercenary quickly moved with his own forces to arrest Zafariyab Khan and reinstate her at Sardhana with support from the Marathas.

Begum Samru soon augmented her army by 6,000 infantry, part of the attempt by the Maratha overlord, Sindiah, to create a large, European-style army capable of resisting that of the East India Company. She did not, however, employ her troops to assist George Thomas when, after being double-crossed by his Maratha allies, he attempted to establish his own kingdom in Haryana to the west of Delhi. Another mercenary force in the Maratha service, that of Perron, smashed Thomas's army in a particularly bloody campaign near Hansi in 1801–1802.

A few months after Thomas's defeat, Sindiah went to war with the British, and most of Begum Samru's troops marched to central India under Saleur, where a quarter of them were annihilated by Major-General Arthur Wellesley's forces at the Battle of Assaye in 1803. Begum Samru surrendered her remaining battalions at Sardhana when the British drove the Marathas from Delhi, and the company confirmed her possession of the *jagir*. She became a staunch ally of the East India Company, which allowed her to retain her army. She led her troops in battle for the last time at the siege of the Jat city of Bharatpur in 1825. After this campaign she retired to Sardhana, where she built a Roman Catholic church and died in 1836.

Begum Samru may be considered one of the most successful mercenary commanders of late-eighteenth-century northern India. Under her leadership the Sardhana army improved greatly in strength, discipline, and efficiency. Begum Samru was also noted for her abilities as an administrator who greatly enhanced the revenues of her *jagir* and created a civil and military infrastructure that formed the basis of the company's government and defensive organization in the Delhi region. She was also active in the cultural life of the nearby imperial capital, where she maintained a large villa, and many Roman Catholic institutions in the region enjoyed her generous patronage. During her life, her adventures and those of her lovers and comrades were celebrated by the Urdu poets of Delhi.

Remarkably, Begum Samru commanded her army while observing the traditional northern Indian custom of purdah, or the seclusion of women. She never appeared before her troops unveiled, traveled in a covered palanquin instead of on horseback, and held conferences with her officers from behind a screen. Throughout her career she was attended by a retinue of twenty to thirty women servants, one of whom she is reputed to have had buried alive as a punishment for treachery.

The most balanced and accurate contemporary account of her career is probably that of James Skinner, a Eurasian soldier in Perron's service, who was the only other mercenary captain to survive the war with the British and retain command of his troops. The best and most recent account of Begum Samru's military leadership and civilian administration is that of Seema Alavi, who has recently published several works on the northern Indian mercenary armies of the late eighteenth century.

Bibliography: Alavi, *The Sepoys and the Company,* 1995; Chand, *The History of Zeb-ul-Nissa,* 1994; Holman, *Sikander Sahib: The Life of Colonel James Skinner,* 1961; Sharma, *The Life and Times of Begam Samru,* 1985; Shreeve, *Dark Legacy,* 1996.

—James W. Hoover

BEN-YEHUDA, NETIVA (born 26 July 1928). Revolutionary, member of the Palmach, Israel, War of Independence (1948).

In 1946, when she was only eighteen Netiva Ben-Yehuda enlisted in the Palmach, an underground military organization. She trained in demolition and bomb disposal as well as in topography and scouting and commanded a sapper unit. She also trained recruits, transferred ammunition, and escorted convoys. Ben-Yehuda related her military experiences in three autobiographical works, thinly disguised as novels, which deal with some of the lighter moments of the **Israeli War of Independence** and reflect her exceptionally witty and controversial personality.

Although the Palmach generally opposed women fighting at the front, as a commander, Ben-Yehuda participated in several battles, performing sabotage operations. On 11 February 1948 she took part in a mission under fire that had a tremendous influence on her psychologically. She and her comrades planted a mine for a busload of Arabs; thirty were killed. Ben-Yehuda later revealed her ambiguity toward warfare. Just before the operation started, she had panicked and for a while had to be dragged on by a friend: "My heart was going to burst. I could not breathe. My legs would not carry me. . . . I just felt like sitting there on the spot . . . pounding temples . . . and I saw something like a red veil while the veins in my eyes were popping brightly." By the time they reached the ambush site, she was physically and emotionally drained. In a letter to her father, Ben-Yehuda admitted that she did not feel any pride in the success of this military operation. According to her account, she performed what was expected of her as a soldier but in fact caused death and injury to living individuals with her own hands. To her, the term "enemy" became a ridiculous one: "When humans commit murder, 'enemies' become suddenly human. . . . On the other hand, the 'enemy' as well kills among us indiscriminately. I received an order and I carried it out. Yet I am confident that there is something evil, very evil [here], a terrible injustice to us and to them, to everyone. . . . Don't worry. I'll continue to follow orders; otherwise we'll all be lost. But meanwhile don't be proud of us. There is no room for joy here. We should not rejoice when those who are hit by our bullets are falling dead."

This ambiguity runs like a thread throughout Ben-Yehuda's autobiographical works. The conflict between peace and war and between love and hate penetrates through numerous tales of heroism and selfless devotion. Ben-Yehuda's contribution to the history of Israel's War of Independence is notable for its integrity and sincerity. The way she depicted events and personalities is unrivaled by other chroniclers of the period.

Bibliography: Ben-Yehuda, *1948*, 1981, *Mibaad La'avotot*, 1985, and *Keshepartza Hamilchama*, 1991.

—Dina Rispman Eylon

BEN-ZVI, RACHEL YANAIT (born 1886, Malin, Ukraine; died 1979, Jerusalem, Israel). Military leader and organizer, Haganah, Israel.

Rachel Yanait's military career included almost twenty years (1920–1940) on the two top governing bodies of the Haganah in Jerusalem; she was the only woman to serve at both levels. She was also a labor and social activist, agronomist, educator, editor and author, and a founding member of the Po'alai Zion labor organization in Russia and of Hashomer, Tenu'at Hapo'alot (Women's Labor Movement), and the Ahdut Ha'avodah labor party and leader of the Haganah (Defense) in Jerusalem. She was married to Itzhak Ben-Zvi, the second president of Israel (1952–1963).

Before Israel gained its independence, Yanait advocated the establishment of a secret self-defense organization, particularly in the Old City of Jerusalem. Apart from administrative work, her military activities included fundraising, recruiting new members (especially women), and strategic planning. She often smuggled firearms hidden on her person. During the day she managed an agricultural school for girls, and at night she participated in the Haganah's secret meetings and military activities.

A stamp issued in honor of Rachel Ben-Zvi. (Library of Congress)

On 18 May 1939, to protest the British White Paper that restricted Jewish immigration, Yanait organized a women's demonstration in Jerusalem. Thousands of Jewish women carried banners against British rule and marched toward the government buildings. When they reached the main street, they were blasted with water from firehoses by the British police, halting the demonstration. Soaked and disappointed, the women resorted to a written petition.

In her book *Hahaganah Biyerushalaim* (The defense in Jerusalem) Yanait passionately described the role she and other women played in the Jewish defense of the city: "The selection procedure of female members was very strict because their activities carried great responsibility and had a positive impact on the male members. They inspired a pure, friendly atmosphere.... They eagerly participated in any enterprise at anytime; whether accompanying male fighters at night... patrolling the dangerous alleys of the Old City, standing watch nightly, or immersing themselves in housework. Their selfless devotion contributed to the wonderful spirit of the Haganah."

Thus Yanait played a leading role in three major movements, "out of which, to a great degree, modern Israel grew." Hashomer and the Haganah evolved into the IDF (Israel Defense Forces), and Po'alai Zion evolved into the present Labor Party of Israel.

Bibliography: Ben-Zvi, *Coming Home,* 1963, and *Derachai Siparti,* 1972; Yanait, Avrahami, and Etzion, *Hahaganah Biyerushalaim,* 1973–1975.

—Dina Ripsman Eylon

BERENICE II EUERGETES (born 273 BCE; died 221 BCE). Ptolemaic queen of Egypt. Accustomed to fighting from horseback.

Evidence, slight as it is, indicates that Berenice II stood in the martial tradition of Macedonian women in the ruling aristocracies of the former parts of Alexander the Great's empire. The Roman writer Hyginus (first century CE) in his *Astronomica* (2.24.2) tells us that when her father Magas, king of Cyrene in modern Libya, and his troops were routed in battle, Berenice mounted a horse, rallied the remaining forces, killed many of the enemy, and drove the rest to flight.

Berenice II had a strong equestrian background, and there are indications that she entered horses in the Olympic Games. Tradition also emphasizes her bravery and even—like her chariot-riding cousin **Berenice Syra**—a certain blood lust. When the young Berenice learned that her first husband, Demetrius the Fair, had taken up with her mother,

Head of Berenice II. (The Art Archive/National Glyptothek Munich/Dagli Orti [A])

Apame, she had him killed in her mother's bedchambers. She stood at the door and instructed the assassins not to hurt Apame, who tried to protect her lover. After the death of her second husband, Ptolemy III Euergetes of Egypt, those assigned to assassinate Berenice proceeded carefully in fear of her "exceptional courage" (Polybius, *Histories* 5.36).

Berenice II was best known in antiquity through the poetry of Callimachus and Catullus for pledging a lock of her hair to Ptolemy in the event of victory in the Third Syrian War (246 BCE). When he returned to Egypt, the lock was gone from the altar on which it had been placed, and the royal astronomer Conon reported that it had been installed as a new constellation near Ursa Major. This made the Lock of Berenice the only war trophy in the Greco-Roman sky.

Bibliography: Hyginus, *Astronomica* 2.24.2; Macurdy, *Hellenistic Queens*, 1975.

—Alexander Ingle

BERENICE SYRA (died ca. 246 BCE). Seleucid queen of Syria. Fought from a chariot.

After the death of her husband Antiochus II Theos, the Seleucid ruler of Syria and other Asian provinces of the former Persian Empire, Berenice Syra conducted an internecine war with Laodice I, an earlier wife of Antiochus. While resident in the capital of Antioch, Berenice discovered that her son had been kidnapped by Laodice's agents. Berenice armed herself, mounted a chariot, and pursued a magistrate of the city, Caeneus, who was implicated in the crime. Her spear failed to find its mark, but she struck him with a rock and drove her horses over his body in the middle of crowds of her own and Laodice's partisans. Afterwards Berenice shut herself up in a stronghold in Daphne with a guard of Gallic mercenaries and waited for aid from her brother, Ptolemy III Euergetes of Egypt, and from Syrian cities that supported her. There she was betrayed and murdered with the connivance of her physician, Aristarchus; several women of her entourage also died while attempting to protect her from the assassins.

Bibliography: Macurdy, *Hellenistic Queens*, 1975; Polyaenus, *Strategematon* 8.50; Valerius Maximus, *Memorable Words and Deeds* 9.10.

—Alexander Ingle

BEST, KATHLEEN ANNIE LOUISE (born 28 August 1910, Sydney, Australia; died 15 November 1957, Melbourne, Australia), also known as Katie. Colonel, nurse, Australian Army Medical Women's Service, Second World War; Director of the Women's Australian Army Corps, 1950–1957.

Kathleen Best was born at Summer Hill, an inner Sydney suburb. She completed training as a nurse in 1933 and worked in hospitals in Sydney before volunteering for the Australian Army Nursing Service Reserve when the Second World War began. In 1940 Best, the youngest matron ever to serve in the Australian army, sailed with the Australian Imperial Force to the Middle East, where she took charge of the 2/5th Australian General Hospital. Best's leadership qualities soon emerged as she and her staff participated in the disastrous Allied attempt to prevent the German invasion of Greece and Crete. During the withdrawal the nursing staff were under enemy fire on several occasions.

Returning to Australia in 1942, Best was placed in charge of the Voluntary Aid Detachments (VADs). In 1943 the VADs were enlisted into the new Australian Army Medical Women's Service, and Matron Best was accorded the rank of lieutenant colonel. A number of government appointments followed, with Best advising on the reabsorption of servicewomen into civilian life in the closing stages of the Second World War.

In 1950 Kathleen Best was appointed founding director of the Women's Australian Army Corps (WRAAC; the prefix "Royal" was granted in June 1951). She was appointed colonel in 1951 and was made an officer of the Order of the British Empire in 1956. When she was appointed as founding director of the WRAAC in 1950, Best confronted the difficult task of raising the new corps from scratch and of overcoming considerable opposition from senior military staff who believed that a women's unit would be neither necessary nor competent. Best organized the WRAAC on the lines of its wartime predecessor, the Australian Women's Army Service. She gained the respect of her corps and was regarded as an inspiring, if rather authoritarian, leader. The WRAAC quickly grew to more than 1,000 members, with regular and part-time units in most capital cities.

Best did not live to see her early achievements consolidated, however. She died of cancer in November 1957, aged forty-seven years. The WRAAC gained permanent status in 1959 under Best's successor, Colonel Dawn Jackson.

Bibliography: Dennis, *The Oxford Companion to Australian Military History*, 1995; M. Lincoln, "Kathleen Annie Louise Best," *Australian Dictionary of Biography*, vol. 13; Urry and Wilcox, *Women and the Australian Armed Forces*, 1994.

—Bobbie Oliver

BHAI, LAXSHMI (born 1827/1835; died 1858; India), also known as Lakshmi Bai. Rani (short for maharani) of Jhansi. Military and political leader, nineteenth-century India. Leading figure in Indian (Sepoy) Mutiny, 1857–1858.

Laxshmi Bhai, the last ruler of the principality of Jhansi, led a military uprising after a British takeover of her homeland. She became a heroine and a symbol of resistance against British rule in India.

Laxshmi Bhai was probably born between 1827 and 1835 in the pilgrimage city of Varanasi, where her Karhada Brahmin father, Moropant Tambe, lived in exile. Moropant, who had served as an advisor to the Maratha court at Poona, joined the deposed Maratha *peshwa*, Baji Rao II, at Bithur (Kanpur) shortly after his daughter's birth. As her mother had died while she was still a child, Lakshmi Bhai was raised and educated along with the male children of other Maratha Brahmin elites at Bithur, and it is here that she is said to have received military training.

In 1842 Laxshmi Bhai was married to the much older Maharaja Gangadhar Rao of Jhansi and probably took up residence in the principality in 1849. Although the new rani of Jhansi bore a son and heir in 1851, the child died after a few months, and the maharajah died in 1853. Lakshmi Bhai's status as a widowed queen and regent liberated her from purdah, the traditional seclusion of high-caste women, and she turned her attention to the administration of Jhansi. She raised and trained an all-female regiment from among her servants and was described at the time by the British political agent at Gwalior as "a woman highly respected and esteemed and ... fully capable of ... assuming the reins of government."

At the time of his death, Gangadhar Rao had named an adopted male relative, five-year-old Damodar Rao, as heir to the throne. Governor-General Dalhousie, however, refused to recognize either the rani or the adopted successor, citing the Doctrine of Lapse, which held that Indian princes who held their states under a grant from the East India Company could not adopt heirs. The state of Jhansi was annexed to British India in May 1854, and the rani was pensioned on 5,000 rupees a month. With assistance from sympathetic British officials at Jhansi, Laxshmi Bhai petitioned London for a change in policy concerning the future of Jhansi, but the government of India refused to restore the state.

> The following item concerned conflict between Russia and Turkey on the Danube River in connection with the Crimean War:
>
> Among 600 Kurdish mounted troops who are at Tulcha [on the Danube] along with 200 Turkish guards hussars, there were many women armed and dressed like the men, except that instead of fezzes, their heads were covered by a red or yellow scarf, while another scarf covered the lower part of the face up to the nose.
>
> —*Russkii Invalid* (St. Petersburg), Issue 2 of 1855, 4 January 1855, quoting the *Neue Preussische Zeitung* (Berlin), translated by Mark Conrad.

At the time of the mass mutiny of the Bengal army in 1857, the British garrison at Jhansi was composed entirely of Indian troops; the contingent joined the rebellion in early June. As elsewhere in northern India, the revolt of the army was followed by civil insurrection and local skirmishing between the great landowners and their family troops. Sixty-three British military and civilian personnel and their family members who lived in Jhansi were massacred shortly after the beginning of the rebellion, an incident from which Laxshmi Bhai chose to distance herself. For a brief period Laxshmi Bhai resumed the government of Jhansi at the request of the British and raised an army of 14,000 troops.

Hard pressed by neighboring princes who had invaded Jhansi, the rani requested British reinforcements. When no assistance had arrived by October 1857, Laxshmi Bhai sought allies among the rebel princes, who by that point desperately required her military support to offset recent defeats at Delhi, Kanpur, and Lucknow. By January 1858 a British force of 4,300 troops under Major-General Sir Hugh Rose had reached Mhow and Indore with orders to capture Jhansi, which controlled the strategic roads to Allahabad and Kanpur and contained a strong fort. The rani of Jhansi seems to have remained uncertain about whether or not to resist the British militarily until March 1858, but she spent this period of indecision in strengthening her armed forces.

Hampered by lack of intelligence and transport, the British army did not reach Jhansi until mid-March and moved hesitantly to invest Jhansi. Under the personal direction of Laxshmi Bhai, the city offered heavy resistance, particularly effective counterbattery fire in which local women assisted the rebel artillery crews. Deploying heavy siege cannon from Madras, Rose was able to destroy several bastions by the end of March, as well as the rebel powder magazine. The insurgent general Tatya Tope attempted to relieve Jhansi with 20,000 troops, but was driven back by Rose's forces. Running short of ammunition after a weeklong bombardment, the British commanders ordered a general assault on the walls of Jhansi. The city was taken on 3 April 1858 despite stiff resistance and house-to-house fighting. Laxshmi Bhai, having retreated to the citadel, escaped the same night with most of her troops and narrowly eluded pursuing British cavalry.

Joining other insurgent leaders at Kalpi, the rani of Jhansi joined in a council of war but was not elected to lead the rebel forces against the advancing British. Although Tatya Tope fought at Koonch upon the advice of the rani, he was compelled to retreat to Kalpi arsenal. The rebels were able to delay Rose's army for several days at Kalpi, but were eventually forced to disperse and regroup astride the route to Gwalior. The insurgent leaders were able to persuade the 8,500-strong garrison of the immensely powerful fortress to join them, against the orders of Maharajah Jayaji Rao Scindia, and obtained access to Gwalior's treasury. Still excluded from field command, however, Laxshmi Bhai separated herself from the other rebel commanders and decided to harass Rose's army with her cavalry. As the British forces approached Gwalior, the rani of Jhansi took up a defensive position near Kotah-ki-Serai on 17 June, where her forces received a charge by the 8th Queen's Irish Hussars. During this attack the rani was mortally wounded, and the rebel army was forced to evacuate Gwalior after light skirmishing. Despite continued guerrilla resistance under Tatya Tope, the remaining insurgents were defeated by the end of 1858.

While the rani of Jhansi was heavily dependent upon advisors and often slow to act in her political affairs, she displayed great independence and adaptability as a military leader. During the mutiny she displayed considerable practical knowledge of warfare and handled troops and resources effectively in both defensive and offensive operations. Her choice of tactics displayed not only considerable daring, but also a knowledge of her enemies' weaknesses and how best to exploit them. She maintained a high level of morale and loyalty among her troops and was one of the most successful insurgent commanders of the Indian Mutiny.

Although vilified by Major-General Rose and many of his officers as "the Jezebel of Jhansi," Laxshmi Bhai had begun to assume a legendary status before her death. Romanticized by many of her British adversaries, as well as by the people of the Bundelkhand region, she later became a symbol of Indian resistance to foreign domination during the political struggle for independence. Today Laxshmi Bhai is celebrated in fiction, poetry, song, and film as both patriot and feminist, and her brief career has recently attracted close scholarly examination.

Bibliography: Kincaid, *Lakshmibai, Rani of Jhansi, and Other Essays,* 1943; Lebra-Chapman, *The Rani of Jhansi,* 1986; Maharashtra (India) Department of Archives and Desai, *Relations of the Rani of Jhansi,* 1990; Mahasveta et al. *The Queen of Jhansi,* 2000; Smyth, *The Rebellious Rani,* 1966.

—James W. Hoover

BIANCA DI ROSSI (died ca. 1253, Bassano, Italy), also known as Bianca della Porta, Blanche of Rossi, Bianca de' Rossi of Parma. Defender, siege of Bassano, Italy (1253).

Bianca di Rossi fought alongside her husband against the forces of the notorious tyrant Ezzelino IV da Romano at the siege of Bassano in 1253. Bianca was a daughter of the powerful di Rossi, the leading Guelf family in Parma, Italy. She married Giambattista della Porta, governor of Bassano. She was apparently a historic figure, but her tragic end, which recalls the ancient rape of Lucretia, inspired legendary retellings in poetry, historical fiction, theater, and opera. Many variants have entered her story, but the basic outline is as follows: after her newly wedded husband fell in battle, Bianca refused to surrender and fought until she was taken prisoner in close combat. Ezzelino made amorous advances to her that she spurned. Determined to defend her honor and that of her homeland, she threw herself from a window in a suicide attempt that resulted in a broken shoulder. After her injury had healed, the tyrant succeeded in forcing himself on her. Outraged and humiliated, she asked permission to once more kiss her husband's body. She placed her head under the heavy cover of his stone sepulcher and then knocked the prop away so the lid fell, crushing her.

Bibliography: Brentari, *Ecelino da Romano,* 1994; Cantù, *Ezelino da Romano,* 1852; Ortalli, "Ezzelino," 1992.

—Gloria Allaire

BIANCA MARIA VISCONTI SFORZA (born 31 March 1425, Abbiategrasso; died 19 or 23 October 1468, Melegnano, Italy). Duchess of Milan. Military organizer, Italy.

Daughter of one great lord and wife of another, Bianca Maria Visconti Sforza often governed territories and directed military enterprises in the course of building an important northern Italian state. She was the only daughter (illegitimate) of Filippo Maria Visconti, duke of Milan. After years of negotiations, on 25 October 1441 she became the consort of Francesco Sforza (later duke of Milan).

Well educated, dignified, energetic, and intelligent, she employed the best humanistic tutors of the day to educate her eight children. She governed capably in the absence of her campaigning husband and accompanied him into exile according to the exigencies of their political fortunes. She did not hesitate to assume military roles as the occasion demanded to help establish and preserve the dynasty.

As her spouse defended his territories against the ongoing expansionism of Venice, she sometimes joined him, riding on horseback "not like a lady but as a valorous captain," according to the contemporary chronicler Campi (Denti, 95). On those occasions when she reviewed the troops, she rode in parade armor accompanied by a troop of liveried Amazons.

In the winter of 1447/1448 the Venetian fleet needed to pass the bridge at Cremona, her dotal city, in order to attack Francesco, who was stationed at Pavia. Bianca Maria headed a squadron of faithful Cremonese against the Venetians. When they disembarked to try to destroy the bridge, she rode out in her parade armor and threw a spear at a Venetian attacker, hitting him in the mouth. Her display of courage rallied the defenders, who drove off the enemy with heavy Venetian casualties. Another time Bianca Maria arrived in Francesco's camp as he was about to abandon a siege due to inclement weather. At her urging, a new bombardment was begun until the resisting city's walls collapsed and the inhabitants yielded.

In 1449, after a three-year war of succession, the influence of her family name on the Milanese as well as their affection for her helped effect the overthrow of the Ambrosian Republic, which had been set up after her father's death.

Portrait of Bianca Maria Visconti. (Perry-Castañeda Library)

In 1450 Bianca Maria and her husband made their triumphal entry into Milan on horseback at the head of the troops, disdaining to ride in the chariot that had been prepared. About two years later, while Francesco was away on the Brescia campaign (ca. 1452), Bianca Maria ordered a military expedition to suppress a rebellion at the castle at Monza. Some chroniclers state that she went there personally to direct the action, but other documentary evidence seems negative on that point. After her husband died in 1466, she demonstrated her political acumen, serving as regent and preserving the duchy for her son Galeazzo.

Bibliography: Collison-Morley, *The Story of the Sforzas*, 1934; Denti, *Storia di Cremona*, 1985; Pizzagalli, *Tra due dinastie*, 1988; Terni de Gregorj, *Bianca Maria Visconti*, 1940; Trease, *The Condottieri*, 1971.

—Gloria Allaire

BOCANEGRA, GERTRUDIS (born 1765, died 1818, Mexico). Soldier, revolutionary, Mexico, War of Independence against Spain (1810–1823).

Gertrudis Bocanegra participated in the 1810 War of Independence in the state of Michoacan, Mexico. She joined her husband and son in supporting the rebellion against the Spaniards. Bocanegra established a spy network of women tracing the movements of Spanish forces. After her husband and son were killed, she joined the forces of her son-in-law and carried out whatever mission was assigned to her. She raised a battalion of armed women and led them until she was arrested. Bocanegra refused to give the names of her fellow rebels even under torture. The Spaniards decided to make an example of her by having her executed by a firing squad on 11 November 1818.

Bibliography: Cotera, *Diosa y Hembra*, 1976; Escobedo, *Galeria de mujeres Ilustres*, 1967; Gugliotta, *Women of Mexico*, 1989; Miles, *The Women's History of the World*, 1989.

—Elizabeth Salas

BOCHKAREVA, MARIIA LEONT'EVNA (born July 1889, Nikolsko, Russia; died?), also known as Botchkareva, Yashka. Commander, First Petrograd Women's Battalion of Death. Captain, infantry, Russia, First World War.

Mariia Bochkareva was the first Russian woman to command a military unit. She was exposed to the military at an early age: her father, a sergeant in the imperial army, fought in the Russo-Turkish War, and at age sixteen Bochkareva married a soldier. The marriage was unhappy, and at the outbreak of the First World War Bochkareva volunteered for service but was rejected by the commander of 25th Reserve Battalion in Tomsk, who suggested that she approach the Red Cross instead. Determined to distinguish herself in battle, Bochkareva petitioned the tsar and was eventually enlisted as a regular. After three months of training, she began front-line duty with the 5th Corps, 28th Regiment of the Second Army, stationed in Polotsk. During this term of service she was decorated for rescuing fifty wounded soldiers from the field. While recuperating from wounds to the arm and leg, Bochkareva worked as a medical sister until her return to the front as a corporal in command of eleven men. On her recovery from an additional wound that had left her paralyzed for four months, Bochkareva once more returned to the front as a senior noncommissioned officer responsible for provisioning a platoon of seventy men. Exhausted by personal injury and disheartened by the demoralization exhibited by the troops, she requested discharge from service in the spring of 1917, after the February Revolution ended the monarchy and brought the Provisional Government to power.

According to Bochkareva, she first recommended the formation of an all-female battalion when President Rodzianko asked if she could offer a solution to the problem of army morale; she believed that a women's battalion would shame the men into more vigorous support of the war effort. Once she agreed to assume responsibility for the recruits' personal conduct, Bochkareva's formal proposal was approved by the army commander in chief, General Brusilov, and on 17 May she approached the war minister, Aleksandr Kerensky. Though army regulations precluded female enlistment, Bochkareva was granted special dispensation by Kerensky to proceed with the formation of an all-female battalion. At a public meeting held in the Maryitskiy Theatre, Bochkareva called for volunteers and was overwhelmed by the response: 1,500 women volunteered that night, and eventually 2,000 pledged themselves to enlist. Bochkareva's volunteers were screened only for poor health, but her authoritarian methods quickly thinned the ranks to about 300 recruits. Hurried training was led by twenty-five male instructors from the Volunskii Regiment of the Petrograd Military District.

Their battalion standard was consecrated at St. Isaac's Cathedral on 25 June 1917, Bochkareva's group was assigned to the First Siberian Corps, and Bochkareva was promoted to the rank of lieutenant. During the July offensive, while attached to the 172nd Division, 525th Kuriag-Daryivski Regiment, Tenth Army, stationed in Beloe, Bochkareva ordered her troops to lead a charge "over the top." Her advance, which was joined by several regimental officers, roused the recalcitrant troops, and during the encounter Bochkareva's troops took 200 prisoners. However, the majority of the Russian soldiers refused to support her attack, and Bochkareva's battalion was forced to withdraw, incurring heavy casualties.

After the collapse of the July offensive, Bochkareva returned to Petrograd, where she was congratulated by Rodzianko and Kerensky on the fine example set by her battalion. Over the summer of 1917 **Women's Battalions of Death** modeled on Bochkareva's were formed in many regions of Russia. Repelled by rumors of lax morality in other units and horrified by the "rouged and wantonly dressed" soldiers she reviewed in the Moscow Battalion, Bochkareva took great pains to maintain discipline among her own troops. During the first days of her battalion's formation she dismissed eighty female officers for frivolous behavior. She was a martinet who believed in rigid hierarchies of authority; at one point she threatened to disband the unit rather than comply with Kerensky's order to introduce soldiers' committees to discuss and vote on issues. Bochkareva's concern with conduct was no less rigorous in the field: while her battalion was trapped by enemy fire in July, she encountered a pair of lovers in the woods "in the act" and claimed that she promptly ran her bayonet through the woman soldier's back.

Like General Kornilov, Bochkareva violently opposed the Bolsheviks and supported Kornilov's plan to restore strict army discipline. Bochkareva's identification with Kornilov's cause proved dangerous, however, as revolutionary activity intensified during the late summer of 1917. Fired on at one point by Russian troops and fearing mob violence, Bochkareva requested a transfer. Though she was promoted to the rank of captain, she disbanded her battalion after the October Revolution. Arrested on two occasions for alleged affiliation with Kornilov, on her release in late 1917 Bochkareva boarded a steamer out of Vladivostok and sailed to America. There she dictated her memoirs (published with her name oddly transliterated as Botchkareva), but apparently returned to Russia later to fight against the Bolsheviks in the **Russian Civil War.** Nothing is known of her subsequent fate.

For her service at Polotsk, Bochkareva was awarded the **Order of St. George,** Fourth Degree. For assistance to the wounded and the capture of German sentries while incarcerated by the Germans during a Russian counteroffensive, she was recommended for the Order of St. George, Second and First Degrees, but, as Bochkareva noted, "being a woman ... received only a medal of the 3rd Degree" (Botchkareva, 131).

Bibliography: Abraham, "Mariia L. Bochkareva," 1992; Botchkareva, *Yashka,* 1919; Bryant, *Six Red Months in Russia,* 1918; Drokov, "Organizator Zhenskogo batal'ona smerti," 1993; Senin, "Zhenskie batal'ony," 1987; Stites, *The Women's Liberation Movement in Russia,* 1978.

—Mary Allen

BOGAT, A.P. (born 1898, Ekaterinoslav, Ukraine; died?). Platoon commander, Budenny's Cavalry, Red Army, Russia, Russian Civil War.

A.P. Bogat wrote the first comprehensive study of women's participation in the Revolution and **Russian Civil War**. She began her military training in the Kharkov underground, which was engaged in struggle against Petliura's Directory. With establishment of Bolshevik control of the region in January 1919, she was named commissar of all Red Crescent hospitals in Kharkov. After the Bolshevik retreat from the city on 24 June 1919, Bogat joined a military force that was operational in the Sumy sector. In her next appointment as commissar in the Sanitary Administration of the 41st Rifle Division, Fourteenth Army, Bogat held responsibility for supervising evacuation of the wounded from front-line regions. In this capacity she participated in battles at Glukhov, Dmitriev, Dmitovsk, Kharkov, Nikolaevsk, and Odessa and in the liquidation of Nestor Makhno's bands. During these engagements Bogat was directly involved in combat on several occasions: in the struggle with Makhno's forces, for example, she captured Makhno's aide-de-camp along with his operational documents. After contracting typhus during the fight for Odessa in 1920, Bogat was taken out of formation, but on recovery enrolled in the 1st Cavalry school in Moscow. In 1922 Bogat graduated and enlisted in Budenny's Cavalry Army, where she served over a period of two years as platoon commander in the 19th Cavalry Regiment and as chief of reconnaissance in the 21st Cavalry Regiment. In 1924 she began studies in military academy, from which she graduated in July 1927. At this point heart trouble forced her demobilization, and after eight years of service in the army Bogat enlisted in the reserves, where she continued to exercise her military skills by providing women's defense training in **Osoaviakhim** organizations.

Bibliography: Bogat, *Rabotnitsa i krest'ianka v Krasnoi Armii,* 1928, *Rabotnitsa i krest'ianka na strazhe SSSR,* 1930, and "V ogne Grazhdanskoi Voiny," 1967.

—Mary Allen

BONA OF LOMBARDY (fl. 1440s–1470s?). Mercenary soldier, Italy, fifteenth century.

A historical figure whose life has given rise to romantic legends, Bona of Lombardy was a mercenary soldier active at the time of the famous Italian dynasties. Little is known of her life. She was apparently a robust peasant girl from Valtellina in the Alps of Lombardy whose muscular physique reflected her hard life. She wore armor and lived in military camps. Her lover Piero Brunoro of Parma loyally served for many years as lieutenant to the great warlord Francesco Sforza. Piero led 2,000 knights in a ceremonial procession for the marriage of Sforza and **Bianca Maria Visconti (Sforza)** in 1441. Bona accompanied Piero throughout the course of his career as a professional soldier. Piero and Bona participated in Sforza's many military campaigns to consolidate his northern Italian duchy.

Piero was unjustly accused of betraying the town of Fabriano in the March of Ancona and was thrown into an Aragonese prison around 1444. For ten years Bona ceaselessly petitioned, in person and through friends, King Alfonso, the kings of France and England, and other European notables until she secured Piero's release. Alfonso finally relented, and Bona and Piero were married. They both entered the service of the Venetian Republic and fought to defend Negroponte against the Turks. Piero was killed, and legend has it that Bona died of grief.

Bibliography: Collison-Morley, *The Story of the Sforzas,* 1934; Pizzagalli, *Tra due dinastie,* 1988; Terni de Gregorj, *Bianca Maria Visconti,* 1940.

—Gloria Allaire

BORDEREAU, RENÉE (born 1766, Soulaines; died 1822, Vezins [Maine-et-Loire], France), also known as "L'Angevine." Soldier, France, 1793–1815.

Of the women known to have fought in the ranks of the "Armées catholiques et royales" during the War of Vendée (1793–1796), like Marie Lourdais or Françoise Desprès, the most remarkable figure is surely Renée Bordereau. Born in a peasant family south of Angers, she may have done some smuggling during her youth, carrying illegal salt between Maine and Brittany. In 1793, as peasants rose in "Vendée" against the **French Revolution,** she followed her father in the riots, during which he was arrested and executed by revolutionaries in December. Dressing in men's clothes and riding a horse, the thirty-year-old Bordereau took part in all the battles of the war: in Nantes, Luçon, Chemillé, Laval, Dol, Antrain, and Le Mans. Fighting at the front line, she commanded about twenty men. She reportedly cut the neck of her own uncle, a republican officer. Bordereau was described by **Marquise de la Rochejaquelein,** who was secretary to her husband in the royalist wars: "There were three or four other women who fought; one belonging to Bonchamps' army followed her father, and the story goes that seeing him fall in an engagement at the Ponts-de-Cé, she was roused to such a heat of vengeance that she slew nineteen men with her own hand. She was of average height and very plain. She was pointed out to me one day at Cholet: 'Look at that soldier with sleeves a different colour from his jacket; it is a girl who fights like a lion.' Her name was Renée Bordereau, known as the Angevin, and she served in the cavalry; she was renowned throughout the army for her extraordinary valour. She is still alive, having fought in all three wars with the most signal courage."

Bordereau went through Brittany during the "Virée de Galerne" in late 1793 and fought continuously from 1794 to 1795. She once managed to kill two opponents at the same time, with a pistol and a sword. After three years of "false peace" she fought again during the small war of 1799. She worked a short time as a butcher in Cholet, but was spotted by the imperial police and jailed from 1809 until 1814 in Angers and in the fortress of Mont-Saint-Michel, until the Restoration gave her liberty. She met King Louis XVIII and dictated her memoirs (she was illiterate) during this year. When she was nearly fifty, she fought again in the war against Napoleon in 1815, then retired in the village of Vezins, where she was commonly dressed in men's clothes. Bordereau died in 1822.

Bibliography: Bordereau, *Mémoires de Renée Bordereau*, 1814; Gilbert, *Brave l'Angevin*, 1976; Rochejaquelein, *Memoirs*, 1933; Martin, *La Vendée et la France*, 1987; Urzureau, "Une amazone vendéenne, Renée Bordereau," 1947.

—Jean-Clément Martin

BOUBOULINA, LASKARINA (born 11 May 1771, Constantinople, Turkey; died 22 May 1825, Greece), also known as "Kapetanissa." Naval commander, Greek War of Independence (1821–1832).

Laskarina Bouboulina was the daughter of Stavrianos Pinotsis, a ship's captain from the island of Hydra. She was married twice: to Dimitrios Vannouzes (1788–1797) and to Dimitrios Bouboulis (1801–1811), both wealthy ship's captains from Spetses, and both killed by Algerian pirates. Bouboulina bore a total of six children.

She successfully defeated Turkish attempts to confiscate her late husbands' vast wealth, took her inheritances, and joined the Filiki Eteria (Company of Friends), the secret organization dedicated to the liberation of Greece from the Ottoman occupation, which had then lasted some 400 years. She had built the largest ship in the Greek fleet of her time (the *Agamemnon*) and maintained this and four more ships, as well as ground troops, all used under her command against Turkish naval and land forces. When the Greek War of Independence formally broke out in 1821, Bouboulina was fifty years old, "handsome as an Amazon," and was nicknamed "Kapetanissa" (lady-captain) by her crew and men. She led her ships in the battles to break the Turkish naval blockade at Nafplion and, fighting alongside her men, led the landing at Myloi. She lost a son during the assault on Kehajambeis. She participated actively in the naval blockade engagement at Monemvassis, the siege at Tripolis, and the campaign and capture of Dramalis. Bouboulina was one of the first freedom fighters to enter the liberated city of Tripolis, riding a white horse, and joined Theodoros Kolokotronis and other military heroes of the revolution in going from village to village throughout the Peloponnesus to keep up the spirits of the population.

A similar figure is Manto Mavrogenous (died 1848), another female hero of the Greek War of Independence. Described as a beautiful woman of Mykonos, she used her wealth to outfit and command two ships that protected the island. She also organized troops and fought in male attire against the Turks. It should be noted that while women like Bouboulina and Mavrogenous became well known for their front-line military actions during the 1821 War of Independence, most Greek women were active in support of their fighting men. At great risk they brought food, water, and ammunition to the front lines, loaded muskets during military engagements, slipped through the lines with messages, and contributed in any way they could to the overall military effort.

Bouboulina lived in liberated Nafplion until 1825, when she returned to Spetses. She died there that same year, the victim of a random shot fired during a family altercation. She was interred at the Church of St. John (Ayios Ioannis), which she had built and owned. Bouboulina's wealth, her willingness to spend it all on the war, and a certain panache in her actions all served to single her out as the epitome of the war's heroines. Bouboulina's fame and military feats were commemorated in historical accounts, legends, and folk songs, and a street in downtown Athens bears her name. Statues of Bouboulina and Mavrogenous can be found in the War Memorial in Athens.

Bibliography: Brown and Safilios-Rothschild, "Greece: Reluctant Presence," 1982; Kordatos, *History of Greece*, 1956; Paparrigopoulos, *History of the Greek Nation*, 1955; *Pyrgos Encyclopedia*, Greek language (Athens: Greece, 1931); Trikoupis, *History of the Greek Revolution*, 1968.

—Diane Louise Chatelan

BOUDICCA (fl. first century CE), also known as Boadicea. Queen of the Iceni, England; led military revolt against Rome.

The date of Boudicca's birth is unknown; she appears on the historical scene in 60 CE upon her accession to the chieftainship of her tribe, the Iceni, in the area of England now called East Anglia. Her husband Prasutagas died in 59, and she inherited his position. It was the inheritance of his wealth, however, that was one of the reasons she became a notable figure.

The Iceni lived under the control of Rome, whose forces under Julius Caesar had first invaded Britain in the middle of the first century BCE. The emperor Claudius had presided over the defeat and subsequent domination of southeastern England, and eleven tribes had sworn loyalty to Rome. Most historians believe that the Iceni under Prasutagas were one of these tribes, for they existed as a client tribe that prospered under Roman rule. Prasutagas was renowned for his wealth and had the foresight (or so he thought) to attempt to bribe the Romans in his will. He left half his estate to Boudicca and their two daughters and half to the emperor of Rome, then Nero. Perhaps if he had not done this, he would not have called attention to his people, but the Roman governor decided that half the wealth was insufficient. The procurator Catus Decianus seized it all, then had Boudicca flogged and her two daughters raped in the bargain. Further, he told the Iceni that gifts they had received from the emperor were not gifts at all, but merely loans that were now due for repayment.

Bronze statue at the House of Commons in London by Thomas Thorncraft, 1902. (Library of Congress)

This high-handed conduct was one of the factors in the revolt that followed. It probably would not have been sufficient, however, to motivate other tribes to ally with the Iceni. The others had suffered at Roman hands as well, though. In order to reduce the expense of maintaining a garrison in Britain, Rome had awarded its retiring veterans land grants, acquired at the expense of local landowners, and many of the veterans had seized other lands as well. In addition, Roman authorities had forced the local population to provide the money for the construction of a large temple, dedicated to the spirit of the emperor Claudius. Although the Romans for the most part exercised much religious tolerance of the beliefs of conquered peoples, forcing them to pay for a temple they would never use angered not only the Iceni but the other major tribe in the area, the Trinomantes. Thus high taxation and a cavalier attitude toward the property rights of the local tribes had already laid a foundation of dislike and mistrust.

Thus when Boudicca called on the tribes for aid, her plea fell on responsive ears. The Iceni armed themselves and prepared for a major campaign. The Iceni and their neighbors had no problem following a woman commander, for women held a role in that society that approached equality; women tribal leaders were not uncommon. It has been suggested that her gender aided in the recruitment of allies, although not because of the brutality she and her daughters had suffered. Some writers suppose that the Druids (priests of the Celtic religion) presented her as an embodiment of Andraste, the Celtic goddess of victory. The Druids also had a score to settle with the Romans, for the governor Suetonius Paulinus was in the midst of campaigning against a Druidic stronghold on the island of Mona (Anglesey), off the northern Welsh coast.

Whatever their motivation, tribes joined Boudicca's rebellion. The main source of information on this campaign comes from the Roman historians Tacitus and Cassius Dio; the latter claims that the rebels numbered some 120,000. They certainly had sufficient force to completely overwhelm and destroy their first target, the town of Camulodunum (Colchester). Tacitus writes that agents in the town had persuaded the Romans that an uprising was unlikely. That, coupled with the fact that they had spent their time and money constructing the temple to Claudius rather than defensive walls, made them easy prey. A handful of soldiers retreated to the temple and held out for two days, but they too were overcome and slaughtered like the rest of the city's population. Legend has it that Boudicca rode in a chariot with scythed wheels, although many historians believe that the British tribes used chariots more for transport than combat. Whatever the details, the rebels' first victory was complete, and they looked toward Londinium (London).

> *A historian on the importance of women's history:*
>
> The elimination of most of the human race from the historical record shrinks our human identity. We don't know fully who we are. We know even less what we might become.
>
> —Elise Boulding, *The Underside of History: A View of Women through Time* (Boulder, CO: Westview, 1976), 4.

The bad news traveled quickly, and Suetonius soon learned of the uprising and the destruction of Camulodunum. He broke off his campaign in Wales and rode with his cavalry for London, leaving his infantry to make its way as quickly as possible. He actually arrived before Boudicca, for it seems that the Iceni spent too much time plundering rather than marching. Still, Suetonius abandoned Londinium as indefensible and retreated northeast toward his advancing infantry. London thus fell to Boudicca's forces as easily as had Camulodunum. Both cities were the sites of awful slaughter, for Tacitus and Dio note that the Iceni and their allies took no prisoners and showed no mercy. This time they marched out of the city quickly after capturing it, intent on finding Suetonius and defeating him. They captured and sacked Verulamium (St. Albans) on the way, but it proved their final victory. Somewhere northeast of that city Suetonius held high ground with perhaps 10,000 men. The exact site of the battle that ensued is unknown, but superior Roman discipline in unit fighting broke the rebel forces, and the resulting carnage was immense. Some 70,000 Romans had died at the hands of Boudicca's army, but the Romans proceeded to inflict 80,000 to 100,000 deaths on the Britons. The massed wagons of families and supplies behind the rebel force hindered their retreat, and the Romans killed every man, woman, and child they could find.

Boudicca escaped but could not rebuild her rebellion. The place, time, and manner of her death are unknown; it is said that she died of disease (Dio) or by her own hand (Tacitus). Her defeat helped the Romans solidify their hold on England, for Suetonius proceeded to exact revenge for the rebellion. The intensity of his vengeance ultimately induced the Roman government to call him home. His replacement was much calmer, and soon the island's economy stabilized and Romanization proceeded with little trouble.

Boudicca is known in England more for her legend than for her actual deeds. She was immortalized by Ben Jonson, as John Fletcher's *Bonduca*, as William Cowper's *Boadicea,* and in Tennyson's poem "Boadicea." It is by the latter name that she is best known to moderns; a statue to her stands by the Thames River. She became the prototypical female warrior to Britain, and Prime Minister **Margaret Thatcher** often found herself compared to the woman who defended her homeland from invaders and presided over the deaths of her enemies.

Bibliography: Cassius Dio, *Roman History,* vol. 8; Dudley and Webster, *The Rebellion of Boudicca* 1962; Fraser, *The Warrior Queens,* 1989; Macdonald, "Boadicea: Warrior, Mother, and Myth," 1987; Tacitus, *The Annals,* 1952.

—Paul K. Davis

BOUILLON, ROSE (born February 1764, France; died ?). Soldier, infantry, France, French Revolution.

Entering the French army in disguise, Rose Bouillon served in 1793 in the Sixth Battalion of the Haute-Saone in the Army of the Moselle until her discharge six months later. When she enlisted in 1793 at age twenty-nine, Bouillon left her two children with her mother in order to serve with her husband, Julien Henri, on the eastern front of the French conflict. During the summer of 1793 her battalion was engaged in heavy combat and her husband was killed. Despite his death, she continued her service. Only after completing the campaign did Bouillon petition for and receive a discharge to return to the care of her children.

On 25 August 1793 the National Convention accorded Bouillon "honorable mention" for her courage in defending France and granted her a pension of 300 livres per year. The representatives of the government also granted pensions of 1,560 livres to each of her children (her daughter until the age of fourteen and her son for life). Her service to France was further recognized in the government-sponsored *Annales du civisme et de la vertu* (1793).

Bibliography: Brice, *La Femme et les armées,* n.d., 321; Dossier XR48, "Bouillon," Service historique de l'Armée, Vincennes, France; France, *Procès-verbal de la Convention nationale,* 27 August 1793.

—S.P. Conner

BRANT, MARY (born ca. 1736; died at Kingston, Canada, 16 April 1796), also known as Molly, Konwatsi'tsiaienni (someone who lends her a flower). Tribal leader, Six Nations Tribe, Canada. Military organizer, British ally, American Revolution.

Molly Brant, sister of the Mohawk chief Joseph Brant, served in a key political and military capacity and is generally known as one of the most important women in Native American history. She was the leading matron of the extensive and important Six Nations Tribe and rendered invaluable assistance to the Crown by encouraging the tribe to keep its alliance with England during the American Revolution. She also provided important information to the Loyalist and British armies during this conflict, playing a pivotal role in the capture of Fort Stanwix and the British victory at the Battle of Long Island. Many historians speculate that her purpose in remaining at Fort Johnson was to be in a position to relay significant information obtained from valuable Native American sources to the Loyalist forces. Molly Brant also significantly aided the negotiations between the British and the Six Nations Confederacy. One contemporary noted that "one word from [Molly] Brant is more taken notice by the Five Nations than a thousand from the white man without exception." Her role during the American Revolution, her ability to influence her tribe, and her work with her brother Joseph in defusing the anger following the announcement of the terms of the Treaty of Paris caused government officials to regard her as a woman of considerable importance to the welfare of the Canadian colony.

Bibliography: Prentice et al., *Canadian Women,* 1988; Robinson, *Mistress Molly,* 1980, and "Molly Brant," 1982.

—Kristine Dawkins-Wright

BRASSEUR, DEANNA (born Pembroke, Ontario, 1954). Major, 416th Tactical Fighter Squadron, Canadian Forces, 1989. Fighter pilot, Canada.

Deanna Brasseur was one of the first three women to earn her wings as a Canadian Forces (CF) military pilot qualified for active duty (**Wendy Clay,** who graduated in 1974, never flew operationally). In 1980 Captains Brasseur, Nora Bottomley, and Leah Mosher graduated from Canadian Forces Flight Training School at Portage. In June 1989, following twelve months of training on CF-5 and CF-18 jet fighter aircraft, Brasseur and Captain Jane Foster became the only two women in the world flying fighters in operational squadrons. Canada was the first country to allow women to fly in a combat role since the Second World War, when the Soviet Union used women pilots. Brasseur was promoted to major in 1989 and posted to National Defence Headquarters, Ottawa, as director of flight safety, in March 1990.

Bibliography: Martinsen, "Canada's First Women Fighter Pilots," 1989; Milberry, *Sixty Years,* 1984; Render, *No Place for a Lady,* 1992.

—Patricia Bowley

BREWER, LUCY (born 1793?), also known as Louisa Baker, George Baker, Lucy West, Eliza Bowen. Marine sharpshooter, United States Navy, War of 1812.

Known only through a popular series of autobiographical pamphlets, Brewer was an ex-prostitute who reputedly served in disguise for three years as a marine sharpshooter on board the USS *Constitution* during the War of 1812. According to the pamphlets published in 1815–1816 by Nathaniel Coverly, Brewer was born in Plymouth County, Massachusetts. At 16 she became pregnant. With no hope of marriage and determined not to bring disgrace on her family, Brewer fled to Boston, where her child was born and soon died; she subsequently worked in a brothel. When the War of 1812 broke out three years later, Brewer determined to leave the brothel and see the world. One of her clients, apparently a lieutenant on board the American frigate *Constitution,* encouraged her to join the crew. Brewer was inspired by the story of **Deborah Samson,** who had served for three years in disguise as a soldier during the Revolutionary War. Brewer disguised herself as a man and, with her lover's assistance to avoid the usual physical examination, enlisted under the name George Baker. Assigned to the fighting troops as a sharpshooter, Brewer participated in the victories over the British frigates *Guerriere* and *Java*. She served on the *Constitution* throughout the war and was discharged in 1815. Brewer returned to her parents' farm and published her experiences in a series of pamphlets, later collected and published as "The Female Marine; or, Adventures of Miss Lucy Brewer." Her later experiences are unknown.

Some authorities dispute that Lucy Brewer ever existed, since there is no independent confirmation of her exploits. Alexander Medlicott, Jr., and Daniel Cohen claim to have checked records throughout Plymouth County and can find no record of a Lucy Brewer (or any of her aliases), and the muster rolls of the *Constitution* do not reveal a George Baker.

Her memoirs, however, were very popular at the time among women and served as an inspiration to other women of what they could accomplish. The United States Marine Corps itself, while producing a report that denied Lucy Brewer's existence, later realized the value of a woman serving in a nontraditional role. Lucy Brewer is now commemorated as the "first girl marine," and a street at Camp Lejune, North Carolina, is named for her.

Bibliography: Cohen, "The Female Marine," 1993; Kittle, "A Female Marine aboard the *Constitution,*" 1980; Medlicott, "The Legend of Lucy Brewer," 1966; United States Marine Corps, *The Legend of Lucy Brewer,* 1957; West, *The Female Marine,* 1966.

—Tim J. Watts

BROWNELL, KADY SOUTHWELL (born 1842, South Africa; died 1915, United States). Daughter of the regiment, 1st Rhode Island Infantry Volunteers ("Mechanics Rifles") and 5th Rhode Island Infantry Volunteers, Union army. United States, American Civil War.

Kady Brownell was born in the Kaffirs territories, South Africa, in 1842, where her father was a colonel in the British army. Upon the death of her mother she was sent to live with relatives in Rhode Island. She married Robert S. Brownell in April 1861, and when her husband enlisted in the 1st Rhode Island "Mechanics Rifles" of Providence three days after their wedding, Kady Brownell announced that she would join him. After initial opposition from the regimental commander, Kady was allowed to go with her husband and was soon named the "daughter of the regiment." Brownell devised a modified uniform for herself, consisting of pants and boots, topped with a knee-length skirt and blouse and cinched with a tasseled sash. She carried a sword and a rifle and practiced using them. Although only five feet tall, she was apparently a steady shot and a credible swordswoman.

It is reported that during the First Battle of Bull Run Kady Brownell, serving as color bearer, held her ground even after a general retreat had begun, and repeatedly placed herself in exposed positions to rally the men in the battles that followed. When the 1st Rhode Island disbanded in August 1861, the Brownells joined the 5th Rhode Island Infantry Volunteers and participated in campaigns in Virginia and North Carolina. While serving as a nurse at the Battle of New Bern, Brownell saw the color bearer of the 6th Rhode Island fall and, taking up the standard, was reportedly hit by an enemy ball as well. Her husband was wounded in the same battle, and Kady Brownell, whose wound was not as severe, chose to remain with him in the hospital. Upon recovery, John Brownell was declared unfit for duty, and the Brownells returned to Rhode Island in 1863

There was some dispute among regimental veterans about whether Brownell was actually on the field at New Bern. Although a journalist wrote in 1862 that Brownell was in the thick of the battle, and Brownell herself claimed the same in newspaper interviews, others said that they remembered her only as a brave nurse, but not a standard bearer. Still, her actions were described in poems and stories, and she eventually received a government pension in recognition of her service as daughter of the regiment. She died at age seventy-two in 1915, followed by her husband in the same year.

Bibliography: Hall, "They All Fought at Bull Run," 1991; Larson, "Bonny Yank and Ginny Reb," 1990, and "Bonny Yank and Ginny Reb Revisited," 1992; Moore, *Women of the War*, 1866; Seeley, *American Women and the U.S. Armed Forces*, 1992.

—Susannah U. Bruce

BRUNEHAUT (born 534, died 613?), also known as Brunhilde. **FREDEGUND** (born 545, died 597). Queens, instigators of war, Frankish kingdoms.

Brunehaut and Fredegund

The punishment of Brunehaut. (Dover Pictorial Archive)

The bloody feud between Brunehaut and her enemy Fredegund has taken on a legendary quality. Brunehaut has been accused of murdering five kings and inspiring decades of war, and she suffered a spectacular death. The equally bloody Fredegund was a military leader and strategist.

Brunehaut and her sister Galswintha were daughters of a Visigothic king. Brunehaut married Sigibert I, Merovingian king of Austrasia (the eastern Frankish kingdom), while Galswintha married Sigibert's brother, Chilperic I, Merovingian king of Neustria (the western Frankish kingdom). Unfortunately, Chilperic's mistress, Fredegund, violently opposed his marriage and had Galswintha murdered; Fredegund had earlier persuaded him to repudiate his first wife, and apparently a second marriage was more than she could bear. Fredegund's agents also assassinated Sigibert; since Fredegund later married Chilperic, she probably hoped to increase her husband's power by doing away with his rivals.

The loss of her husband and her sister drove Brunehaut to revenge, and she began a bloody feud that lasted for forty years. Brunehaut fought for the next three generations to secure the power of her sons, grandsons, and great-grandsons. She was an ardent champion of centralization of power in the hands of the monarchy, of tax reform, and of maintenance of Roman roads and monuments. Her efforts on behalf of Merovingian autocracy antagonized the local nobles, who rallied behind Fredegund's son Chlotar II.

Fredegund is described as being fierce, cruel, and adulterous, with a special affinity for political assassination. She was also a woman of considerable ability who was absolute mistress of her husband's passions and policies for two decades. She was actively involved in the military affairs of Neustria. For example, she devised a tactic involving the use of camouflage: retainers concealed themselves with masses of branches and foliage, then went out ahead of the Neustrian line as a kind of "moving woods." This allowed the soldiers to make a surprise attack; Fredegund's ingenuity caught her enemies unawares and resulted in victory for the Neustrian army over the superior combined forces of Austrasians and Burgundians.

The wars sparked by the bitter enmity of Brunehaut and Fredegund continued after the deaths of Chilperic and then of Fredegund in 597 and were the cause of prolonged military conflict and much bloodshed. Brunehaut continued to fight for her descendants, but around 613 she and her great-grandsons were captured. The men were strangled, while the elderly Brunehaut was tortured for three days, then tied by her hair, one arm, and one leg to the tail of a wild stallion and dragged to her death.

Bibliography: Bachrach, *Liber historiae Francorum*, 1973; Dill, *Roman Society in Gaul*, 1926; *The Origins of France*, 1982; Riché, *The Carolingians*, 1993; Wallance-Hadrille, *The Long-Haired Kings*, 1962.

—Maurice Webb

C

CANDACE. Rather than a particular personality, Candace was a title in ancient Ethiopia referring to the queen mother; the term *qere* was used in designating the ruler.

The most notable Candace was the one mentioned in passing in the Bible, whose prime minister met with the apostle Philip. That probably was not the same Candace mentioned in 7 BCE by the Roman historian Strabo, who described her as the queen, living in the royal city of Napata although the capital city was Meroe. Strabo's Candace may have been named Amanirenas, and Strabo described her as "a masculine sort of woman, and blind in one eye."

Apparently Candace Amanirenas ordered a raid against a Roman outpost in southern Egypt, returning with loot, prisoners, and stolen statues. The Roman consul Petronius pursued, and as he approached Napata, Candace sent envoys offering to return her prisoners. Petronius ignored this offer and launched an assault against Napata that razed the city, though Candace escaped. He withdrew his forces to Premnis and lodged a garrison in fortifications there, then returned to Alexandria.

Upon hearing news that Candace was leading a force against Premnis, Petronius returned and reached the fort before she did. Rather than attack the newly reinforced garrison, she again sent envoys hoping to negotiate with Julius Caesar. Petronius sent them on to Alexandria, where they successfully negotiated with Caesar, who not only forgave them for their aggression but returned tribute they had earlier surrendered.

Bibliography: Fage, *The Cambridge History of Africa*, 1978; Strabo, *Geography*.

—Paul K. Davis

CARROLL, ANNA ELLA (born 1815; died February 1894). Pamphleteer, military commentator, and writer, United States, American Civil War (Union).

Anna Ella Carroll wrote political pamphlets for the Lincoln administration during the **American Civil War.** She claimed to have devised the Union's plan to invade the Confederacy via the Tennessee River, which proved a successful military strategy. There is strong evidence that others conceived of the Tennessee River plan before Carroll, and there is no proof that Carroll actually presented her plan to the Union government. However, she had many supporters, and some historians and feminists later glorified her alleged contribution to Union military strategy. Carroll's case has received far more attention than the real contributions of women like **Jennie Hodgers, Sarah Rosetta Wakeman,** and Dr. **Mary Walker** (see also American Civil War).

Anna Ella Carroll. (Reprinted from Carroll, *The Great American Battle*, New York and Auburn: Miller, Orton & Mulligan, 1856)

The legend of Anna Ella Carroll of Maryland is one most students of the American Civil War run into sooner or later. Carroll was a mid-nineteenth-century political pamphleteer and writer who worked for Millard Fillmore and other politicians, and for the American, or Know-Nothing, Party in the 1850s. In 1861, when the war began, she wrote letters to newspapers supporting Abraham Lincoln and eventually won a verbal contract from the War Department to produce pamphlets in support of the president's actions. In November 1861 she allegedly came up with the Union's strategy to invade the Confederacy in the west by going up the Tennessee River instead of down the Mississippi River. The Tennessee and its companion river, the Cumberland, were not strongly fortified and provided the Union with a route of invasion that would drive the Confederates out of a largely untenable position in Tennessee and relieve the Union loyalists, particularly in the eastern part of that state. Furthermore, if Union gunboats were damaged in any action on the rivers, they would float downstream, north, into Union territory. The plan was good commonsense military strategy.

Determined to milk her agreement with the War Department for all it was worth (Carroll's finances were always precarious), she went to Tennessee early in the war. She talked to a riverboat pilot named Charles Scott, returned to Washington, and drew up a map of operations for the Tennessee campaign. According to Carroll, she then met with the assistant secretary of war, Thomas A. Scott, with whom she believed she had an agreement, and presented her plan.

Unbeknownst to Carroll, however, plans by the military were already well under way to follow the route she suggested. General Ulysses S. Grant had captured Smithville and Paducah at the mouths of the rivers and was awaiting gunboats to venture up the rivers into Confederate territory. By February, when Grant and Admiral Andrew Foote captured Forts Henry and Donelson on the rivers, then drove toward the Confederate rail junction at Corinth, Mississippi, Carroll was elated. It was clear to her that the Union had used her plan—obviously so, since they had done exactly what she had told Scott to tell them to do. Convinced of the value of her military strategy, Carroll presented a bill to Congress shortly after the war's end for strategical services rendered. The Congressional Committee on Military Affairs turned her down flat. Carroll was enraged at her treatment and certain of two things: she had saved the Union from destruction, and the only reason she was not being properly rewarded was because she was a woman and a civilian.

In order to convince Congress to pay her, Carroll reminisced about her wartime contributions in numerous pamphlets and petitions. At first, it was a simple-enough story. In April 1865 Carroll wrote a letter to the *National Intelligencer* in Washington, D.C., to tell the American public that it should render heartfelt thanks to Captain Charles Scott of Tennessee, a riverboat pilot, for informing the War Department of the efficacy of invading the Confederacy by using the Tennessee River instead of the Mississippi. General Grant used the plan, and Tennessee fell to Union forces.

But over the course of the next thirty years Carroll reconstructed the past, changing details of the events surrounding the creation of the Tennessee Plan, and inflating her role so she could convince the public and Congress that she deserved recognition as a war hero. By 1868 Carroll had devised the plan, not Scott. By 1876 the plan had meant the salvation of the Union, and Carroll deserved $250,000 and a major-generalship from a grateful people. By 1885 Carroll had explained to a naïve and childlike Grant the strategic importance of the rivers, a Grant who never knew till then why he had invaded Tennessee. By her death in 1894 Carroll had convinced a generation of suffragists that she had been denied recognition for her military strategy because of her sex, and they had called for a statue of her in the halls of Congress.

The problem lies first in whether a plan in Carroll's handwriting ever went to Thomas Scott in the War Department, and second, if such a plan was presented, whether it actually influenced military decision making. There are no War Department records to confirm that Carroll's plan existed, although Scott's testimony does seem to indicate that he saw something presented by Carroll. Carroll herself certainly believed (as did her followers) that she was capable of devising strategy; that is one reason why she got so much support, and why supporters saw the whole affair as a believable case of the military discriminating against a civilian and a woman.

The legend Carroll told, and her story as used or interpreted by others, highlights the expectations of gender-defined behavior in the mid- to late-nineteenth century as well as the time frame within which other writers told or interpreted her story. This woman convinced many important people of the validity of her version of the events surrounding the Tennessee Campaign of 1862 and presented alleged verification by public men such as Senators Benjamin Wade and Cassius Clay and Secretary of War Edwin Stanton. Carroll's "disremembering" of an accurate past provides a case study not only of the difficulties of using reminiscences as a primary source, but also of the desire of many Americans to find heroines in military history and to latch onto the first likely subject, accurately or not, as representative of the many nameless women who were there.

Carroll alleged in her pamphlets (about ten in all over a period of twenty years) that she must have been the first to come up with the plan and thus deserved recognition. But evidence in the *Official Record of the War of the Rebellion* and nineteenth-century newspapers such as the *New York Times* and the *Washington National Intelligencer*, as well as in papers of the leading military and political figures of the period, indicates that the military strategy was not exclusively Carroll's and in fact predated Carroll's map.

If Carroll did not submit a plan that taught the Union how to invade Tennessee, then how did she convince numbers of important people that she did? In both the *Official Record* and the *Times* the persons suggesting the strategy were civilians. Given that the military is a closed system, notorious for ignoring or downplaying the contributions of civilians, it was reasonable for Carroll's defenders to suggest that the army would not have paid any attention to suggestions of military strategy printed in the newspapers. In addition, there is correspondence in the *Official Record* between various officials in the War Department and civilian boat builder James Eads (who seems to have been the first to point out the importance of controlling the Tennessee and Cumberland Rivers), but given the general confusion that was part of the early months of the war, there is no proof that Eads's advice was actually used by the military. Grant usually took full credit himself for coming up with the Tennessee plan or cited his superior, General Henry Halleck. If Eads was denied recognition because of his civilian status, then it was plausible that Carroll might have suffered a parallel treatment. Proving Carroll's claim invalid thus cannot rest simply on the military records and newspapers.

Carroll supported her claim with statements from Senators Clay and Wade and also from Lincoln, Stanton, and even Grant himself. However, these statements, and Carroll's use of them, are suspect. To begin with, Carroll was not averse to lying about herself and her background: she always claimed a close kinship to Charles Carroll of Carrollton, for instance, though genealogical research indicates that if her family was in fact related to that revolutionary patriot, it was most likely as a black sheep. As far as her military claim to fame, Lincoln was long dead before she cited him, and Stanton had also died before corroboration was printed. Cassius Clay was in Russia in 1861 when Carroll formulated and presented her strategy, so his support for her claim is dubious. That leaves Benjamin Wade and General Grant. But Wade's most supportive letter, quoting Stanton's alleged deathbed praise of Carroll, was not printed until after Wade's death. From Grant there is only a letter to his friend Elihu Washburne in 1862 when the hero of the campaign was being sought, pointing out to Washburne that General Henry Halleck and Grant himself had thought of the invasion strategy throughout the fall of 1861, prior to Carroll's trip to Tennessee.

The degree to which Carroll reinvented her past regarding the Tennessee strategy can be seen most clearly in an article written by a supporter of her claim. H.R. Shattuck interviewed Carroll in 1885, when she was seventy years old and a cause célèbre among suffragists as an example of men's disregard for intelligent and nontraditional women. Carroll reminisced to her interviewer regarding a meeting she supposedly had with General Grant where she explained the importance of the plan to him. "He was like a child," she said (Shattuck, 408). "He asked me all about it . . . and until I described my plan to him, and showed him how success came from its consummation, I am convinced he had no conception that it was this that had brought him to victory." This unlikely alleged naïveté of Grant, added to the fact that no corroboration of this meeting exists, leads any historian to question the validity of Carroll's tale.

All the evidence seems to indicate that Carroll's claim to have been solely responsible for the Union's Tennessee invasion was wildly exaggerated, and she was not an unsung heroine of the Civil War. Thus the extent to which her life continues to be cited as an example of women's actions in the military sphere is truly remarkable. Carroll's story will not go away. She had letters of support from Susan B. Anthony and General E. S. Bragg, columns of fundraising pleas for her care in the leading suffragist magazines, and theater, radio, and television presentations on her life and work. In 1987 William Safire's novel *Freedom*, which covered the first two years of the Civil War, used Carroll as a major character and her claim as a major focus. Since then, at least five additional pieces on Carroll have come out, the latest retelling as a biographical introduction to an edited version of her magnum opus for the American Party, *The Great American Battle*, edited by James Whiskas. Carroll survives as a legend because of her gender. She represents to Americans an example of the ways in which historians have ignored women in the military.

Carroll said that Lincoln and Stanton would not publicly acknowledge her work because of her sex. Carroll's supporters argued the same; they have focused on the discrimination against Carroll and, by extension, against many women throughout the centuries, rather than on the facts in the case. Rebuttals of Carroll's claim based on official records, documents, and evidence have been ignored. Carroll's story survived not because of overwhelming evidence for its truth, but because of strong sympathy for an "unsung heroine."

In examining Carroll's claim as a historical problem, historians can follow a complete historiographical creation of a myth, from start to current status, through numerous incarnations in many modes of presentation, with the added level of gender analysis. Carroll serves as a case study of the difficulties in separating legitimate argument from dismissals of accomplishments based on gender expectations. She knew that her claim to have designed military strategy was something outside the limits of proper womanly behavior. A nineteenth-century woman active in the military in any form was suspect, but Americans knew that women had accomplished amazing feats during times of war. Thus Carroll's claim seemed plausible, as did the alleged discrimination against her. Carroll's legend survived for the same reason. The truth became irrelevant as the symbolic value of Carroll's story fulfilled the need of her proponents to have some woman restored to a place of honor in the field of military history. Unfortunately, they chose the wrong woman.

Bibliography: Coryell, "Anna Ella Carroll and the Historians," 1989, and *Neither Heroine nor Fool*, 1990; Shattuck, "Anna Ella Carroll: The Originator of the Tennessee Campaign," 1885; Snyder, "Political Strategist and Gadfly," 1973; Williams, "The Tennessee River Campaign and Anna Ella Carroll," 1950.

—Janet L. Coryell

CARTIMANDUA (fl. 43–70). Queen of the Brigantes, a Celtic tribal state in northern Britain.

Cartimandua is one of only six native British chieftains whose name is known. It is significant that another of these chieftains, a contemporary named **Boudicca,** was also female. Our only literary source, the Roman historian Tacitus, writes in the *Annals* that "it was indeed usual for Britons to fight under the leadership of women" (14.35) and in the *Agricola*, "Britons make no distinction of sex in their appointment of commanders" (16).

Tacitus says that "Cartimandua ruled the Brigantes in virtue of her illustrious birth" (*Histories* 3.45), indicating that she was descended from a royal family whose male line had died out. Surviving coinage supports this interpretation. While little is known of her realm, scholars agree that Cartimandua reigned over a confederation of tribes, although the extent of her authority is not clear. The political role of her first consort, Venutius, is debatable; was he a coruler, with responsibility for the northern portion of Brigantia? Tacitus is silent.

The Roman emperor Claudius invaded Britain in 43 CE, seeking both security from cross-Channel pirates and the glory of outdoing his illustrious ancestor Julius Caesar, who had settled for a reconnaissance in force in 55 BCE. By 47 CE the Brigantes were allies of the Romans, protecting the northern flank of the new province, Britannia. It is believed that one of the eleven unnamed British kings said to have submitted to Claudius on a triumphal arch dated to 52 was that of the Brigantes, probably Cartimandua herself.

Tacitus first mentions Cartimandua in connection with her loyalty to the Roman conquerors. In 51 Caratacus, a British chieftain hostile to the Romans, fled to her court after a defeat in battle with the invaders. Seeking an alliance with the Brigantes (many of whom opposed Cartimandua's Romanophile policy), Caratacus was instead captured by Cartimandua and sent in chains to the Romans. This act was motivated by her profitable alliance with the Romans, who had already helped her put down a "disturbance" around 48.

At this point Cartimandua's story, as told by Tacitus, begins to resemble a modern soap opera. In 69 (scholars disagree on the chronology) the queen and her consort, Venutius, had a falling out. Venutius was apparently seen as the heir of Caratacus's mantle and thus as the leader of anti-Roman forces in Britain. Cartimandua tried to forestall trouble by capturing Venutius's relatives. Tacitus states that this "enraged the enemy, who were stung with shame at the prospect of falling under the dominion of a woman" (*Annals* 12.40).

Cartimandua then deposed Venutius, replacing him as "partner of her bed and throne" with his armor bearer, Vellocatus. Tacitus, apparently outraged by this immorality, declares that "by this enormity, the power of her house was at once shaken to its base." The spurned Venutius now led an invasion of Cartimandua's territories. Faced with external invasion as well as internal revolt, the queen felt compelled to request aid from her Roman allies, who sent her auxiliary infantry and cavalry. "After fighting with various success, they contrived to rescue the queen from her peril. Venutius retained the kingdom, and we [Romans] had the war on our hands" (*Histories* 3.45). Cartimandua thus lost her throne and, having been a Roman ally, apparently lived out her days as a Roman subject. The Romans successfully subdued the Brigantes by 79.

Cartimandua, though less famous than Boudicca, was the more important figure, especially to the Romans. From the alliance Cartimandua had gained wealth and increased military power over her rebellious subjects, while the Romans enjoyed a quiet border to the north. Indeed, Tacitus's failure to mention the Brigantes in connection with Boudicca's revolt in 61 indicates that Cartimandua kept her subjects from joining it. Her loyalty saved the Roman regime in Britain from total collapse. The alliance was thus vital to both parties.

However, the failure of Cartimandua's marriage and the resulting civil war doomed the alliance and triggered Roman military intervention in Brigantia. Tacitus blames her "lust and savage temper," subscribing to what Antonia Fraser calls the Voracity Syndrome: Cartimandua's sexual appetite for Vellocatus ultimately destroyed her military capability. Renaissance commentators on the queen's career also criticized her perceived moral failings. Without more complete evidence, it is impossible to conclude whether Cartimandua deposed Venutius merely out of lust for another man, or whether their political disagreements were the real problem. Had she retained Venutius as her husband and allowed him to continue in his opposition to Rome, the outcome would have been much the same. In any event, Cartimandua's troubles cost the Romans much, but they obviously saw her allegiance as valuable.

Bibliography: Dudley and Webster, *The Roman Conquest of Britain, A.D. 43–57*, 1965; Hanson and Campbell, "The Brigantes: From Clientage to Conquest," 1986; Richmond, "Queen Cartimandua," 1954; Tacitus, *The Agricola*, 1970 and *The Annals*, 1952; Webster, *Rome against Caratacus*, 1982.

—Curtis F. Morgan

CATHERINE OF ARAGON (born 16 December 1485, Alcala de Henares, Spain; died 7 January 1536, Kimbolton, Huntingdonshire, England). Queen of England, military organizer and leader.

Undated portrait of Catherine of Aragon. (The Art Archive/Galleria degli Uffizi Florence/Dagli Orti [A])

The youngest child of Isabella of Castile and Ferdinand of Aragon, Catherine was married in 1509 to King Henry VIII of England. Catherine is best known as the first of Henry's six wives; it was his efforts to dissolve the marriage and England's consequent break from the Church of Rome that account for most of Catherine's historical fame. However, she also contributed importantly to the most decisive English military success in the period.

When Henry led a lavishly mounted expedition against France in 1513, he named Catherine to be regent of England in his absence. Along with the usual responsibilities of governance, she gave direction to the continuing efforts to supply Henry and the army in France. She undertook the defense of England against the customary attack from Scotland that could be expected in the king's absence. King James IV of Scotland, although married to Henry's sister, adhered to the traditional Scottish alliance with France and in August invaded an England stripped of its best soldiers and equipment. Catherine organized the collection of monies, supplies, and soldiers. For her field commander she commissioned the seventy-year-old earl of Surrey, who hastened to the crumbling north. Catherine meanwhile worked to create a reserve, and in the early days of September she rode out of London to lead them northward. They soon learned that on 9 September the English and Scots had fought at Flodden, where the Scots had been utterly defeated, losing their artillery and perhaps as many as 10,000 men, including great lords and James IV. Catherine had contributed mightily to the victory, and it contributed to her popularity with the English even when later she was divorced by England's king.

Bibliography: Mattingly, *Catherine of Aragon*, 1941; Scarisbrick, *Henry VIII*, 1968; Wernham, *Before the Armada*, 1972.

—Marvin A. Breslow

CATHERINE II OF RUSSIA (Ekaterina Alekseevna) (born 21 April 1729, Stettin, Pomerania; died 6 November 1796, St. Petersburg, Russia), also known as Catherine the Great; born Sophia Augusta Fredericka, princess of Anhalt-Zerbst. Empress of Russia.

The famous ruler Catherine II of Russia is not usually seen as a military figure, although her lengthy reign (1762–1796) is much celebrated or condemned for great territorial expansion attained mostly by war. Catherine was vividly aware of the military repute of her renowned predecessor Peter the Great and aspired to be his female equivalent in her own era. She read widely and knew about many other warrior rulers, both female and male. Indeed, Russia was dominated by female monarchs in this period: from Peter's death in 1725 until the end of Catherine's own reign in 1796, men ruled for only four years. Catherine had no personal combat experience, but achieved fame as a perceptive politician, able administrator, coordinator of high strategy and diplomacy, and patron of military success in many theaters of action.

Catherine (born Sophia) grew up a Prussian subject and daughter of a Prussian soldier and administrator. Her marriage to Grand Duke Peter Fedorovich, a cousin and the future Emperor Peter III of Russia, was facilitated by King Frederick the Great of Prussia, the preeminent military hero and statesman of the age, whom she met several times, corresponded with, and saw as a role model.

According to her autobiographical sketches, young Sophia showed tomboyish traits as she enjoyed rough-and-tumble games. Later she became an accomplished equestrienne, an essential and much admired skill in Russia with military implications. Like her Russian predecessors Anna Ivanovna and Elizabeth, she was an avid hunter and skilled marksman. Presumably she learned some rudiments of military life from her soldier father and other male relatives and, after she arrived in Russia in 1744, from her fiancé and husband, who was enthralled with military matters and even trained his clever consort in the manual of arms. Catherine had seventeen years to prepare for her role as empress. Her husband would not become Emperor Peter III of Russia until December 1761.

Upon presentation to the imperial Russian court of Empress Elizabeth, she was awarded the Order of Saint Catherine, which Peter the Great had instituted in honor of his second wife, Catherine I (1684–1727), who had accom-

---

On cross-dressing among the rich, famous, and powerful in the eighteenth century:

During the 18th century, Russia was dominated by female monarchs. Between the death of Peter the Great in 1725 and the end of the century, Russia was ruled by empresses for all but four years.... [Empress Elizabeth held masquerades every Tuesday, according to the memoirs of her daughter-in-law, Catherine II]. Catherine tells us that the empress Elizabeth ordered all the men "to be dressed as women and all the women as men.... The men wore whalebone petticoats, the women the Court costume of men." Cross-dressing on a regular basis like this reinforced the fact that women controlled the court.

Masquerade balls were in vogue throughout Europe, and it was not uncommon at these events for men to dress as women and vice versa. But for the monarch to decree that every participant cross-dress was unprecedented. Moreover, according to Catherine, the Empress also ordered every guest to appear without a mask, a practice that was completely different from other masquerades.... By eliminating the masks, Elizabeth was significantly altering the meaning of the masquerade.

—Gary Kates, *Monsieur d'Eon Is a Woman* (New York: Basic Books, 1995), 74.

panied him on several campaigns and European sojourns. When Sophia converted to Russian Orthodoxy, she was given the name of Catherine in honor of Catherine I.

An interesting aspect of Empress Elizabeth's court was weekly masquerades, where, according to Catherine's memoirs, the empress ordered all the men to dress as women and the women as men. Perhaps this served to emphasize the power of women in the Russian court. When she became empress, Catherine held such masquerades on a less frequent basis. She seems to have enjoyed appearing in male disguise and fooling the members of her court.

In her years at Elizabeth's court Grand Duchess Catherine became cognizant of the social and political role of the imperial Guards, and she witnessed the army's ceremonial role on numerous occasions at St. Petersburg, Moscow, Kiev, and elsewhere. From her martinet husband she heard constant praise of Frederick the Great, particularly during the Seven Years' War (1756–1762/1763), in which Russia and her allies brought Prussia to dire straits. During this conflict Catherine began to follow political and military affairs closely in view of Elizabeth's faltering health and the rivalry of court factions over succession to the throne. Yet when Elizabeth vowed to lead her armies in person, Catherine scoffed at such empty bravado.

Catherine carefully cultivated support among various statesmen and among the Guards military units in particular. From about 1760 she became intimately involved with Colonel Grigory Orlov, a guardsman famed for battlefield exploits, good looks, and wide contacts. He and his four brothers played important parts in the conspiracy that deposed Peter III and placed Catherine on the throne via the coup of 28–29 June 1762. Dressed in a green Guards uniform and mounted on the stallion Brilliant, the new empress led her forces from St. Petersburg to Peterhof to obtain her husband's submission and abdication. (This scene was memorialized by a huge painting by Vigilius Ericksen later displayed at Peterhof and much reproduced in prints.) Arrested and imprisoned at Ropsha, Peter III was murdered a week later. It is not known whether Catherine sanctioned his death, though she professed to be horrified by the deed and recognized that she would be seen as culpable in any event. Many guardsmen and other soldiers were rewarded for their role in her accession. At her coronation in Moscow in September 1762 and during the early years of her reign she undertook reforms in the army and attended summer maneuvers at Krasnoe Selo in 1765. During her journey to the Baltic provinces in the summer of 1764 she made short sea cruises at Reval and Rogervik and on one occasion was seen to display some apprehension while boarding a ship. Unlike Peter the Great, Catherine had no love for the sea and rarely ventured on shipboard the rest of her reign.

Catherine soon initiated an active foreign policy in Poland and Kurland, based on bribery in the first instance but also backed by the threat of force. Russian troops entered Polish territory in support of the candidacy of Stanislas Augustus Poniatowski for the elective kingship in 1763, provoking Polish resistance that eventually flared into civil war in the late 1760s and spilled over into hostilities with the Ottoman Empire in 1768. Catherine's consultative style took institutional shape in 1768–1769 with the formation of the imperial council, which she chaired for the rest of her reign. The Polish and Turkish conflicts escalated into a long and costly war that made Russia vulnerable to Austrian and Prussian pressure for a three-way partition of Poland in 1772–1773. Against the Ottoman Empire, however, Catherine's forces won notable victories on land and sea while occupying Moldavia and Wallachia, the Crimea, and Middle Eastern ports and threatening Constantinople itself. Even so, the conflict engendered costly diversions: a massive plague epidemic in 1770–1771 and the frightening Pugachev Revolt in 1773–1774. In July 1774, when Pugachev's motley legions suddenly seized Kazan and appeared to threaten central Russia, Catherine vowed before her council to lead the counterattack in person. She was talked out of her folly by Grigory Potemkin, her new favorite, and calmed down within days at news of peace with the Turks. The peace of Kuchuk-Kainardji of 10 July 1774 was celebrated elaborately in Moscow in 1775 with rich awards to field marshals Rumiantsev-Zadunaiskii, Dolgorukii-Krymskii, and Orlov-Chesmenskii.

These triumphs boosted Russia's international prestige, as symbolized by Russian mediation of the War of the Bavarian Succession (1779), its leadership of the Armed Neutrality of 1780, and visits from Catherine's cousin, King Gustavus III of Sweden (1777 and 1783), and by Emperor-King Joseph II of Austria (1780 and 1787). Despite threats of renewed war with Turkey, Russia succeeded in annexing the Crimean khanate in 1783, a triumph that advertised the benefits of the Russo-Austrian alliance and the so-called Greek Project aimed at ejecting the Turks from Europe. The importance of southern expansion was underlined by Catherine's famous Tauride tour to the Crimea in 1787, part of the silver anniversary of her reign. She reviewed the army and the new southern fleet throughout this highly publicized triumphal progress, while Potemkin staged spectacular receptions, including an "Amazon" detachment in one instance. If she hoped that the Turks would be intimidated, the opposite occurred with their declaration of war in August 1787, setting off a series of conflicts that stretched into the 1790s.

Under Potemkin's leadership Russia in alliance with Austria defeated the Turks in the south, but the conflict was complicated by a war with Sweden in 1788–1790 that brought the smell of gunpowder to St. Petersburg and led Catherine to proclaim herself ready to fight the Swedes in the streets—the closest she ever came to actual combat. A war scare with England and Prussia and Potemkin's sudden death in 1791 robbed the triumph over the Turks of much luster in Catherine's eyes. She also became seriously concerned about the revolution in France and partly justified the second partition of Poland in 1793 as her part in defeating "Jacobins." Catherine delighted in Field Marshal Suvorov's triumph over the Polish rebels in 1795 and Valerian Zubov's invasion of Persia in 1796. Her dispatch of Suvorov westward against revolutionary France was canceled by her sudden death from a stroke. Though she never explicitly accepted the sobriquet "the Great" during her lifetime, it became common after her death and is closely associated with territorial expansion and victory in war.

Bibliography: Alexander, *Catherine the Great,* 1989; Catherine II, *Memoirs of Catherine the Great,* De Madariaga, *Russia in the Age of Catherine the Great,* 1981, and *Catherine the Great,* 1990.

—John T. Alexander

CAVANAUGH, KIT (born 1667; died 7 July 1739, London, England), also known as Christian Davies, Christopher Welsh, Mother Ross, "the female soldier." Foot soldier, 2nd Dragoons, England.

Disguised as a male soldier, Kit Cavanaugh fought in the Nine Years' War and the War of the Spanish Succession. For thirteen years she managed to keep her gender secret despite getting wounded twice, being imprisoned, and sharing quarters with male soldiers. Only when she was critically wounded for a third time was her true identity disclosed. Even then she did not completely forgo her military life.

Cavanaugh donned male disguise and enlisted in King William III's forces at Flanders in 1694. Her alleged motive for this unusual decision was to track down her husband, who had abandoned her by joining the army. Almost immediately after the Battle of Landen Cavanaugh was captured and imprisoned by the French. Upon her release she joined the 2nd Dragoons, a Scottish unit under Lord John Hay, until the Peace of Ryswick in 1697.

Five years later Cavanaugh reenlisted to fight the Franco-Bavarian troops in a southeastern campaign across the Netherlands into Germany. During this campaign Cavanaugh finally encountered her husband, but convinced him to let her pass as his brother rather than as his wife. The charade ended when Cavanaugh was critically wounded at the Battle of Ramillies in the Spanish Netherlands in 1706. Although she was dismissed from service and forced to dress once again in women's clothes, Cavanaugh remained attached to camp life by accompanying her husband and then, after he died, by marrying two other soldiers in quick succession.

From the 1640 broadside entitled "The valorous acts performed at Gaunt, by the brave bonny lasse Mary Ambre":

When Captain couragious whom deeth could not daunt,
Had roundly besieged the City of Gaunt,
And manly they marched by two and by three,
But the formost in battell was Mary Ambree. . . .

The skie she then filled with smoak of her shot
And her enimies bodies with bullets most hot
For one of her owne men a score killed she
Was not this a brave bonny Lasse, Mary Ambree? . . .

In this womans praises Ile here end my Song
Whose heart was approved in valour most [strong]
Let all sorts of people what ever they be
Sing forth the great valours of Mary Ambree.

—Dianne Dugaw, *Warrior Women and Popular Balladry, 1650–1850* (Cambridge: Cambridge University Press, 1989), 37–39.

Following the Peace of Utrecht in 1712, the duke of Argyle presented Cavanaugh to Queen Anne, who awarded her a small pension for her service. At her funeral soldiers fired three grand volleys in her honor.

Bibliography: Defoe, *The Life and Adventures of Mrs. Christian Davies*, 1928; *Dictionary of National Biography*, "Christian Davies," 1950; Miles, *The Women's History of the World*, 1989.

—Susanna Calkins

CENTRAL WOMEN'S SCHOOL FOR SNIPER TRAINING. Founded 21 May 1943, Veshniaki, region of Amerevo, to train women snipers for service in the Red Army. Soviet Union, Second World War.

In December 1942 the Komsomol Central Committee issued a decree mobilizing Komsomol girls into women's sniper-training courses that had been established in the Vseobuch (Universal Military Training) Central Sniper School in November. In May 1943 the People's Commissariat of Defense reorganized these courses into the Central Women's School for Sniper Training. With an initial enrollment of 1,120 students who had passed stringent health and eyesight tests, the school commenced operation on 7 June under the command of Nora Pavlovna Chegodaeva, a graduate of the Frunze Military Academy and a veteran of the **Spanish Civil War**. Later that month two battalions and a separate company of sniper instructors were formed. Courses consisted of classroom instruction in weapons handling, use of telescopic sights and binoculars, observation, the art of camouflage, and preparation of a foxhole. In the field students were sent on long marches with full military equipment, and on the shooting range they developed their marksmanship. Physical training was also provided to prepare students for hand-to-hand combat and fighting with a bayonet. According to a former school instructor, in shooting competitions held versus students from the men's school, the women's more responsive, "tender hands" helped them to regularly outperform their male comrades. In September the school transferred from its summer camp in Amerevo to winter quarters in the settlement of Silikatnyi near Moscow, and by January 1944 the first 121 graduates, equipped with monogrammed custom-made rifles, were dispatched to the Karelian, First and Second Baltic, and Belorussian Fronts. By the end of the war the school had trained 1,885 snipers who were credited with the extermination of 11,280 Germans. Well-known graduates from the school or sniper courses who were awarded the **Hero of the Soviet Union** included Aliia Molgagulova, Tatiana Kostyrina, **Natalia Kovshova, Maria Polivanova,** and Tatyana Baramzina. On 24 January 1944 the school was invested with the **Order of the Red Banner**.

Bibliography: Krivich, *Zhenshchiny-snaipery Leningrada*, 1966; Nikiforova, *Rozhdennaia voinoi*, 1985; *Snaipery*, 1976; author interview with Nina Lobkovskaia, 1991.

—Mary Allen

CEZELLY, FRANÇOISE DE (born ca. 1555, Montpellier, France; died 14 October 1615, Montpellier, France), also known as Constance of Ciselli. Governor of Leucate during the French Wars of Religion.

When Leucate was under siege by members of the Catholic League, with Spanish support, Françoise de Cezelly took the place of her husband, the governor of the town, who had been taken prisoner by the enemy. She refused to surrender the town in exchange for the life of her husband.

Françoise was the daughter of the director of the Bureau of Accounts of Montpellier and niece of the governor of Leucate. In 1577 she married Jean Boursier de Barry, who succeeded his uncle-in-law as governor of Leucate and was a former member of the Catholic League now in the service of King Henry III. In July 1589 Jean de Barry fell into the hands of the duke of Joyeuse, one of the leading members of the league. The league set siege to Leucate, but Françoise took over her husband's duties, and, according to Michaud, "dressed like an Amazon, with a pike in her hand, she drove back the besiegers and rendered all their efforts useless." Joyeuse then demanded the surrender of the town in exchange for her husband's life. Françoise refused; concerning her husband she exclaimed, "I will not redeem, through unworthy cowardice, a life that it would be shameful to enjoy." Jean de Barry was executed in August 1589, and his body was returned to Leucate. In retaliation, the garrison wanted to execute a prisoner of war, but Françoise refused. As a reward for her actions, the new king, Henry IV, offered her a pension and maintained her in the office of governor until her son was able to succeed her in 1610. Her son in his turn vigorously defended Leucate and refused to surrender it to the Spanish in 1632 and 1637.

In a square in Leucate a monumental statue still recalls the memory of Françoise de Cezelly. She is shown standing erect, with a pike in her hand, as Michaud described her.

Bibliography: Cazals and Fabre, *Les Audois,* 1990; Michaud, *Biographie universelle,* 1813; Roman d'Amat, *Dictionnaire de biographie française,* 1959; Vaissete, *Histoire générale de Languedoc,* 1889.

—Marie-Thérèse Lalaguë-Guilhemsans (trans. by Valerie Eads)

CHAPUY, REINE (born 12 May 1776, France), also known as Chapuis. Volunteer in the French Revolutionary cavalry, 1793.

Like her five brothers before her, Reine Chapuy served in French armies; in 1793 she spent nearly a year as a cavalry soldier. At age seventeen Chapuy left Versailles to join the 25th Cavalry Regiment (24th Cavalry Regiment in the reorganized army) and, according to her peers, served with distinction in combat. Although Chapuy desired to make the military her career, she was discharged after ten and a half months because of the 1793 decree that ousted women from military service (see **French Revolution and Napoleonic Era, women soldiers in**). She petitioned vehemently, but to no avail, that "the arms of a woman serve as well as those of a man when the blows are directed by honor, a thirst for glory, and the certainty of annihilating the tyrants who oppose France." Recognized with an "honorable mention" in the proceedings of the National Convention of France in December 1793, Chapuy received no pension for herself, but her parents were rewarded with 300 livres in annual aid.

Bibliography: Brice, *La femme et les armées*, n.d.; Cère, *Mme Sans-Gêne*, 1894; France, *Procès-verbal de la Convention nationale*, 5 ventôse an II.

—S.P. Conner

CHARPENTIER, MARIE (born 1751, Paris). The only woman recognized for her role in the siege of the Bastille, 14 July 1789; gendarme in the Gendarmerie nationale of France, 1792–1793, French Revolution.

Marie Charpentier was the only woman to receive a certificate for bravery as a "Vainquer [sic] de la Bastille" when that jail fell in Paris on 14 July 1789, the date generally viewed as the symbolic beginning of the **French Revolution.** There she was reputed to have been within the first ranks of the besiegers who entered the bastion. Later, in August 1792, she was commissioned in the 35th Division of the Gendarmerie nationale of France (the national police force), where she served with distinction until 1793, when she was discharged. Meanwhile, her husband Joseph Hanserre died in the French Caribbean colonies, leaving her with two children and few resources.

Although Charpentier had received a pension of 200 livres and recognition by the National Assembly of France for her valor at the Bastille, she petitioned the French government for her widow's pension and for additional assistance based on her service to France. According to her dossier, the injuries received as a gendarme (traumas to her abdominal area from the butt of a gun) were insufficient to consider her infirm. In a letter based on the 1793 exclusion of women from the military (see French Revolution, women soldiers), the official who replied to her entreaties noted that in any case, "her sex is in no way appropriate to service in the war."

Bibliography: Dossier XR48, "Charpentier," Service historique de l'Armée, Vincennes, France; France, *Procès-verbal de la Convention nationale*, 2 fructidor an II.

—S.P. Conner

> *A French female soldier protests the 1793 decision of the National Convention to expel all women from the army:*
>
> Catherin Pochetat, an actress with a distinguished military record, did not intend to be unceremoniously cast aside with the other "femmes inutiles." She had manned the cannons on 10 August, and she had been wounded in the campaign of 1792 at Jemappes. Along with other petitions on her behalf, the twenty-three year old soldier presented her case to the Convention: "Legislators, the law which requires women to retire from military service—how can it be applicable to a person who has served without slackening since the beginning of the campaign? What? This disgraceful exclusion will be the thanks I receive for the blood I have shed for France? I will be run out of the French armies—me—who has pursued the Austrians as many times as they have fled before us." Pochetat's pleas, however, brought no sympathy for her request. Because of her wound, she was discharged and granted a pension of 300 livres for life.
>
> —Susan P. Conner, "Les Femmes Militaires: Women in the French Army, 1792–1815," *Proceedings of the Consortium on Revolutionary Europe, 1750–1850* 12 (1982): 297.

CHECHNEVA, MARINA PAVLOVNA (born 15 August 1922, Protasovo, Orel Region, USSR; died 12 January 1984, Moscow, USSR). Pilot, author. Soviet Union, Second World War.

Marina Chechneva was born in the village of Protasovo in the Maloarkhangelsk District of the Orel Region and spent her first twelve years in this bleak area of northern Russia. In 1934 her family moved to Moscow. Four years later, at the age of sixteen, Chechneva enrolled in an **Osoaviakhim** flying club, where she learned sport flying. She dreamed of becoming a professional pilot, an ambition that was encouraged by her father, abhorred by her mother, and supported by Valeriia Khomiakova, one of the air club's instructor pilots and later a famous fighter pilot.

Chechneva longed to become a fighter pilot, but settled for a job as an instructor pilot at the Central Flying Club in Moscow from 1939 to 1941. After the Germans invaded the Soviet Union in June 1941, the club was evacuated to Stalingrad. In October 1941 Chechneva was permitted to join the new women's aviation group formed by **Marina Raskova**.

After completing training, Chechneva was assigned to the **46th Guards Bomber Regiment,** which flew the frail Po-2 biplane. She became a squadron commander in 1942. Chechneva completed 810 combat sorties and dropped 115 tons of bombs. She was considered one of the best and bravest pilots in her regiment and often was assigned the most dangerous missions, including daytime reconnaissance flights. She once flew 18 sorties in a single night; this was possible because of the long winter nights in Russia and the relatively short sortie duration of the Po-2. However, few male Po-2 pilots achieved the maximum sortie rates that Chechneva and others in the 46th attained.

Chechneva remained in the air force after the end of the war, serving in ground-attack aviation with a Soviet unit in Poland. She had her first child during this period and continued flying until she was demobilized in 1948. She and her pilot husband returned to Moscow, and after he was killed in an accident the next year, Chechneva devoted herself to aerobatic performances. In 1957 her health forced her to retire from flying; thereafter she served with DOSAAF (the successor to Osoaviakhim) and as an editor with an aviation magazine.

In the 1960s Chechneva completed graduate work in history and wrote a dissertation on Soviet women who had served in the Second World War. She also wrote at least six books and published many articles about her wartime experiences and the women with whom she served. Her books include *Samolety ukhodiat v noch'* [Aircraft go out into the night] (Moscow: 1961), *Boevye podrugi moi* [My fighting women friends] (Moscow: 1968), *Nebo ostaetsia nashim* [The sky remains ours] (Moscow: 1976), and *"Lastochki" nad frontom* ["Swallows" over the front] (Moscow: 1984). She died in 1984 at the age of sixty-two.

Bibliography: Chechneva (see bibliography); Cottam, *Women in War and Resistance,* 1998.

—Kazimiera J. Cottam

CHEN, abbreviation of Cheil Nashim (women's corps). 1949–present, Israel. Administrative organization overseeing women in the Israel Defense Forces (IDF).

For decades Israel was the only nation that conscripted women into military service. The involvement of Jewish women in military organizations dates back to prestate Israel (see **Israel: War of Independence, women's military roles**). From the early days of the twentieth century women participated in underground Jewish military organizations. When these were incorporated into the IDF, women lost all combatant roles that they had earlier shared with their male comrades. This was the result of three major sociopolitical factors: (1) female soldiers who served as noncombatants during the Second World War in the British **ATS** (Auxiliary Territorial Services) and later joined CHEN outnumbered the former female members of the Palmach who wanted to retain a combat role; (2) the religious parties in the Israeli coalition governments fiercely opposed women's conscription for moral reasons; and (3) gender inequalities in Israeli society were reflected in the Israeli army. Once the Israeli army was formed, a decision was made to establish a separate jurisdictional unit for women. This unit, CHEN, supervises the introductory training of female recruits and female officer training and oversees judicial issues pertaining to female soldiers.

CHEN was formally inaugurated as part of the IDF by the Defense Service Law of 8 September 1949, which broadly defined the regulations of conscription of Jewish women into the army. CHEN is an administrative organization, operating in much the same way that the **WAC** or **WAVES** formerly operated in the American military, with responsibilities for recruitment and assignments, but without operational control. The Hebrew word *chen*, coincidentally, means "charm." CHEN is commanded by a member of the General Staff. In 1949 the rank of the commander of the Women's Corps began as colonel but in 1986 changed to brigadier general. The commander offers her recommendations in matters regarding women, such as housing, health, and training, to the chief of staff. A CHEN officer serves in each military unit and sees that these recommendations are followed.

Compulsory military service is regarded by Israelis as a period in life when adolescents become adults, and as a melting pot for both men and women from different backgrounds. Research indicates that among women there is a significant correlation between background and military status. It is also reported that the military experience influences the developmental course of young women, altering their demeanor and self-image. Women who serve in the army demonstrate an increase in self-confidence and seem to aspire to a professional career after their discharge. Nonservers, on the other hand, cling to traditional roles and have lower self-esteem.

Military service is compulsory, and single women must enlist when they reach the age of eighteen. The initial period of conscription was twenty-four months, but it was eventually reduced to twenty months (men are conscripted for thirty months but usually serve thirty-six months). Reserve service is obligatory for childless women up to age thirty-four. Married women and mothers are exempted, and marriage or pregnancy can be grounds for separation from service; however, the largest group of women routinely exempted are Orthodox women (about 25 percent of Jewish women qualify for automatic exemption based on religious belief). Many women are exempted, as personnel requirements do not require Israel to draw upon more than about 60 percent of its pool of eligible women (about 90 percent of eligible men are drafted). Conscription percentages vary over time; in 1993–1994 the conscription rate for eighteen-year-olds was 70 percent for women and 83 percent for men. Women recruits tend to have higher test scores than men, and surveys show that at the end of their service nearly three-quarters of all women felt that they were overqualified for the jobs to which they were assigned. Until recently, upon conscription a female soldier underwent gender-segregated basic training run by CHEN that continued from one to three months, depending on her assigned vocation. She was then transferred to her assigned unit. From then on she is supervised by the general military units.

Traditionally, women serve in clerical and secretarial positions, technical and mechanical trades, educational and cultural positions, and operational noncombatant roles. These may vary periodically according to existing military needs. As of 1988, women served in less than 40 percent of the general military positions, although they were widely dispersed, with women in virtually all military units. Women also serve in the career army (Kevah), where egalitarianism is more often enforced. Still, analysts like Dafna Izraeli conclude that "within the military most men and women are treated so differently that one might say that they experience very different militaries."

Women are supposed to be evacuated before any unit goes to combat. However, this is sometimes difficult to accomplish, and women still occasionally find themselves in military action. In the 1956 Sinai Campaign there were three women C-47 pilots (Yael Rom, Yael Finkelshtein, and Rina Levinson) who flew paratroopers into action. During the Yom Kippur War in 1973 the surprise attacks on Israeli forces resulted in the death of three female soldiers at forward posts and many others wounded.

Changes occur periodically. Since 1977 women have served as instructors in combat-related training; for example, many women became tank instructors and jumpmasters. In several jobs, such as instructor in the tank unit, the female soldier is now allowed to join her male trainees in the field. In March 1997, for the first time since its establishment, IDF inaugurated joint basic training for men and women. However, this training period is reported to have been easy and relatively short.

The popular image of an Israeli female soldier holding a machine gun in her arms has long been an anachronism, though one that has come full circle in the twenty-first century. Though female soldiers play an important role in the IDF, they have not served in any combatant role since their brief experience in the War of Independence (1948). However, as in other nations, Israel is opening the door to combat roles for women; in connection with this, the role of CHEN has been reduced since 1997. In May 1996 the first all-female border police unit graduated from basic training. The border police, a paramilitary unit of the national police, formed a unit of twenty-five women, assigned all the same tasks as the men, except for patrolling the West Bank, where the border police shares responsibilities with the army. While the border police are not part of the IDF, the women's unit set a precedent. In November 1997 the Israeli Supreme Court opened the way for women to compete for slots in combat aviation, and several women have recently graduated as combat pilots and navigators. In January 2000 the Knesset opened all branches and services to women; women must meet the existing qualifications for combat positions, and these roles will entail longer service commitments. It remains to be seen how many women will volunteer and qualify for these new roles.

Bibliography: Binkin and Bach, *Women and the Military*, 1977; Bloom, "Israel: The Longest War," 1982, and "Women in the Defense Forces," 1991; Bloom and Bar-Yosef, "Israeli Women and Military Service," 1985; Izraeli, "Gendering Military Service in the Israel Defence Forces," 2000; Swirski and Safir, *Calling the Equality Bluff*, 1991; Williams, *The Israel Defense Forces*, 1989.

—Dina Ripsman Eylon

CHEVREUSE, MARIE DE ROHAN-MONTBAZON, DUCHESS OF (born December 1600, France; died 12 August 1679, Gagny France). Involved in a number of plots under Louis XIII; one of the heroines of the Fronde under Louis XIV (France).

For most of her life Madame de Chevreuse was at war against her king and her country, allying herself with the enemies of France and drawing those around her into conspiracies against her sovereign. At twenty-two she married her second husband, Claude de Lorraine, duke of Chevreuse, son of Henry of Guise, head of the (Holy) League during the Wars of Religion. She was intelligent and energetic; it was said of her that she was capable of governing a kingdom. A tireless conspirator, still active at the beginning of the Wars of the Fronde in the 1640s, she was gradually eclipsed by the rising influence of the **duchess of Longueville.**

She was quick to immerse herself in both amorous and political intrigues. She was one of the originators of the affair with the duke of Buckingham (1625) made famous by Alexandre Dumas in *The Three Musketeers*. Braving the animosity of Louis XIII and Cardinal Richelieu, she conspired with Huguenots, the English, Savoyards, and important Frenchmen as well in trying to bring about Richelieu's death. Fearing for her life following the failure of the plan and the execution of her lover Chalais (1626), she fled to Lorraine. There she continued plotting against France; the king, judging Madame de Chev-

Portrait of Marie de Rohan-Mountbazon Chevreuse. (Perry-Castañeda Library)

reuse to be less dangerous in Paris, allowed her to return to court. However, she immediately resumed her cabals, scheming with conspirators in Lorraine and Spain, and was exiled on a number of occasions. In 1637 she was forced to make a hasty flight to Spain disguised as a man and hotly pursued by agents of the king. She crossed into England, where she rallied the enemies of Richelieu, then moved on to Flanders, where she was implicated in a conspiracy led by the count of Soissons, who was killed in battle in 1641.

Reentering France after the death of Louis XIII in 1643, she threw herself into the "les Importants" conspiracy, encouraged an assassination attempt against Cardinal Mazarin, and became involved again in schemes with the English and the Spanish. Forced to flee England, she was intercepted by Cromwell's fleet and detained for a short time. She then traveled to the Netherlands, where she took advantage of the initial disorders of the Fronde to return to France in 1649. Favorable for a while to Mazarin, she later allied herself with the princes, obtaining the exile of the cardinal, then later reversed her position again. Both sides in turn took advantage of her talents as a negotiator, but as she grew older, she gradually played a less important role in public life, ceasing all political activity with the death of her daughter in 1652.

Hearing her name while on his deathbed, Louis XIII found the strength to exclaim, "That is the devil." Her intrigues had so shaken the kingdom that Mazarin remarked that "France was not calm except when she was not there." Madame de Chevreuse ended her life living in religious seclusion at a Benedictine convent. Though not strictly speaking a military leader, Madame de Chevreuse instigated actions that led to military conflict.

Bibliography: Batiffol, *La duchesse de Chevreuse,* 1913; Cousin, *Madame de Chevreuse,* 1886; Roman d'Amat, *Dictionnaire de biographie française,* 1959.

—Marie-Thérèse Lalagüe-Guilhemsans (trans. by Annette Parks)

CHI ZHAOPING (fl. 21 CE, Pingyuan, now in Shandong Province, China). A local leader in a peasant revolt during the last years of the Former Han dynasty (206 BCE–25 CE).

Not much is known about Chi Zhaoping except that she was from Pingyuan and knew how to interpret the *Classic on the Playing Blocks,* apparently a manual of the ancient Chinese game *bo* (playing blocks). In 21 CE she gathered several thousand people in strategic places of the lower part of the Yellow River to fight the government of Wang Mang (45 BCE–23 CE), who usurped the Han dynasty for fifteen years. There was no record about the outcome.

Bibliography: Ban Gu, *The History of the Former Han Dynasty,* trans. Homer H. Dubs, vol. 3 (Baltimore: Waverly Press, 1955), 414–415.

—Sherry J. Mou

CHIOMARA (c. 189 BCE). Queen of Galatia. Prisoner of the Roman army, Asia Minor.

When a Roman army led by Gnaeus Manlius Vulso stormed and captured the Galatian stronghold in Asia Minor, the Galatian king, Ortiagon, managed to escape, but his wife, Chiomara, was captured along with a band of women. She was raped by her guard, a Roman centurion, yet lived to get her revenge. The centurion, who is described in the sources as a crass brute, agreed to release her in exchange for money and allowed her to send one of her fellow prisoners to her kinsmen to make arrangements for payment. At the meeting where the exchange was to take place, while the centurion was examining the ransom money, Chiomara gave a signal to her comrades and had him beheaded. She wrapped his head in her clothing and took it to her husband. At their reunion her husband was surprised at this "unfeminine deed" and said, "Being faithful is a fine thing," to which she replied, "It is a better thing that only one man who has slept with me should be alive."

Bibliography: Livy, 38.24.2; Pauly, A., et al., "Ortiagon," *Real-Encyclopädie der klassischen Altertumswissenschaft* (Alfred Druckenmuller Verlag, 1988); Polybius, *Histories* 21.38.

—C.A. Hoffman

CHRISTINE DE PISAN (born ca. 1365, Italy; died ca. 1430, France), also known as Christine de Pizan, de Pezano. Born in Italy, lived and worked primarily in France. Wrote influential military treatises during the Hundred Years' War.

Christine de Pisan was a humanistic writer and scholar of the late fourteenth and early fifteenth centuries whose interests included the practice and ethics of warfare. Her classic book on the subject, *The Book of Deeds of Arms and of Chivalry,* was widely studied by military leaders of her age and retains lasting importance as an exposition of important military ideas of the period.

Christine was the daughter of the scholar Thomas de Pezano, who held the chair of astrology at the University of Bologna (1345–1356). In 1368 the family moved to Paris to join him at the court of Charles V, to whom he had become an advisor. In 1380 Christine married Etienne de Castel, a university-trained man who became a royal secretary. The couple moved in educated chancellery circles with men who became France's first humanists. Christine's literary ca-

Christine de Pisan

Christine de Pisan. (The Art Archive/Bibliothèque Nationale Paris/The Art Archive)

reer followed the untimely death of her husband in 1389 when she was about twenty-five. Left with three children to support, she became almost certainly the first professional female author of the Middle Ages. Initially she worked as a copyist, but continued to educate herself in the classics and began to write original works. Among the best known of her many works are several that comment on the social situation of contemporary women: *The City of Ladies*, the *Book of Three Virtues*, and her correspondence concerning the misogynist nature of Jean de Meung's portion of the *Romance of the Rose*. In her famous work *The City of Ladies* she cites some examples of military women as role models, including the **Amazons** and **Zenobia,** and even envisages them all in her ideal city, defending it against male enemies. The image of a woman "under siege" by men was a fairly common metaphor in the art and literature of the Middle Ages, but Christine had an interest in the practical experiences of women under siege, as well as the symbolic imagery.

Christine also wrote about political and military affairs, inspired by the events of the ongoing Hundred Years' War. *The Book of Deeds of Arms and of Chivalry*, composed in 1408–1409, was probably commissioned by the duke of Burgundy for the dauphin of France. It draws heavily on earlier works—most obviously the classical works of Vegetius, Frontinus, and Valerius Maximus. Indeed, Christine was attempting to popularize these works among men-at-arms who did not read Latin.

The Book of Deeds of Arms and of Chivalry (*Le livre des faits d'armes et de chevalerie*), as its title implies, deals with strategy and the undertaking of warfare in a practical sense, but reflects a contemporary concern that moral law be considered alongside necessity and above blood lust. Christine describes the circumstances of lawful and just war, to be waged only by sovereign rulers and in defense of a just cause, that is, against usurpation and tyranny and not out of aggression or for revenge. She outlines the qualities of the ideal leader (based heavily on humanistic thought) and discusses the raising, training, and disciplining of an army in ethical as well as military terms.

The Book of Deeds of Arms and of Chivalry was clearly influenced by other important works of military thought. Indeed, Christine believed that Honoré Bonet (or Bouvet) appeared to her in a dream and invited her to pick fruit from his own *Tree of Battles* (ca. 1387), and her final chapters, dealing with the rules and diplomacy of warfare, are heavily indebted to Bonet. In matters of tactics Christine draws most from classical writers, especially Vegetius, most notably using almost word for word his passages on drawing up an army for battle. However, Christine's book is far from wholly derivative. She adeptly applies Vegetius's stratagems to the circumstances of medieval warfare, for example, taking into account technological advances. In ethical matters she argues with Bonet, most notably refusing to accept that prisoners may be slain in battle. She also draws on conversations held with anonymous acquaintances. These discussions inform passages that address, in the most practical terms, the specific circumstances of medieval warfare: for example, organization within a feudal army, the positioning of campsites, and siege techniques and the selection and use of contemporary siege engines.

It is in Christine's discussion of sieges that we see reflected her interest in the position of women. In this period women, though generally excluded from the battlefield, often found themselves besieged in castles or walled cities and at times were even responsible for their defense. Christine stresses that the wives and widows of nobles should see that their castles are well provisioned with supplies that keep well, such as corn and salt pork, and, above all, that women should know the laws of arms, what to expect in a siege, and how to repel attackers. Indeed, far from portraying women as passive in warfare, Christine urges them to auxiliary action with the example of the women of ancient Rome who cut off their hair to make bowstrings during a siege.

The survival of numerous manuscript copies of *The Book of Deeds of Arms and of Chivalry*, the earliest dating to 1434, attests that this was a popular and influential work. A decorated manuscript of it was thought a suitable present for **Margaret of Anjou** on her marriage to Henry VI of England in 1445. The treatise first went into print, to be read by a wider audience, in 1488. Henry VII of England had its popular diffusion in mind when he commissioned a printed English translation from William Caxton. It appeared in 1489 or 1490, and its twenty extant copies indicate a high circulation. (The modern edition of Caxton's translation and a discussion of its manuscripts are contained in A.T.P. Byles, ed., *The Book of Fayttes of Armes and of Chyvalerye*, London, 1932.) However, Christine was not always given credit for her work. A significant group of manuscripts exists in which all references to her authorship have been omitted, feminine pronouns have been changed to masculine, and Christine's invocation to Minerva, as an Italian woman calling on the Roman goddess of war, has been deleted. The lasting importance of the work is also seen in its influence on later works; for example, *Le Jouvencel* by the experienced soldier Jean de Bueil is just one example that shows the influence of Christine di Pisan.

Christine's interest in military affairs must be seen in the context of the troubled period of history in which she wrote. After she had finished the treatise, France was plunged into a new period of strife and uncertainty by civil war between the Armagnac and Burgundian factions, the insanity of King Charles VI, and the English victory at Agincourt in 1415. When the Burgundians seized Paris in 1418, Christine and her son Jean, secretary to the dauphin, had to flee. In this period she wrote much on the misery caused to the French people by the war, as well as nonpartisan letters to those whom she blamed for the discord. Her last work, however, a poem written in 1429 celebrating the victory of **Joan of Arc** at Orléans and the coronation of Charles VII, reflects a new optimism in her adopted country at the end of the Hundred Years' War—a war whose social consequences had greatly shaped her writing and whose course she herself had sought to influence.

Bibliography: Christine de Pisan, *The City of Ladies*, 1521; McLaughlin, "The Woman Warrior," 1990; Willard, "Christine de Pisan's Treatise on the Art of Medieval Warfare," 1970, and *Christine de Pisan*, 1984.

—Claire Taylor

CIMBRIAN WOMEN (ca. first century BCE). Fought and provided support to male warriors. Germanic tribe, Roman frontiers.

Florus reports that the Cimbri, a Germanic tribe, were forced to migrate when their homeland became flooded. When the Romans denied their request for a grant of land in which to settle, they went on the offensive. Several Roman generals were defeated before a final encounter occurred south of the Alps on the left bank of the Po at Vercellae (also known as the Raudian Plain). The Roman army under Marius defeated the Cimbrian forces in the field, then proceeded to attack their base, where the women and children had been left. The women defended themselves by forming the baggage train into a barricade and fighting fiercely from that position against the Romans. When they were unable to come to acceptable terms with Marius, they killed their children and committed suicide rather than be captured and sold into slavery. Plutarch's account differs from others, claiming that after the defeat of the main Cimbrian force, many Cimbri retreated to their base of operations; here, he says, the women attacked and slaughtered their own retreating warriors as they arrived, before killing the children and committing suicide. Sources agree that the armed Cimbrian women fought against the Romans, and they committed murder/suicide to avoid capture, but the extent of these activities is unclear. A great number of the Cimbri were killed in the fighting, as many as 140,000, but some sources say that 60,000 survivors were in fact captured.

Bibliography: Florus, *Epitome* 1.38; Orosius, *Historiae adversum paganos* 5.16; Plutarch, *Life of Marius*, 23–27.

—C.A. Hoffman

CLARKE, MARY ELIZABETH (born 3 December 1924, Rochester, New York, United States). Last director of the Women's Army Corps (1975–1978); longest-serving woman in the army (thirty-six years). First army woman promoted to major general (1 November 1978); first woman to command a major army post (Fort McClellan, Alabama, 1978–1980).

In August 1945, at age twenty, Mary Clarke entered the army and attended the last Second World War **WAC** (Women's Army Corps) basic training class at Fort Des Moines, Iowa. She later said that she had enlisted "to do something for the war effort" and intended to serve for the duration plus six months. But when the war ended and a male commander told her that she couldn't make it through the officers' training program, she decided to stay in the army to prove him wrong. Clarke graduated from WAC Officer Candidate School (OCS) in September 1949; over the next two decades her assignments included recruiting, staff positions, and commander of the WAC Training Battalion.

Retired Women's Army Auxiliary Corps members Colonel Bettie J. Morden (*left*), Brigadier General Elizabeth Hoisington (*center*), both of Arlington, Virginia, and Major General Mary Clarke (*right*), of Jacksonville, Alabama, cover their hearts during a wreath laying ceremony at Iowa's World War II memorial, July 20, 2000, in Des Moines, Iowa. From 1942 to 1945 about 72,000 women passed through Des Moines, site of the WAAC training center at Fort Des Moines. (AP Photo/Charlie Neibergall)

In September 1972 she assumed command of the WAC Center and School at Fort McClellan, Alabama. In 1975 Clarke was selected as director of the WAC on the basis of her leadership, executive abilities, and public relations skills. She officially became director on 1 August 1975 and was promoted to brigadier general the same day. During General Clarke's thirty-three-month tenure as WAC director, many major changes in the role of army women occurred: women were admitted to the service academies, the first women graduated from Army ROTC programs, they were required to undergo mandatory weapons training, the WAC Center and School was closed, and women began attending OCS and basic training in integrated classes with men.

General Clarke was concerned that with the end of the WAC, women would

> **NEW ARMY STUDY UNDERMINES OLD TABOO**
>
> *Through a regimen of regular jogging, weight training, and other rigorous exercise, more than 75% of the 41 women studied were found fit for traditional male duties associated with the military. Before training, less than 25% of the women were capable of performing the tasks. All but one of the females were civilian volunteers and none had previously adopted a routine of strenuous physical activity. The women included lawyers, mothers, students and even bartenders. Several recently had children and thought the training would put them back in shape. They were unaware that their performance may have toppled one of the last citadels of bias against women in both the military and society.*
>
> *The 24-week training study began in May 1995 with women spending 90 minutes a day, five days a week building themselves up for endurance tests. They ran a 2-mile wooded course wearing a 75-pound rucksack and performed squats holding a 100-pound barbell on their shoulders. Nationally-certified trainers oversaw the conditioning. Improvement of over 33% was noted by scientists who wrote the report....*
>
> *Nearly concurrently with this test, the Ministry of Defense in Great Britain was conducting the same kind of study! The Sunday Times of London reported on December 3, 1995, that "by using new methods of physical training, women can be built up to the same levels of physical fitness as men of the same size and build."*
>
> —J. Michael Brower, *Armed Forces Journal International* May 1997: 13.

no longer have anyone to represent them or their interests. She drafted several plans to establish the position of a senior military spokeswoman to present the women's point of view on policy issues. Each plan met with great resistance within the army hierarchy, primarily because it was seen as conflicting with the army's attempt to fully integrate women.

The ceremony to disestablish the WAC was held in April 1978; the WAC was legally disestablished on 20 October 1978. General Clarke noted in her parting speech that "this action today in no way detracts from the service of WACs who have been pioneers—in fact it honors them.... The significance of the abolishment of Office of the Director ... is the Army's public commitment ... to the total integration of women in the United States Army as equal partners." Clarke's next assignment was commanding general of the Military Police/Training Center of Fort McClellan, the first time a woman had been selected to command a major army post. She was promoted to major general on 1 November 1978. She commanded Fort McClellan from May 1978 to August 1980. She then served as director of the Human Resources Development Directorate in the Pentagon until her retirement on 31 October 1981. Clarke had served thirty-six years, a record for women in the army. She was awarded the Distinguished Service Medal at her retirement ceremony. After retiring, Clarke served on the Defense Advisory Committee on Women in the Services (DACOWITS) and as a member of the Presidential Commission on the Assignment of Women in the Armed Forces in 1992.

Clarke was again in the news in 1997 when the army decided to close the WAC Museum at Fort McClellan, Alabama, along with the base itself. WAC veterans and supporters had provided half a million dollars in private contributions in the early 1970s to pay for the building the museum occupied, and had donated more than 5,000 artifacts. "We built this ourselves, from the first stone to the last," said Mary Clarke.

Bibliography: "Clarke, Mary Elizabeth," *Who's Who in America*, 51st edition (New Providence, NJ: Reed Elsevier, Inc., 1996); Morden, *The Women's Army Corps*, 1990; "Onetime Enlisted Woman Retires as a General," *New York Times* 31 October 1981: 53.

—Vicki L. Friedl

CLAY, WENDY (born 1943, Fort St. John, British Columbia). Surgeon general, Canadian Forces. Medical doctor, master of health science (M.H.Sc.), Fellowship in Community Medicine (FRCPC), Canadian Forces pilot.

Wendy Clay's military career was a series of firsts for women in the Canadian Forces (CF). She was the first female medical cadet officer in the CF, the first woman to go through the subsidization plan since the Second World War (graduated 1967), the first female flight surgeon (1969), and the first female base surgeon (Moose Jaw, 1971). Clay was the first woman to complete the six-month advanced aviation medicine course at the RAF Institute of Aviation Medicine, Farnborough, England (1974). As a lieutenant colonel (1977), she was director of the CF Medical Assessment and Training Division and senior medical officer of the Canadian peacekeeping contingent in Suez. In 1989, as brigadier general, she became one of Canada's first female general officers, with Brigadier General **Sheila Hellstrom** and Commodore Laraine Orthlieb. She was promoted through the ranks to become the first female surgeon general (1995), the highest rank ever held by a woman in the CF and possibly in the Commonwealth.

Despite her achievements in the medical field, Clay received more publicity for her role in aviation. In July 1972 Wendy Clay became the first woman in the CF to graduate from basic pilot training. In August 1974 she earned her military wings, but because women were not yet accepted into combat roles, she never flew operationally. Two themes influenced the progress of Clay's military and medical career: she combined her training as a military pilot and her specialization in aviation medicine to better understand medical problems faced by pilots, and she accepted the publicity that followed when she earned her wings as a necessary burden so that more women would be admitted to pilot training.

Bibliography: Cardwell, Mark, "Major-General Practitioner," *Medical Post* 31.3 (1995): 10; Render, *No Place for a Lady*, 1992.

—Patricia Bowley

CLEOPATRA VII (69–30 BCE). Queen philopator and pharaoh (51–30 BCE).

Cleopatra VII was the final Ptolemaic ruler of Egypt—the last in the line descended from the Macedonian general Ptolemy, who had taken control of Egypt after the death of Alexander the Great two and a half centuries earlier. The Ptolemies had been allies of Rome for most of that period, struggling to maintain independence as Rome tried to increase its control in the region. Unfortunately, Cleopatra ended up on the wrong side in the Roman civil wars; after her death Egypt became the personal province of the Roman emperor.

Cleopatra's father, Ptolemy XII, paid tribute to the Romans to keep them from interfering in Egyptian affairs. When he died in 51 BCE, Cleopatra came to the throne of Egypt at age seventeen. Following long-standing tradition, she was to be co-ruler and wife to her younger brother Ptolemy XIII, who was

Cleopatra VII

Undated portrait of Cleopatra VII. (Perry-Castañeda Library)

under the influence of the local governor Pothinus. It was Pothinus who influenced Ptolemy XIII to at first support Pompey in his civil war against Julius Caesar, then have Pompey murdered when he arrived in Egypt. Ptolemy also took Pothinus's advice to rule alone, without the aid of Cleopatra.

Cleopatra refused to give up her throne without a fight. She escaped to Syria and raised an army with which to invade Egypt and reassert her claim. As she gathered her forces on the Egyptian frontier, Julius Caesar arrived in Egypt and mediated an agreement between brother and sister. Caesar with a small force of soldiers moved into the capital and began to collect tribute. Despite a thirty-year age difference, Cleopatra and Caesar began an affair that produced a son, Caesarion. Ptolemy and Pothinus fomented a rebellion against Caesar, but this was thwarted when Caesar received reinforcements from Asia Minor; Ptolemy was killed in the fighting. Cleopatra proceeded to marry another, even younger brother, Ptolemy XIV, and was acknowledged by Rome as queen of Egypt and a client of Rome.

Caesar returned to Rome and established himself as emperor. He did not disguise his association with Cleopatra; in fact, he immortalized her with a bronze statue in the Temple of Venus Genetrix in Rome. Cleopatra followed Caesar to Rome in 46 BCE, with her son and her brother, and lived with Caesar until his assassination in 44 BCE. Unable to negotiate Roman legitimacy for their son, she returned to Egypt, where on the fifteenth birthday of her brother/husband, she had him killed rather than share power with him.

Cleopatra ruled independently, a tributary of Rome, supporting no one (or perhaps at one time or another everyone) in the civil war that ensued after Caesar's death. Octavian's victory brought to Egypt (41 BCE) Mark Antony as commander of Roman forces in the East, and he soon was as smitten with Cleopatra as Julius Caesar had been. He ultimately rejected his own wife (Octavian's sister), married Cleopatra and legitimized their three children, and began to challenge the authority of Octavian in the eastern Roman dominions. Octavian declared war—not on Antony, but on Cleopatra.

In 32 Antony and Cleopatra took some 80,000 men to Greece, from whence Antony planned to attack Italy, with Cleopatra providing much of his navy and supplies. Octavian brought an almost equal force of soldiers to the region around Actium, in western Greece, to fight Antony. He also had a large fleet to fight Antony's armada if need be. It was at sea that the great battle took place on 2 September 31 BCE. Antony commanded the right wing of his fleet, while Cleopatra commanded the Egyptian naval contingent in reserve. Octavian's commander Agrippa outmaneuvered Antony's heavy ships with faster, lighter cruisers that harassed Antony's forces into disarray and cornered his fleet in the Gulf of Ambracia. At Actium, outside the gulf, Antony tried to break out, but was unable to do more than batter a small hole in Octavian's line. Cleopatra took advantage of that gap in the enemy's force and commanded her ships to try to run for Egypt. Antony abandoned his men and ships to catch up to her, and his forces surrendered to Octavian a week later.

Antony and Cleopatra were unable to withstand Octavian's invasion of Egypt, and by 30 BCE Octavian declared Egypt to be a possession of Rome. Antony fell on his sword. Cleopatra spent the weeks after Actium sending messages to Octavian in an attempt to negotiate and retain some measure of her power. Octavian gave her little mercy when he seized Alexandria; after all, she had caused the humiliation of his sister and fomented civil war. Perhaps he was also motivated by practicality and ambition; by seizing the treasury of Egypt for himself, Octavian became wealthier than the state of Rome. Octavian used the money to pay off his army, ensuring its continued loyalty. He planned to display Cleopatra during his triumphal march through Rome, where she was much hated; the idea was more than her pride could bear, and Cleopatra committed suicide at the age of thirty-nine. The Roman poet Horace eulogized her thus: "Serene of countenance she looks at the stormed fortress / And calmly stretches forth her hand / To seize the asp. Undauntedly / She allows the venom to pervade her body. / Rather than be borne to Rome on a swift cruiser, / be led in triumph through the streets / Here perished a proud and brave woman."

Cleopatra was much more a power politician than a military leader, though her actions had far-reaching military effects. She did command the forces gathered in Syria during her challenge to Ptolemy XIII. She also was in command of the reserve contingent at Actium, but did not participate in combat. The Egyptian defeat has often been attributed to the early withdrawal of a supposedly cowardly Cleopatra from the battle scene, but this claim is now discredited by most historians. At the geopolitical level, Cleopatra was a pivotal player. If her brother/husbands had dominated, it seems likely that Rome would have gained control of Egypt even sooner. As it was, Egypt was the prize that lay much of the foundation for the establishment of empire in Rome. As the personal possession of the emperor, Egypt provided necessary funding for the Caesars, and Octavian's victory led not only to the fall of Egypt as an independent power, but to the domination of the West over the East for centuries to come.

Bibliography: Grant, *History of Rome,* 1997; Mainzer, *Caesar's Mantle,* 1936; Rappoport, *History of Egypt,* 1904; Volkmann, *Cleopatra: A Study in Politics and Propaganda,* 1958.

—Paul K. Davis

CLOELIA (fl. ca. 508 BCE). Prisoner of war, Italy, revolt of Rome against the Etruscans.

One of the heroines of the early Roman Republic, Cloelia was held up in later tradition as an example of courage and patriotism. During the war that followed the expulsion of the Etruscan kings from Rome, the Romans gave Cloelia as a hostage to the Etruscan king Lars Porsenna. According to Livy, she escaped across the Tiber River, leading a group of other young Roman women. The Romans returned her to the Etruscans, in accordance with the treaty, but Porsenna then released her in recognition of her bravery and allowed her to take back other hostages as well.

In its broadest details, the story may be true; Cloelia's family was well known in the early Republic, and there is no inherent implausibility in the tale. There are many variations and developments in the tradition; those versions that have Cloelia riding a horse across the river are unlikely. This embellishment probably arose from the identification of Cloelia and the subject of an equestrian statue on the Via Sacra, which must date from a later period.

Bibliography: Livy, 2.13; Ogilvie, R.M., *A Commentary on Livy, 1–5 Books* (Oxford: Clarendon Press, 1965), 267–268; Pauly, A. et al., *Real-Encyclopädie der classischen Altertumswissenschaft* "Cloelius" 13 (Stuttgart, A. Druckenmuller, 1964–1975).

—Josephine Crawley

CLOTILD (fl. 589–590, Poitiers, Gaul), also known as Chrodieldis, Chrodielde. She claimed to be the daughter of King Charibert I (king of Paris, 561–567). Nun at the Nunnery of the Holy Cross (Abbaye Sainte-Croix), Poitiers. She rebelled against her abbess and raised an army against her. **LEUBEVÉRE** (fl. 586–590, Poitiers, Gaul), also known as Leubovera. Abbess of the Nunnery of the Holy Cross (Abbaye Sainte-Croix), Poitiers. Raised troops to defend her abbey.

Only a little is known of Clotild and Leubevére. Although they are certainly not the only example of military women in the Kingdom of the Franks, they may represent a unique occurrence in Frankish history: that of women taking up arms against each other. The primary source of this historical event is Gregory of Tours's *History of the Franks*. The conflict was not a major rebellion or outright war, but was important enough for Childebert II, king of Metz and Burgundy (575–595), to become involved. The events took place between the winter of 589 and November 590.

Both women were nuns at the same abbey. Both probably came from higher economic and social classes. Clotild claimed to be the daughter of a king. Leubevére, the abbess, was undoubtedly well-to-do and apparently also was related to royalty, possibly Guntram, king of Orléans and Burgundy (561–592). Beyond this, nothing is known of their early lives.

The origin of the conflict went back to the founding of the abbey by Saint Radegund in 557 CE. Radegund had obtained a piece of the true cross for her nunnery. When she died in 587, the care and protection of the convent was given to Maroveus, bishop of Poitiers (584–590), first at the request of Abbess Agnes and then officially by King Childebert II. Maroveus, however, was not popular among all the nuns, as he appears to have resented having to take responsibility for the nunnery.

Around 586 an unnamed nun of the Holy Cross escaped over the wall and took refuge in the Church of Saint Hilary. There she presented to Bishop Gregory a list of charges against her mother superior, Agnes. These charges are not specified by Gregory in his *History*, but he believed them to be false. The nun was "pulled up into the nunnery again by ropes" and later, by her own will, was enclosed in a cell to become a "recluse nun."

In 589 Clotild brought charges against her abbess, Leubevére, who had succeeded Agnes. Gregory says that Clotild "pretended to be" the daughter to King Charibert I (king of Paris, 561–567) and made the complaint that the abbess treated the nuns like servants rather than as the royalty that many of them apparently were. With her cousin Basina, also a nun of Holy Cross and daughter of King Chilperic (king of Soissons, 561–584), and about forty other nuns, Clotild left the nunnery to go to Tours to see Bishop Gregory. It was apparently her intention to have the abbess expelled and herself made abbess of Holy Cross. Gregory advised her to ask Bishop Maroveus, the official protector of the convent, to settle the matter. Clotild refused to do this; she also refused to return to Poitiers and told Gregory that she intended to take the matter before the kings.

Undated painting of Clotid. (Library of Congress)

Despite Gregory's warning that she might be excommunicated by the bishops' council, Clotild went to King Guntram. Upon her return to Tours, she found that several of the nuns had married. Clotild and company returned to Poitiers and took up residence at Saint Hilary, refusing to return to Holy Cross while Leubevére was still abbess. To protect themselves, the nuns hired what Gregory describes as "burglars, murderers, adulterers and criminals of all sorts." The recluse nun, having escaped from her cell, also joined Clotild. At Saint Hilary Clotild was visited by the bishops' council (Bishops Gundegisel of Bordeaux, Nicasius of Angoulême, Safarius of Perigueux, and Maroveus of Poitiers), who tried to persuade the nuns to return to Holy Cross. The nuns declined, and the "mob of ruffians" physically attacked the bishops, who made a hasty and bloody retreat. Clotild then took control of the estates belonging to the convent (Holy Cross), threatening to throw the abbess over the wall, given the opportunity.

By 589 King Childebert II ordered Count Macco (apparently of the region) to suppress the "revolt." The bishops themselves were uncertain about how to proceed. Bishop Maroveus wanted to administer communion to the nuns and speak to them but was opposed by Bishop Gundegisel. Childebert sent a priest, Theutar, to negotiate, but Clotild declined to speak to him until the ban of excommunication was lifted. This, in turn, the bishops refused to do. By the winter of 590 the rebellion seems to have lost its appeal to many of the nuns, primarily, it seems, due to a lack of fuel that made the cold unbearable. Some returned home, some went to other convents, and a few remained with Clotild, including her cousin Basina. In an apparently desperate moment Clotild ordered her thugs to break into the convent and take the abbess by force. The thugs, however, kidnapped Prioress Justina by mistake. Discovering their error, they sent her back to the convent, broke in again, and this time did kidnap the abbess. They also looted the convent. Clotild then moved into Holy Cross as its abbess. Abbess Leubevére was held at Basina's residence.

Bishop Maroveus then threatened not to say Easter Mass in Poitiers if Clotild did not release the abbess, effectively turning the community against her, but to no avail. Clotild's hired thugs, possibly led by Duke Childeric the Saxon, then went on something of a rampage. They committed murder and assaults, all, it seems, at the orders of Clotild. She also ordered her "hired assassin" to kill Leubevére if any attempt was made to rescue her. However, an official of King Childebert managed to free the abbess and take her to Saint Hilary. All this proved too much for Basina, who attempted to reconcile with her abbess, which they did briefly; but when one of the abbess's servants killed one of Basina's, the abbess and the nun were again on opposite sides. Kings Childebert and Guntram now called for an immediate end to the rebellion, granting Count Macco permission to use force if necessary. Clotild's thugs prepared to defend the convent, but Macco's forces proved stronger. Clotild, probably seeing her forces about to be overrun, took the piece of the true cross for which the convent was named and threatened revenge on anyone who touched her. Macco's forces ended the revolt by capturing the nuns. Childeric the Saxon, it should be noted, was ordered executed by Childebert for his crimes, but choked to death on his own vomit before the sentence could be carried out. Clotild and Basina were excommunicated. Clotild was later restored to communion but never returned to the convent; she lived in a villa provided for her by Childebert.

Bibliography: Gregory, Bishop of Tours, *The History of the Franks* 9.39–10.20; Jones, A.E., Jr., "A Biographical Catalogue of Named Persons from the History of the Franks of Gregory of Tours (ca. A.D. 260–591)," Ph.D. dissertation, University of South Carolina, 1992.

—Stefanie Buck

COCHRAN, JACQUELINE (born 1910?, Florida; died 7 August 1980, Indio, California). Colonel, United States Air Force Reserves; director of women pilots, WASP. United States, Second World War.

Jacqueline Cochran was one of the most accomplished female pilots in the United States. After Amelia Earhart, Cochran was also the best-known woman pilot in the United States before the Second World War. Born in poverty, she worked as a beautician, married the wealthy Floyd Odlum, and with his sponsorship became a private pilot who set numerous world records in the 1930s. Cochran and Amelia Earhart were the first women to compete in the Bendix Trophy Race in 1935, a prestigious transcontinental race. In 1938 Cochran became the first woman to win the race. She served as director of the Women's Flying Training Department (WFTD) and **WASP** from 1942 to 1944. In 1948 Cochran was commissioned a lieutenant colonel in the Air Force Reserve, retiring as a colonel in 1978. After the war Cochran returned to sport flying and in 1953 became the first woman to break the sound barrier. In 1960 Cochran was the first woman to fly at twice the speed of sound and the first to land a jet on an aircraft carrier; in 1962 she became the first woman to complete a transatlantic flight in a jet. She became the first woman elected to the American Aviation Hall of Fame in 1971.

Undated portrait of Jacqueline Cochran. (Library of Congress)

Cochran made a vivid impression on all who knew her. Former WASP Marian Hodgson described Cochran as "pretty, ambitious, ruthless, generous, and totally fearless." Overall, she said, Cochran made her uncomfortable: "I had the feeling that she would head toward her goal like the proverbial steam roller, heedless of obstacles in her path and oblivious to squashed forms that lay in her wake. Still, if that was how she achieved her visionary schemes, maybe that was what it took." Many WASP members credit Cochran as a pathbreaker and believe that they would never have had a chance to fly without her. For example, former WASP Katherine Landry Steele says, "Cochran was a gutsy, tough woman who crawled her way to the top. In my opinion, she is the only woman in the United States who could have put this program into action."

Jacqueline Cochran had influence in high places. She was invited to the White House several times by the Roosevelts, and although she was not one of their intimate circle, her connection to President Roosevelt and the First Lady greatly enhanced her power. In a 1939 letter to First Lady Eleanor Roosevelt, Cochran suggested that a women's air corps be formed for noncombat military flying jobs in order to release male pilots for combat duty. Cochran was adamant that women pilots should have their own organization, and that they should be commanded by a woman as well. She believed herself to be the best person for this job. A letter of introduction from the president gave Cochran an entrée into the War Department in July 1941.

When the Army Air Forces (AAF) dragged its feet about admitting women pilots, Cochran decided to sponsor a group of female American volunteer pilots who would fly with the British **ATA,** thereby proving the capability of American women. Cochran's ultimate goal (in which she failed) was to establish a permanent women's air corps as part of the Army Air Forces. In the fall of 1942 Cochran was made director of the newly created Women's Flying Training Detachment (WFTD), and by August 1943, after engineering a merger between the WFTD and the **WAFS,** she was named director of women pilots with administrative control over all female pilots in the AAF.

No WASP could be hired, fired, or reassigned by any using agency without Cochran's approval—an unusual administrative requirement within a military organization. Furthermore, Cochran was "authorized to consult directly with the commanding generals of the several commands and air forces"—again, an unusual authority for anyone operating within a military structure. In effect, Cochran had established herself as a free agent, operating independently of the military chain of command. The Air Staff later characterized her leadership as "aggressive control which was out of keeping with the traditional role of a staff officer"; Cochran objected to this description. She defended her actions, stating that "the Women Pilot program was something entirely new, dealt with something that crossed command lines and was experimental in nature."

AAF Regulation 40-8 (21 December 1943) spelled out the authority and responsibility of the director of women pilots. Many routine administrative matters involving women pilots, including resignations, transfers, and accident reports, as well as any news releases or publicity matters, had to be coordinated with the director of women pilots. This system created additional friction between Cochran and the using agencies, especially the Ferrying Division. Cochran energetically opposed any attempt by other persons or agencies to supersede her ultimate authority over women pilots, which resulted in frequent conflicts (see WAFS).

Cochran rejected suggestions that the WASP should be part of the **WAC,** just as the AAF was technically part of the army. Not a woman to mince words, in her memoir *The Stars at Noon* Cochran called WAC commander **Oveta Culp Hobby** "the woman I loved to hate" and noted that they "had knock-down, drag-out fights" over jurisdiction. Cochran told AAF General Hap Arnold, "[The WASP] will become part of the Women's Army Corps over my dead body." Firsthand observers frequently noted power struggles between Cochran and Nancy Love (see WAFS), primarily arising from Cochran's determination to establish herself as the final authority over women pilots. Captain James Teague, who attended meetings with Cochran in 1942 when the WAFS and WFTD were still separate groups, noted in a memo to his commander that Cochran "made it quite clear that she considered herself the only person who could efficiently be in charge of the Women Ferry Pilots," although that group had already been formed under Nancy Love's command. Teague warned that "Miss Cochran will take inch by inch and try to move in on us. I don't believe I am exaggerating the extent to which she will go." Like the German test pilot **Hanna Reitsch,** Cochran never seemed to regard another woman as her peer when it came to flying skill or leadership.

During the battle to militarize the WASP in 1944, the complicated situation in Congress was exacerbated by the all-or-nothing stance taken by Cochran on the proceedings. In July and again in August 1944 Cochran recommended that WASP be militarized, but if that could not be achieved, then the program should be disbanded altogether. Several newspapers dubbed her position as an "ultimatum." When the militarization bill failed, the WASP program was disbanded, although the war was far from over and there was still a need for trained ferry pilots.

Cochran's assessment of the WASP, stated in her final report, was that women could meet the same standards and qualifications as male pilots, were unhampered by unique physical or medical problems, and had a proven flying record for flight safety and graduation rates comparable to those of men in the same sort of work. Moreover, she stated that young American women could provide "an effective women's airforce of many scores of thousands of good dependable pilots." But thirty years later Cochran testified to military boards that she regarded the idea of allowing women into the USAF as pilots as a waste of time, since few women would make a career in the military. She noted that most of the WASP members had simply gone home and raised families after the war, rather than pursuing a career in flying. (She did not note that it was virtually impossible in the immediate postwar years for women to take up careers in aviation.) When congressional hearings were held in 1977 to determine whether the surviving WASP members should be given veteran status, Cochran did not attend, perhaps for health reasons. Cochran's stance against the militarization of female pilots aroused confusion and animosity among some female pilots and veterans.

A controversial and vibrant woman, undeniably one of the great pilots of the twentieth century, Jacqueline Cochran was both adored and abhorred; there seems to be little middle ground in people's response to her. Like many women who broke the rules of their time, she was strong, determined, and successful on the basis of both skill and willpower and created both devotees and enemies in the process.

Bibliography: Arnold, *Global Mission,* 1949; Cochran, "Final Report on Women Pilot Program," 1945, and USAF Oral History Interview K239.0512-940, 11–12 March 1976; *The Stars at Noon,* 1954, Cochran and Brinley, *Jackie Cochran,* 1987; Hodgson, *Winning My Wings,* 1996; Marx, "Women Pilots in the Air Transport Command (Revised)," 1945; Pennington, "Women and Military Aviation in the Second World War," 2000.

—Reina Pennington

CORBIN, MARGARET COCHRAN (born 12 November 1751, Pennsylvania; died ca. 1800, Highland Falls, New York), also known as "Captain Molly," "Dirty Kate"; sometimes confused with "Molly Pitcher" (Mary Ludwig Hays McCauley). Artillery gunner, member of the army's Invalid Corps, United States, Revolutionary War.

> George Washington issued orders relating to women. His 28 December 1782 General Orders, for example, specified that "on every fifteen men actually in a regiment, there shall be allowed a draught for sixteen rations, so as to supply the women of the regiment or corps, that is to say the rations drawn may exceed the number of Noncommissioned and privates one fifteenth."
>
> —John C. Fitzpatrick, *The Writings of George Washington from the Original Manuscript Sources, 1745–1799* (Washington, DC: U.S. Government Printing Office, 1931).

Margaret Corbin became a soldier in the Continental army during the Revolutionary War, and after the war she continued as a member of the army's Invalid Corps stationed at West Point. She was compensated by the state of Pennsylvania for her military service and became the first woman in American history to receive a congressional soldier's pension. While many women followed their husbands into the Continental army, Corbin's military service was unusually well documented, and she was given official status as a soldier in her own right.

Margaret Corbin was born on the Pennsylvania frontier to Scots-Irish parents. After her father was killed and her mother abducted in an Indian raid in 1756, Corbin was raised by an uncle. In 1772 she married John Corbin, a Virginian. When he enlisted as a private in the First Pennsylvania Regiment of Continental Artillery under Captain Thomas Proctor at the start of the Revolutionary War, Margaret went along. There is some indication that she enlisted under her own name.

During the Battle of Fort Washington on 16 November 1776 the Corbins trained their gun on attacking Hessian soldiers and on British ships on the Hudson River. John Corbin was killed during the fight, and Margaret assumed full control of the weapon. Corbin was herself wounded in the battle by British grapeshot that tore her chest and nearly severed her arm. Fort Washington was surrendered to the British, but Margaret Corbin was not taken prisoner of war. Probably because she was a woman, and therefore was not perceived by the British as a regular soldier, she was allowed to leave New York City.

The Pennsylvania Executive Council granted Margaret Corbin thirty dollars on 29 June 1779, and on 6 July 1779 the Continental Congress granted her military half pay and a yearly suit of clothing for life in recognition of her wounds received "whilst she heroically filled the post of her husband who was killed by her side." Corbin then enlisted as the only woman in the Invalid Corps at West Point, where she joined other disabled soldiers. The Invalid Corps had been organized in 1777 to provide guards for West Point, recruiting officers, and occasional garrison troops. In 1782 Margaret Corbin married another member of the Invalid Corps, whose name is unknown. Also in 1782 Corbin petitioned Congress to have her pension increased and was denied. She did manage to gain access to a full monthly rum and whiskey ration, despite the general policy of the Commissary at West Point to deny women access to liquor. When she submitted a bill for the 257 gills of whiskey that she had been denied under the policy over the years, the army ordered that she be paid for the missed rations.

Margaret Corbin was discharged from the Invalid Corps in April 1783 as the corps was disbanded at the conclusion of the Revolutionary War, but she continued an unofficial connection to West Point. She remained in the West Point area for the rest of her life, where she was regarded as a respected if eccentric figure who was frequently observed wearing her old artillery jacket. Corbin was buried in an unmarked grave, but on 4 July 1926 she was reinterred at West Point, where a monument was dedicated in her honor.

Bibliography: Blanco, *The American Revolution, 1775–1783,* 1993; De Pauw, "Women in Combat," 1980; Engle, *Women in the American Revolution,* 1976; Evans, *Weathering the Storm,* 1989; Hall, *Margaret Corbin, Heroine of the Battle of Fort Washington,* 1932.

—Sarah J. Purcell

CORNUM, RHONDA L. SCOTT (born 31 October 1954, Dayton, Ohio, United States). Colonel, United States Army; Ph.D., M.D. Flight surgeon, United States, Gulf War.

Rhonda Cornum was the only female officer to become a prisoner of war during the Persian Gulf War. Her 1992 autobiography *She Went to War,* which told the story of her experience, was selected by the *New York Times* as one of the most notable books of the year and was placed on recommended reading lists such as that of the Commandant of the United States Marine Corps.

Cornum graduated from Cornell University with a B.S. degree in microbiology and genetics in 1975, the same year in which her daughter Regan was born (the product of a first marriage that ended a few years later). Cornum went on to earn a Ph.D. in nutrition and biochemistry in 1980, also from Cornell. She joined the army after being recruited at a medical conference in Atlantic City. After her first assignment at Letterman Army Institute of Research, Cornum entered the military's Uniformed Services University of the Health Sciences in Bethesda, Maryland, in 1982. While attending medical school, Cornum met and married air force officer Kory Cornum, and together the two trained as flight surgeons, learning to fly helicopters. After graduating from medical school on 17 May 1986, Rhonda Cornum held a variety of assignments, including an internship in general surgery at Walter Reed Army Medical Center and chief of the Physical Exam Section at Lyster Hospital, Fort Rucker, where she was named Fort Rucker's Flight Surgeon of the Year in 1988 and Flight Surgeon of the Year for the army in 1990. She was working in the Biomedical Applications Research Division at the United States Army Aeromedical Research Laboratory, Fort Rucker, when she deployed to Saudi Arabia in August 1990.

In Saudi Arabia thirty-six-year-old Major Cornum served as a flight surgeon to the 2-229th Attack Helicopter Battalion, which was attached to the 101st

Airborne Division. She says that she put up a sign in her aid-station tent reading, "Suffering Is Stupid, but Whining Is Worse," which was one of her philosophies of life. During the long months of waiting in Saudi Arabia another woman officer had suggested that separate women's quarters be established on the base. However, Cornum says, "I argued that putting the women in a separate area was a terrible idea. The commanders wanted to enhance unit integrity and cohesiveness, so segregating people according to sex was obviously not the way to do it" (*She Went to War*, 27–28).

On 27 February 1991, the last day of the four-day ground war in the Persian Gulf, Major Cornum and seven crew members were sent on a Blackhawk helicopter mission to rescue a downed and injured air force pilot from behind enemy lines. About thirty minutes into the flight the helicopter received ground fire from Iraqi antiaircraft guns and crashed in the desert, killing both pilots, one of the door gunners, and two infantrymen. Cornum and two others survived. Cornum was badly injured, with two broken arms, torn ligaments in her right knee, and a bullet wound in the shoulder. She and the other survivors were quickly taken prisoner by Iraqi soldiers. On her first night of captivity Cornum was sexually abused (though not raped) by an Iraqi guard. Other guards treated her with relative consideration, although medical care was minimal. When she finally was taken to a hospital, the medical staff treated her well. Cornum and her fellow prisoners were released on 6 March 1991, after eight days of captivity. During the release process she met fellow POW Captain Bill Andrews, the air force pilot her crew had tried to rescue. Cornum noted, "I never thought we had wasted the lives of our crew members trying to rescue him. Every one of us would go again if we had the chance" (*She Went to War*, 110).

After her release Cornum testified before Congress on her treatment while imprisoned and on general issues pertaining to congressional concerns about American women taken as prisoners and the role of women in combat. Although much was made in the press about the fact that one of Cornum's guards molested her, she herself does not regard her treatment as a POW as any worse than that of her male counterparts: "Sexual abuse was one form of torture, just like being beaten . . . most of us were tortured in one way or another," and "being a POW is a rape of your entire life" (*She Went to War*, 173, 203).

Cornum is a strong advocate of gender-integrated service; in her book she writes, "People talk about male bonding in the military and how female soldiers supposedly will disrupt unit cohesion. Real bonding, however, goes far beyond whether the people involved are two men or two women or one of each. . . . going to war with a unit, risking your life with them, builds an intimate and intense relationship. The soldiers don't all have to be men for that to happen." She also notes that although many people ask whether her POW experience affected her views on women in combat, "I think exactly what I have always thought, that everyone should be allowed to compete for all available jobs, regardless of race or gender" (*She Went to War*, 84, 197).

Cornum graduated from the USAF Air Command and Staff College in June 1992 and completed a residency in urology. Her subsequent assignments included assistant deputy commander for clinical services (DCCS) and staff urologist at Eisenhower Medical Center at Fort Gordon, Georgia, and commander, 28th Combat Support Hospital at Fort Bragg, North Carolina (July 2000–March 2001). Then Colonel Cornum began a six-month assignment as commander of Task Force Med Eagle (TFME) at Eagle Base, Bosnia and Herzegovina, followed by an assignment as a student at the United States Army War College, Carlisle Barracks, Pennsylvania. Colonel Cornum's awards include the Expert Field Medic, Air Assault, and Airborne badges, a Bronze Star, an Air Medal, and for her Gulf War service, a Distinguished Flying Cross and the Purple Heart.

Bibliography: Cornum, *She Went to War*, 1992; Francke, *Ground Zero*, 1997; Perry, Joellen, "Rhonda Cornum: The Iraqi Army Couldn't Quash Her Fighting Spirit," *U.S. News and World Report* 20–27 August 2001: 30–32; Stiehm, *It's Our Military, Too!*, 1996.

—Vicki L. Friedl

CRUSADES, WOMEN IN, 1096–1291. Eastern Mediterranean and western Europe.

For nearly two centuries the Christians of western Europe sponsored a series of military expeditions against Muslims to take control of the region in the Middle East regarded as the traditional Holy Land. In his speech at Clermont in southern France in 1095, Pope Urban II called for the First Crusade, and it was extraordinarily successful. Jerusalem was taken in 1099, and by 1118 most of Syria and Palestine was under Christian control. This led to the formation of the Crusader States, the kingdoms established in the newly conquered territory. The Second and Third Crusades continued the fight for control of the Holy Land; the fall of Edessa in 1144 sparked the Second Crusade, the fall of Jerusalem in 1187 sparked the Third. The Fourth Crusade of 1204 became a Latin invasion of a fellow Christian state, Byzantium, culminating in the conquest of Constantinople. The fall of Acre in 1291 marked the practical end of the Crusades in the Middle East, though the crusading impulse continued to spawn grandiose plans to retake the region for Christianity.

An unknown number of women took part in the Crusades in a variety of roles. In many cases wives accompanied their husbands on campaign. For example, the wives of Baldwin of Boulogne and Raymond of Toulouse accompanied their husbands on the First Crusade, as did **Emma FitzOsbern. Eleanor of Aquitaine** is the most famous example; she went on crusade with her first husband, Louis VII of France, but there has been no attempt to determine what portion of the army of the Second Crusade she raised from her large duchy. In the Second Crusade there were reportedly two bands of women "armed as Amazons and mounted on horseback," a French party with Eleanor of Aquitaine and a German one with Emperor Conrad III. These women were likely noblewomen with their own followings of troops, provided at their own expense. Eleanor's widowed daughter, **Joanna of England,** accompanied her sister-in-law, Berengaria of Navarre, on part of the Second Crusade.

For every woman of the high nobility participating in a crusade, however, there were scores who were the wives of commoners and the lesser nobility. The accounts of the period often mention the presence of these women. Any medieval army on the march attracted camp followers and hangers-on of both sexes, but it is impossible to determine accurate numbers. William of Tyre, writing nearly a hundred years after the fact, described the plague in Antioch during the First Crusade and claimed that "women in particular were the victims in this epidemic, nearly fifty thousand perishing in a few days" (1.299). Fifty thousand is a number that cannot be verified, but the implication is that a significant number of women had accompanied the crusaders. Some historians have dismissed these women as being mainly prostitutes and "dead weight"; other recent research into camp followers indicates that most women who followed armies were not prostitutes. However, the conflation of women's roles into sexual stereotypes may account for the description of any woman engaging in sexual activity, even legal and common-law wives, as a "whore." The fanciful descriptions (and obvious enthusiasm) of the Muslim chronicler Imad ad-Din concerning the activities of these prostitutes during the Third Crusade was not echoed on the Christian side.

Some prostitutes were undoubtedly attendant and created a unique problem by their presence in a "holy" campaign. Several significant defeats (the second Battle of Dorylaeum, for example) were ascribed to the "licentiousness" of the crusaders. During the eight-month siege of Antioch in the First Crusade the crusaders' leaders came to believe that their lack of success was a result of debaucheries by the soldiers; they ejected all women from the camp, including the wives and nuns who were present. After thus showing their religious devotion, the crusaders achieved a victory over the city, which was regarded as a divine reward. However, they immediately returned to their former habits of consorting with women, resulting (according to the chroniclers) in the arrival of a Muslim army that besieged them within Antioch. Again they did penance and were rewarded with a victory over the besieging army.

Women were always present within the major crusading armies and within the Crusader States as wives, camp followers, and support personnel. Many of these women performed important logistical functions. In addition, women often assisted the soldiers. For example, during the First Crusade women played a role in the victory at Dorylaeum (1097) by carrying water to men who were maintaining a standing shield wall. These women were under enemy fire as they moved back and forth to the front lines. In several instances women are mentioned bringing water and food to men during the fighting.

This occurred at Nicaea during the First Crusade, during the siege of Jerusalem, and during the Battle of Damietta in the Fifth Crusade (here they also brought stones for use as missiles). Some women assisted more directly, performing tasks that traditionally fell within the male sphere. According to the French chronicler Ambroise, who was present at the siege of Acre during the Third Crusade, a woman working to fill in the moat was slain by a Muslim arrow. As she died, she asked her husband to use her body to help fill the moat, which was done.

Some women went even further, taking up arms against the infidel. William of Tyre describes women fighting at Jerusalem during the First Crusade: "Even women, regardless of sex and natural weakness, dared to assume arms and fought manfully far beyond their strength" (1.362). Several Muslim writers record similar instances of Christian women taking up arms; Ibn al-Athir wrote, "Among the prisoners were three Frankish women who had fought from horseback and were recognized as women only when captured and stripped of their armor." His account can be compared to that of Beha ad-Din, a historian in Saladin's service, who wrote, "Someone told me that he had seen four women engaged in the fight, two of whom were taken prisoners." Beha ad-Din further relates the finding of the bodies of two women on the battlefield of Acre, as well as the story of a "green mantled" female archer who wounded several Muslims before being killed. Her bow was taken to Saladin, where the historian presumably saw it. All of these Muslim accounts concern the actions at Acre during the Third Crusade. While it is possible that the same story is being repeated by several of the Muslim writers (Ibn al-Athir certainly used Beha ad-Din's work as a source for his larger history), it seems possible that a female warrior or warriors participated actively in this action.

Bibliography: Brundage, "Prostitution, Miscegenation, and Sexual Purity in the First Crusade," 1985; Comnena, *The Alexiad*, 1969; Foucher de Chartres, *A History of the Expedition to Jerusalem, 1095–1127,* 1972; Gabrieli, *Arab Historians of the Crusades,* 1969; Nicholson, "Women on the Third Crusade," 1997; Peters, *The First Crusade: The Chronicle of Fulcher of Chartres and Other Source Materials,* 1998; William of Tyre, *A Middle English Chronicle of the First Crusade,* 2001.

—Paul Westermeyer

CYNANE (Greek: Kunane) (born 357 BCE; died 320 BCE), also known as Cynna (Kunna). Macedonian princess and military leader.

Cynane came from the most famous military family of her age, which included her father Philip II of Macedon, her half brother Alexander (later called the Great), and her stepmother **Olympias**. Her own mother Eurydice I (Audata) was an Illyrian noble who also may have had experience in battle. Cynane was famous for her military knowledge and gave her daughter **Eurydice II** (Adea) a military education. Cynane commanded armies during the

Macedonian expansion northward into the Balkans in the 4th century BCE. She played an active role in fighting, almost certainly from horseback. She is credited with defeating an Illyrian army and killing an Illyrian queen in individual combat.

About three years after Alexander's death and the fragmentation of his empire, Cynane crossed the Strymon River in Paeonia with a mercenary army for the purpose of forcing the marriage of her daughter Eurydice to the king of the moment, Philip Arrhidaeus. When her general Alcetas confronted her, she reproached him for betraying the royal line. The ancient Greek writer Polyaenus relates that Alcetas's troops were awestruck by the sight of the daughter of Philip II; when Alcetas had Cynane assassinated, they mutinied. The dynast Perdiccas allowed Eurydice to marry Philip Arrhidaeus. However, three years later Olympias led an army into Macedon against Philip and Eurydice in what the Greek historian Duris called the "first war between women," and Cynane's daughter was killed. Thus Alexander's half sister and his niece died much like himself: at a young age and in a militant and violent manner.

Bibliography: Arrian, *History of the Successors of Alexander* 1:20–24; Macurdy, *Hellenistic Queens*, 1975; Polyaenus, *Strategematon* 8.60.

—Alexander Ingle

D

DADESHKELIANI, PRINCESS KATI (fl. ca. 1915, Georgia, Russia), also known as Prince Djamal Dadeshkeliani. Cavalry, ambulance company, Russia, First World War.

Princess Kati Dadeshkeliani was a member of the ruling family of Svanetia, a small province in the territory of Georgia. She entered the Russian imperial army during the First World War after a series of family tragedies left her in a state of severe depression (her father was murdered, her sister committed suicide, and her newlywed husband was killed in combat during the first weeks of the war). A family friend, Colonel Khogandokov, suggested that she join the fighting to get her mind off her troubles. With the help of the colonel, who secured her a military passport in the name of Prince Djamal Dadeshkeliani, she joined the Grand Duke Mikhail's Caucasian "Savage Division," disguised as a young male officer. Khogandokov taught her the rudiments of military etiquette, and she seemed to have little difficulty emulating male behavior.

The princess initially served as an aide-de-camp to Khogandokov and as a cavalry courier, assigned to regimental headquarters in Czortkow, Galicia. Her lifetime experience with horses allowed her to fulfill these duties with relative ease, despite the danger of her missions, which required riding long distances under heavy fire. She soon became bored with the tasks of a courier, however, and requested to be sent to the trenches. Khogandokov granted her wish, despite his reservations about a woman's ability to withstand battle conditions. Dadeshkeliani spent a long week in mid-1915 engaged in front-line fighting, but her experiences in the trenches quickly proved disenchanting: the princess lost her romantic conception of war and suffered disillusionment with her own abilities. Despite her successes with other "male" occupations, she ascribed her poor performance and intense fear in the trenches to inherently female weakness. The princess served the remainder of the war in company ambulances and was promoted to chief medical assistant. She felt much more comfortable in this capacity and had little problem performing these functions even in the most dangerous situations on the front lines. Over the course of her three-year term of service "Prince Djamal" was awarded two St. George's Crosses (**Order of St. George**). In August 1917 the entire regiment was ordered to Petrograd (presumably to defend the city), where Dadeshkeliani rejoined her family and resumed her female identity, thus ending her military career.

Bibliography: Dadeshkeliani, *Princess in Uniform*, 1934; Wheelwright, *Amazons and Military Maids*, 1989.

—Laurie Stoff

DAHOMEY, WOMEN IN ARMY OF. Kingdom of Dahomey, West Africa, eighteenth and nineteenth centuries.

Women made up nearly one-half of the armed forces of the West African kingdom of Dahomey in the nineteenth century. Nearly all were members of a larger institution, the palace/bureaucracy of the king, which eventually included as many as 8,000 persons, of whom all were women except a few eunuchs. The epithet "Amazons" was used derisively by European observers to describe the women soldiers. Women armed with muskets were first seen by European visitors to Dahomey in 1727. During the eighteenth century a small portion of the palace women were armed and trained to serve as the king's bodyguard. Succession struggles were invariably violent as eligible princes allied with powerful women within the palace to win the throne. Since physical control over the palace was a prerequisite to claiming power, the armed women in the palace played crucial roles in the violence that accompanied virtually every succession. By late in the century their numbers were estimated at between 150 and 500.

Les Amazones à L'Exercise from *Le Dahomey,* by Edouard Foa (Paris: A. Hennuyer, 1895).

Following his coup in 1818, King Gezo maintained the female bodyguard as a military center and added to his military strength an army of women that included parallel right and left wings. Recruits were drafted from families in the kingdom or drawn from among young female war captives who were brought up into military life within the palace. By the 1840s the women's army was better equipped, disciplined, and trained than that of the men. Missionary Thomas Birch Freeman commented that Dahomey's women soldiers "excel in martial appearance any body of native troops I have ever seen," while John Duncan said that they were "all well armed, and generally fine strong healthy women" and that "their appearance is more martial than the generality of the men; and if undertaking a campaign, I should prefer the females to the male soldiers of this country." In the mid-nineteenth century Sir Richard Burton (the famed explorer who searched Africa for the origins of the Nile River) provided the best-known Western account, commenting that he believed that there was little difference between the sexes in many African peoples; that in fact women seemed just as strong as men, and even stronger and larger than men in the case of Dahomey. Burton speculated that the ferocity of the women soldiers was a result of their sexual abstinence.

> Every people have had their female warriors.
>
> —Voltaire, Pierre Augustin Caron de Beaumarchais, Jean-Antoine-Nicolas de Caritat Condorcet, and Jacques Joseph Marie Decroix, *Oeuvres Completes De Voltaire*, Vol. 17, 70 vols. ([Kehl]: Impr. de la Sociâetâe littâeraire-typographique, 1785), 130.

The women's and men's forces had a parallel organization of companies and regiments. Each was commanded by officers of the same sex as the soldiers. The vast majority of the women warriors were infantry, armed with trade guns and short swords. Valor in warfare was rewarded with titles and appointment to high office in the kingdom. In the mid-nineteenth century, for example, a woman named Yewe was official royal host to British visitors as well as commander of two regiments. She had been promoted for successfully scaling the stone ramparts of a fortified town.

The women soldiers numbered 5,000 to 6,000 prior to a disastrous 1851 attack on the Yoruba city of Abeokuta (Nigeria), when they lost 1,000 to 2,000. In the early 1860s an epidemic of smallpox and a second fruitless attack on Abeokuta thinned their ranks to around 3,000. In the 1890s the women warriors were in the front ranks of the troops that defended Dahomey from an invading French colonial army. Decimated by the superior firepower of the French, the army of Dahomey (male and female soldiers alike) was disbanded when the kingdom fell to the French in 1894.

Bibliography: Almeida-Topor, *Les Amazones,* 1984; Bay, *Wives of the Leopard,* 1998; Burton, *A Mission to Gelele, King of Dahome,* 1864; Duncan, *Travels in Western Africa,* 1847; Edgerton, *Warrior Women,* 2000; Law, "The 'Amazons' of Dahomey," 1993.

—Edna G. Bay

DARRAGH, LYDIA BARRINGTON (born 1729, Dublin, Ireland; died 28 December 1789, Philadelphia). Patriot, United States, Revolutionary War.

Lydia Darragh gained information about the plans of the British high command during the Revolutionary War and went to the American lines to inform them of a planned attack on the patriot army camped at Whitemarsh during the winter of 1777. Her story is interesting as an illustration of the way some women followed their patriotic beliefs, even when that meant violating the bounds of acceptable behavior within their community.

Lydia Darragh, a midwife and nurse, had emigrated to America with her husband William shortly after their marriage in November 1753 and had joined the Philadelphia Monthly Meeting of Friends. During the British occupation of Philadelphia in 1777–1778 the Darraghs lived in Loxley Hall on Second Street directly opposite the headquarters of General William Howe. From this vantage point Darragh was able to overhear conversations and obtain occasional bits of information on the activities of the British. She passed on this information to her eldest son Charles, who was serving as an officer with the revolutionary forces.

According to family history, British officers sometimes requisitioned one of the Darraghs' back chambers for use as a council room. On the evening of 2 December 1777 Lydia Darragh eavesdropped on one of these meetings and learned of a planned surprise attack on General George Washington's army at Whitemarsh. If the first published account of her wartime exploit is accurate, the forty-eight-year-old Darragh did not inform her husband of her intention to carry a warning to the American forces, but acted on her own. She acquired a pass to leave the city on the pretext that she was going to get flour at the Frankford mills about five miles away. After depositing her flour sack at the mill, however, she continued on toward Whitemarsh until she encountered an American officer, Lieutenant Colonel Thomas Craig, on the road. Darragh delivered the crucial information to Craig, then walked back to retrieve her filled flour sack before recrossing British lines into Philadelphia. On the night of 5 December Howe marched 12,000 soldiers out of the city to storm Washington's camp, but without the advantage of surprise, he was unable to dislodge the Americans from their entrenched position and returned within three days.

Darragh was not actively involved in the war except for this incident, although it may be assumed that she continued to support the efforts of the revolutionary army. After the revolution she was widowed and kept a store on Second Street until her death in 1789. There are no known contemporary records that confirm the details of Darragh's revolutionary activities. Her obituary (*Pennsylvania Mercury*, 2 January 1790) makes no reference to the events of the war. Darragh would have had reason not to publicize her service to the patriot cause during her lifetime. As a practicing Quaker, she was bound to adhere to the sect's pacifist principles. Her son Charles was dismissed from the Society of Friends in 1781 for his participation in the war. Darragh herself was censured by the Friends for a period after her husband's death in 1783, but apparently later reconciled, as she was allowed to be buried on Quaker ground. The first published account of Darragh's wartime activities appeared in 1827 ("American Biography," *American Quarterly Review* 1 [March 1827]: 32–34), and one of her descendants wrote about her revolutionary activities in 1899.

Bibliography: Darrach, 1899; Dexter, *Colonial Women of Affairs*, 1931; Ellet, *The Women of the American Revolution*, 1848–1850; Leonard, *All the Daring of the Soldier*, 1999.

—Donna DeFabio Curtin

DE LA TOUR, FRANÇOISE MARIE JACQUELIN (born 1602, France; died 1645, Fort La Tour, New Brunswick, Canada). Defender during civil wars in 1640s Acadia, Canada.

Françoise Marie Jacquelin was born in France in 1602. In 1625 she traveled to Nova Scotia, where she married Charles de La Tour and assisted in the establishment of Fort La Tour. She was known as an excellent hunter, skilled with firearms, who frequently traveled throughout the New World to advance her husband's trading business and to secure supplies and recruit soldiers. A male contemporary referred to her as "a remarkable woman or an uncommon man." Life was prosperous until the ambitious Jesuit Seigneur d'Aulnay Charnise became aggressive in his designs on Fort La Tour. During the increasing hostilities that arose between the inhabitants of Fort La Tour and Charnise, Madame de La Tour assumed the role of military commander in her husband's absence. In one engagement she commanded a portion of their fleet of three ships and numerous small craft, chasing Charnise from the mouth of the St. John River. Her forces valiantly defended the fort with cannons, blowing Charnise's ship out of the water and throwing his soldiers from their scaling ropes and ladders. Meanwhile, her husband was outlawed and forced into hiding. Madame de La Tour's defense of the fort had been so successful that Charnise's hatred was now transferred to her with increasing venom. Four months later he renewed his attack upon the fort. For four days she led the defense of the fort with forty-five soldiers but was defeated. Following her capture, Madame de La Tour was forced into a halter that held her head rigid, and she was forced to watch as her soldiers were hung one by one. De La Tour herself was sent to Port Royale as a captive, where she died a few weeks later.

Bibliography: Greer, Allan, "La Tour, Françoise-Marie de Saint-Etienne de," *The Canadian Encyclopedia*, vol. 2 (Edmonton: 1985), 944; Moody, Barry, *The Acadians* (Toronto: 1981).

—Kristine Dawkins-Wright

DEBORAH (fl. early twelfth century BCE). Old Testament, possibly apocryphal figure; prophetess, judge over the Hebrews. Military leader and strategist in conflict between Hebrews and Canaanites.

Deborah

After the slaying by Ehud of Eglon, King of Moab, the Moabites are subdued. Deborah the Prophetess encourages Barak to attack Sisera, Captain of the Army of Jabin, King of Canaan. From the Maciejowski Bible, France (probably Paris), 1250, MS. M.638, f. 12. (The Pierpont Morgan Library/Art Resource, NY)

The books of Joshua and Judges in the Old Testament describe some of the conflicts between the Hebrews and local tribes in the "Promised Land" of Canaan. The Hebrews had no single leader over their twelve-tribe confederation, but successive judges rose to prominence to settle disputes between individuals and tribes. Deborah was one such judge, the fourth described in the Book of Judges. The Hebrews, according to the biblical account, had turned away from God and had been delivered to enemies until they became wise enough to return to God. After twenty years of oppression at the hands of the Canaanite king Jabin of Hazor, Deborah apparently grew tired of hearing the complaints her people leveled against the Canaanites and decided (or was divinely inspired) to overthrow the Canaanite yoke.

Deborah called upon Barak of the Hebrew tribe of Naphtali to lead a revolt. Barak hesitated, either through fear or through a lack of belief in her prophetic abilities. He would only call forth volunteers if she would cocommand; she agreed but warned him that the glory for defeating Jabin's general, Sisera, would go to a woman.

The Hebrews had long feared Jabin and Sisera, for they fought in iron chariots. The Hebrews had not yet learned the art of ironsmithing, so they had to fight with inferior bronze weapons when they had any weapons at all, for the Bible says that the Canaanites had banned Hebrew ownership of weapons. Barak called out 10,000 men to fight, according to Deborah's instruction, and gathered them on Mount Tabor, which is above Megiddo along the Plain of Esdraelon and the river Kishon. It is likely that Deborah's plan to fight Sisera was timed to coincide with seasonal rains, for the Kishon would have risen and the ground would have been muddy. The chariots would have been mired and useless, making them easy targets for a swarming Hebrew force. The size of the Canaanite force is not revealed, but they would probably have lost heart upon seeing the chariots, the strength of their army, unable to fight.

As Deborah had predicted, Sisera fled as his army was being slaughtered, escaping to the camp of Heber the Kenite, with whom his people had cordial relations. Assured of safekeeping by Heber's wife Jael, he gratefully accepted some milk and lay down to sleep. Jael, loyal to the Hebrews, killed Sisera by driving a tent peg through his temple, then went outside to await Barak's arrival. The Bible relates that the Hebrews returned to their God for forty years, but apparently returned to their evil ways after Deborah's death. She receives no other mention after the Song of Deborah celebrates the victory over Sisera in Judges 5.

Deborah stands out as one of the few women to rise to prominence in Hebrew history, the only one listed as a judge and military leader over the Hebrews. Although her historical existence cannot be proven, her story is interesting for the almost matter-of-fact way in which a woman is shown to exercise military leadership and acumen, against all the traditions of her culture. Her example would provide an inspiration in centuries to come; for example, during the Renaissance the Italian artist Artemisia Gentileschi immortalized the victories of Deborah (as well as **Judith**).

Bibliography: Asimov, *Asimov's Guide to the Bible,* 1968; Bible, Judges 4 and 5; Lamsa, *Old Testament Light,* 1964.

—Paul K. Davis

DELAYE, MARGOT (fl. second half of the sixteenth century, Montélimar, France), also known as Marguerite Delehaye. Partly legendary heroine of the siege of Montélimar, France, Wars of Religion.

It is not certain that Margot Delaye ever existed. Did a different woman, bearing a different name, accomplish the exploits attributed to her? Or did popular memory group a number of exploits owing to several different women under this name? The question is still open.

What is known is that in May 1570 Montélimar was besieged by an army under the Huguenot Louis of Nassau, and women participated vigorously in the defense of their town. According to tradition, Margot Delaye was credited with disabling Louis by striking him over the head with a cast-iron cooking pot. Delaye was herself seriously wounded during the fighting; she lost an arm. The defense of Montélimar was successful, and the enemy army had to give up on taking the town.

A woman named Marguerite-Catherine Ponsoye, called Gandonne, was probably the origin of the legend. According to town records, she was "disabled in her right hand" on the ramparts and for this received compensation from the consuls. (Perhaps she later married a Delaye.) Another woman, Catherine Arnaud, was "disabled likewise in the right arm" in the course of the siege. Since a Margot Delaye does not appear in the records, it seems that an amalgamation was made of the deeds of Ponsoye, Arnaud, and some others, because the women wounded on the walls were so numerous that the consuls were unable to compensate them all. At the scene of these exploits the consuls erected a statue to the glory of Margot Delaye. This statue, mentioned in the *Dictionnaire des Gaules,* still existed in 1766, but the inscription was already illegible. By the nineteenth century it had disappeared. However, today in the castle of Montélimar one can admire a large tableau by François Grellet (1880) showing Margot Delaye on the ramparts preparing to topple a load of stone rubble on an attacker.

Bibliography: Baumont, Stephane, *Histoire de Montélimar* (Toulouse: Privat, 1992); Coston, baron de, *Histoire de Montélimar et des principales familles qui ont habité cette ville,* vol. 2 (Montélimar: Bourron, 1883); Expilly, *Dictionnaire,* 1766; Roman d'Amat, *Dictionnaire de biographie française,* 1965.

—Marie-Thérèse Lalaguë-Guilhemsans, trans. Valerie Eads

DIONISIA DE GRAUNTCOURT (fl. mid-twelfth century, Northamptonshire, England), also known as Dionisia of Clapton, Dionisia, Denise.

During the reign of King Stephen (1135–1154), while she was still a girl, the lightly armed Dionisia de Grauntcourt overthrew a mounted soldier (*miles*) with a blow of her spear and took home his war horse (destrier) as a trophy. The story is preserved in the records of the estate of Clapton, which her family held at that time. No information is given about the circumstances under which the encounter took place, whether during a battle or a sporting contest, or if it had any connection to the civil war fought between King Stephen and his cousin, the empress **Matilda,** but Northamptonshire generally held for Stephen.

Bibliography: King, Edmund, ed., *A Northamptonshire Miscellany,* Northamptonshire Records Society Publications 32 (1983), 8, 45.

—Valerie Eads

DIX, DOROTHEA LYNDE (born 4 April 1802, Hampden, Maine; died 17 July 1887, New Jersey State Hospital at Trenton). Nurse, United States, American Civil War.

Dorothea Dix, at her own suggestion, was appointed by the U.S. secretary of war in 1861 as superintendent of female nurses—the highest nursing post of the **American Civil War** and the only major federal appointment won by a woman. Her primary responsibilities included organizing and recruiting a corps of nurses to serve with the Union army. She is credited with recruiting and training around 6,000 nurses for the Northern troops.

Dix, a schoolteacher and writer of children's books and devotional verse, is best known for her work as a social reformer who helped found thirty-two mental hospitals and worked to improve conditions for the insane. When the

Civil War was imminent, Dix was nearly sixty years old, but neither her age nor recurring illness kept her out of the war effort. In 1861, shortly after the Civil War began, Dix arrived in Washington, D.C., ordered a plain black silk dress that would become her uniform, and prepared for her part in the war. Using her connections in Congress and the president's cabinet and offering to work without pay, Dix secured an official appointment for herself. Then, without waiting for approval, she began pulling together teams of female volunteers. Her standards for nurses said more about social practices in the nineteenth century than about the rigors of wartime nursing; for example, she stated that "all nurses are required to be plain looking women. Their dresses must be brown or black, with no bows, no curls, no jewelry, and no hoop-skirts." Ostensibly fired in 1863, Dix continued to work on her own (she had never drawn a salary from the government)—often a hardship due to the fact that she had alienated many of the nurses who worked under her. Although Dix did not become a hero of the war in the mold of Clara Barton or "Mother" Bickerdyke, she did work for the war effort and as superintendent of female nurses until the office was discontinued in 1866.

Dorothea Lydia Dix, 1868. (Hulton/Archive by Getty Images)

Bibliography: Dorothea Lynde Dix manuscripts, Houghton Library, Harvard University, Cambridge, MA, and Library of Congress, Washington, DC; Gollaher, *Voice for the Mad*, 1995; Snyder, *The Lady and the President*, 1975.

—Gayle Veronica Fischer

DOLGORUKAIA, PRINCESS SOFIIA ALEKSANDROVNA (born ?; died?, unk, Russia). Pilot, Russia, First World War.

Princess Sofiia Dolgorukaia was one of the last known women to receive a pilot's certificate from a prerevolutionary Russian flying school. Educated in France and known as a bit of a daredevil, Dolgorukaia was an avid automobile racer who had already made a name for herself in that sport by participating in the St. Petersburg to Kiev auto race. Though she started her flight training in June 1910 under the famous Delagrange at Chartres, France, Dolgorukaia received her pilot certificate (No. 234) from the Imperial Russian Aeroclub Flying School at St. Petersburg only on 5 June 1914. At the outbreak of war in August 1914 Dolgorukaia, like many other patriotic women, offered her services to her country. However, it was not until the first revolution in February 1917 that women were encouraged to take active service by the Provisional Government's minister of war, Aleksandr Kerensky; at that time

Dolgorukaia was given permission to serve with the 26th Corps Air Squadron as a military observation and reconnaissance pilot.

Bibliography: Smithsonian Air and Space Museum, archival materials: "LB71768" and "Robertson, Russian Women in WWI"; White, "Gossamer Wings," 1991.

—Christine White

DRUCOUR, SEIGNEURESS ANNE-MARIE AUBERT DE COUSERAC

(born ?; died October 1762, France). Defender, Canada, French and Indian War.

Anne-Marie Aubert de Couserac was the wife of Augustin de Boschenry de Drucour, the French governor of Ile Royale (Acadia). During the British siege of Louisbourg in 1758 Madame Drucour fired artillery rounds each day throughout the five-week battle and provided aid to the wounded. Anne-Marie and Augustin had arrived in Acadia from France in 1754. Augustin, a career military officer, assumed command shortly before the ongoing war with Britain entered its final stage, which was to include the invasion of Acadia. In 1758 a massive force under Jeffrey Amherst landed near Louisbourg, forcing the French forces, outnumbered in terms of manpower by a nine-to-one ratio, to retreat into the fortress-town. The siege lasted from 19 June to 26 July. French resistance was determined but at times squandered good opportunities to inflict serious damage on the superior British forces. Though Augustin Drucour ordered the French ships in the harbor to open fire on the British, only one did so with any effectiveness. French soldiers did occasionally emerge from the fortress, but the overall consequences of these patrols could have been little more than morale building for the people of Louisbourg. It was in that spirit that Anne-Marie Drucour fired three cannon shots per day at the British positions. Following the surrender of Louisbourg, Anne-Marie Drucour provided care to those wounded during the battle. She and her husband were permitted to leave for England on August 15. Anne-Marie eventually returned to France, where she died in October 1762.

Bibliography: Arsenault, Bona, *Louisbourg, 1713–1758* (Quebec: Le Conseil de la vie française en Amerique, 1971); Downey, Fairfax, *Louisbourg: Key to a Continent* (Englewood Cliffs: Prentice-Hall, 1965); Fortier, John, "Boschenry de Drucour, Augustin de," *Dictionary of Canadian Biography,* vol. 3 (Toronto: University of Toronto Press, 1974).

—William Henry Foster

DUCHEMIN, MARIE ANGÉLIQUE JOSÉPHE

(born 1772, northern France; died 1859, Paris, France), also known as the widow Brûlon. Sutler and soldier, French revolutionary armies, 1792–1799. First female recipient of French Legion of Honor, 1851.

Marie Duchemin, who served for seven years in seven campaigns, first enlisted in the 42nd Infantry Regiment, which became the 1st Battalion of the

83rd Demi-brigade of the French Armies. She later served in the 2nd Battalion of the 84th Demi-brigade.

Duchemin joined two brothers and her husband in service with the revolutionary army; her family from Dinan (Côtes du Nord) had a tradition of military service. Duchemin worked first as a sutler, then became a soldier after her husband's death. Some sources say that she fought first in male disguise, but was then permitted to remain in service as a woman. She was promoted to corporal (some sources indicate further promotions to the rank of sergeant). During various campaigns in Corsica and Italy Duchemin was twice wounded in the leg and injured in both arms. Nonetheless, she continued serving until 1799, when she was discharged and granted aid based on the severity of her injuries.

Her later years were plagued by difficulties with French officials. Since women had been officially ousted from the military in 1793, prior to her discharge (see **French Revolution and Napoleonic Era, women soldiers in**), Duchemin's promotions were deemed invalid, and she was denied entry into Les Invalides, the veterans' hospital in Paris. She continued to petition and to collect testimony from her former commanders. Finally, in 1822 Duchemin became the first woman admitted to Les Invalides, where she was granted the honorary rank of second lieutenant. "Her conduct had been irreproachable as a soldier," remarked one of her references, "and she showed a strength of soul uncommon among her sex."

Among her most significant decorations was the Legion of Honor, conferred by Napoleon III in 1851 for her service to the First French Republic. She was the first woman in French history to receive this highest governmental honor.

Bibliography: Brice, *La femme et les armées*, n.d.; Dossier "Duchemin" and portrait, Archives de la Légion d'Honneur, Paris, France; Dossier XR48, "Brûlon," Service historique de l'Armée, Vincennes, France.

—S.P. Conner

DUERK, ARLENE B. (born 29 March 1920, Defiance, Ohio, United States). Director of the United States Navy Nurse Corps (1970–1975). First woman to achieve flag-officer rank in the United States Navy (1 June 1972).

Admiral Arlene Duerk's naval career spanned the years from the Second World War to the Vietnam War. She served as a navy nurse in many different jobs, including educator, recruiter, senior nurse, chief of nursing services, and director of the Navy Nurse Corps. Her career was highlighted by her selection as the navy's first woman admiral.

After graduating from high school, Duerk trained at the Toledo School of Nursing, graduating in 1941. She was commissioned as an ensign in the Navy Nurse Corps in 1943, serving at stateside hospitals and on the hospital ship *Benevolence* in the Pacific. Duerk remained in the Naval Reserve when the Second World War ended and completed a B.S. degree in nursing in 1948. In

1951, during the Korean War, the navy recalled Duerk to active duty, and two years later she transferred from the Naval Reserve to the Regular Naval.

Duerk worked at a large variety of assignments after the Korean War. In 1958 she was promoted from lieutenant to lieutenant commander and in 1962 was promoted to commander, serving in the Philippines, Japan, and the United States. On 1 July 1967 she was promoted to the rank of captain and was selected as director of the United States Navy Nurse Corps in May 1970. As director, she was in charge of more than 2,000 navy nurses, as well as civilian personnel and paramedical personnel in thirty-nine navy hospitals and dispensaries all over the world. Captain Duerk became the first female admiral in the United States Navy on 1 June 1972. She later commented, "Being the first of anything has its responsibilities. I'm more than an officer. I'm a symbol, for women in the Navy and women in the military. Women thinking of careers like mine can know that . . . the ultimate is possible." Admiral Duerk continued as director of the Navy Nurse Corps until her retirement on 1 July 1975.

Bibliography: *Current Biography;* Holm, *Women in the Military,* 1992.

—Vicki L. Friedl

Undated portrait of Nadezhda Andreevna Durova. (Library of Congress)

DUROVA, NADEZHDA ANDREEVNA (born September 1783, Ukraine; died 21 March 1866, Elabuga, Russia). Married surname Chernova, also known as "Aleksandr Sokolov," "Aleksandr Andreevich Aleksandrov," "Kavalerist-devitsa" [cavalry maiden]. Soldier and then officer in Russian light cavalry regiments, Napoleonic Wars, 1807–1816.

Nadezhda Durova served in three different regiments and saw action against the French in East Prussia in 1807, the invasion of Russia in 1812, and the European campaign of 1813–1814. She was the only woman to hold the Russian Cross of St. George (**Order of St. George**) for valor before the early twentieth century. Her major contribution to military history, however, came as much from the pen as from the sword: in 1836–1841 she published remarkable journals and sensational fiction drawn from experiences of the nine years she had served in the Russian cavalry two decades earlier. In particular, two autobiographical works drawn from her sporadic diaries, *The Cavalry Maiden* (1836) and *Notes* (1839), are classic accounts of the alarms of the Napoleonic Wars interspersed with what Tolstoy called the "obligatory and blameless idleness" of peacetime military life.

In an introductory memoir to *The Cavalry Maiden* Durova attributed her attraction to military life to her upbringing. Durova recalled that until her fifth birthday her father, a hussar captain, used his orderly as a nanny who amused her by "taking me into the squadron stables and sitting me on the horses, giving me a pistol to play with and brandishing his saber." When her father retired to become mayor of Sarapul in the Urals, Durova came under the supervision of her mother, who was determined to feminize her tomboy daughter, whose favorite toys were make-believe weapons. Her father undercut his wife's threats and punishments by giving the adolescent Durova a spirited horse, Alcides, and a Cossack-style uniform to ride in. Throughout her writings Durova takes seven years off her age, portraying herself as a virgin of sixteen when she joined the army. In fact, at eighteen, perhaps acceding to family pressure, she had married and given birth to a son before leaving her husband and returning home. In September 1806, when her mother's failing health threatened to trap her as lifetime caretaker to her father, son, and younger siblings, Durova quit the female sphere for good.

With hair cut short, wearing her uniform, and riding Alcides, Durova persuaded a Cossack troop to let her travel with them to Grodno, Lithuania, where in the spring of 1807 Russian cavalry regiments were preparing to renew the war against Napoleon, who had taken most of Poland the year before. Officers seeking recruits had no incentive to inquire closely into the childish appearance or lack of credentials of young volunteers, and Durova was accepted as a cadet in the Polish Horse [Konnopol'skii] regiment. In Durova's first battle at Guttstadt, East Prussia, that June, she portrays herself as both bewildered and fascinated by the spectacle. There she performed the feat that won her the St. George Cross by preventing two French dragoons from hacking down an unhorsed Russian officer. Her rescue of a soldier at Friedland two weeks later, however, brought reprimand rather than reward: her commander told her that her "bravery was scatterbrained and . . . [her] compassion witless." Fearing further foolhardiness, he sent her to the wagon train "in order to preserve a brave officer . . . in times to come," but it was the worst punishment

---

Entries from the diary of Nadezhda Durova:

June 1807. Friedland. Over half of our brave regiment fell in this fierce and unsuccessful battle. Several times we attacked, several times we repulsed the enemy, and in turn we ourselves were driven back more than once. We were showered with caseshot and smashed by cannonballs, and the shrill whine of the hellish bullets has completely deafened me. Oh, I cannot bear them! The ball is a different matter. It roars so majestically at least, and there are always brief intervals in between. After some hours of heated battle, the remnants of our regiment were ordered to pull back a little to rest. I took advantage of it to go watch the operations of our artillery, without stopping to think that I might get my head torn off for no good reason. Bullets were showering me and my horse, but what do bullets matter beside the savage, unceasing roar of the cannons? An uhlan from our regiment, covered in blood, with a bandaged head and bloodied face, was riding aimlessly around the field in one direction or another. The poor fellow could not remember where he was going and was having trouble keeping his seat in the saddle. . . .

[August 1812]. Kutuzov has arrived! Soldiers, officers, generals—all are ecstatic. Misgivings have given way to calm and confidence. Our entire camp is ebullient and imbued with courage. . . .

Borodino. The 26th. A hellish day! I have gone almost deaf from the savage, unceasing roar of both artilleries. Nobody paid any attention to the bullets which were whistling, whining, hissing, and showering down on us like hail. Even those wounded by them did not hear them; we had other worries!

—Nadezhda Durova, *The Cavalry Maiden: Journals of a Female Russian Officer in the Napoleonic Wars,* trans. Mary Fleming Zirin (London: Angel, 1988), 48, 142–143.

Durova could imagine. The campaign ended with the Treaty of Tilsit, and Durova entered the peacetime routine of regular drills and menial chores. Billeted separately from peasant soldiers, she was able to maintain her masquerade as a man. During this period she was left inconsolable by Alcides' death in a freakish accident: he accidentally impaled himself on a fence post while frisking.

In December 1807 Alexander I, alerted to the presence of a woman in his army by her father's appeal to locate her and send her home, ordered Durova to be brought to St. Petersburg for a personal interview. Deeply moved by her impassioned plea to remain in the cavalry, Alexander granted Durova a commission as cornet in the Mariupol' Hussars and the glory and burden of bearing his name: "You will call yourself . . . Aleksandrov . . . and I will never forgive you for even the shadow of a spot on it." Perhaps regretting that impulsive gesture, he turned all further dealings with Durova over to his war minister. Rumors about a woman officer soon spread, but the absolute monarch had declared Durova an honorary man, and apparently nobody ever dared question her right to serve. She continued to dress as a man and did not advertise the fact that she was female, which undoubtedly made her daily life in the army much easier.

From January 1808 to April 1811 Durova's regiment was assigned to guard the newly defined Russian borders with Napoleon's Duchy of Warsaw and Austrian Galicia. A long narrative Durova inserted into *Notes* indicates that by 1811 her position in the Mariupol' Hussars had become untenable because of the stubborn infatuation her colonel's daughter had conceived for the enigmatic "Aleksandrov." Durova asked to be transferred and was assigned to the Lithuanian Uhlans, serving in the same region. After the Napoleonic invasion in June 1812 her largely Polish regiment was assigned to the rear guard and clashed repeatedly with troops on the French side that included Poniatowski's Poles. Durova's official service record lists "active combat" throughout the period of Russian retreat—at Mir, Dashkova, Smolensk, Gzhatsk, and Kolotsk Monastery. At Borodino, the Russians' last-ditch stand before Moscow, she was struck in the leg by a cannonball. Durova herself remarks on the irony of her change in attitude over the five years since her first battle: suffering from pain and cold (careless with money, she rarely managed to be properly equipped and warmly clothed), she was only too happy to head for the once-detested wagon train. On 29 August 1812 she was promoted to lieutenant.

After the Russians surrendered Moscow to the French in early September, Durova sought out the commander in chief, Mikhail Kutuzov, told him "all," and asked to be assigned to his staff. Before long he sent her home to Sarapul to recuperate from the contusion of her leg. She missed Napoleon's headlong flight from Russia that winter, and her hopes of a brilliant career with Kutuzov were dashed by his death in April 1813. She rejoined the Lithuanian Uhlans in western Russia and was assigned to pasture the emaciated regimental horses. Judging by the ecstatic tone of her description, that idyllic summer was the happiest she had known since Alcides' death. Durova's journals skip lightly over her participation in mopping-up actions in Poland and Bohemia and the siege of Hamburg in 1813–1814. Her last two years in the cavalry were marked by growing frustration. She enjoyed life in Germany and grumbled about having to return to Russia to take charge of a squadron quartered in a muddy village near Vitebsk. Her account of detached duty in Petersburg in the winter of 1815 indicates how antipathetic the post-Napoleonic army under Aleksei Arakcheev's rigid discipline was to her rebellious spirit. In 1816 she retired as a staff captain and apparently returned home.

In 1836 Durova emerged from provincial obscurity when, perennially short of money despite a government pension, she sold her account of the 1812 retreat to Aleksandr Pushkin's magazine *Contemporary*. Throughout her retirement Durova signaled masculine independence by wearing military-style jacket and boots, bobbing her hair, and referring to herself in masculine terms. Her writings are in feminine voice, however, and a few of her stories are narrated by a woman cavalry officer. In 1841, apparently disillusioned with her literary celebrity and frustrated in hopes of making a substantial income as an author, Durova abruptly disappeared. She spent the rest of her life in Elabuga, where she was known as an amiable, animal-loving eccentric. At her death in 1866 she was buried with military honors. Her posthumous reputation has been a good barometer of Russian attitudes toward "woman's place": Durova has been glorified in wartime and ignored when state and society have chosen to emphasize women's nurturing role. The first uncensored edition of *The Cavalry Maiden* since 1836 appeared in the Soviet Union in 1983, the 200th anniversary of Durova's birth.

Bibliography: Durova, *Zapiski Aleksandrova (Durovoi): Dobavlenie k Devitse-kavalerist*, 1839, and *The Cavalry Maiden*, 1988; Zirin, "Nadezhda Andreevna Durova," 1994.

—Mary F. Zirin

E

EARLEY, CHARITY ADAMS (born 1918, Kittrell, North Carolina, United States). Lieutenant colonel (ret.), WAC (Women's Army Corps). United States, Second World War.

A member of the first **WAC** officers training class, Charity Adams distinguished herself as commander of the only unit of African American women to serve overseas in the Second World War, the 6888th Central Postal Directory Battalion in Europe. Like many other Wacs, she entered military service with advanced education and work experience. She was a high-school valedictorian, graduate of Wilberforce University, and math teacher before signing up during the WAC's early recruiting effort. In addition to commanding the 6888th, Adams helped to investigate complaints of racism and worked to raise the morale of African American Wacs.

Adams left the service in 1946; her later career included serving as dean of students at Tennessee A&I University and Georgia State College. She married Stanley Earley, a doctor, and had two children. She published her memoir, *One Woman's Army,* in 1989 and still travels occasionally for book signings. Charity Adams Earley remained active in addressing racial issues in civic affairs, education, social services, and business practices. In November 2000 the eighty-year-old veteran traveled to Washington, D.C., and from her wheelchair joined President Clinton in breaking ground for the World War II Memorial on the National Mall. She lives in Dayton, Ohio. (See also **WAC, African Americans in.**)

Bibliography: Earley, *One Woman's Army,* 1989; Johnson, *Black Women in the Armed Forces, 1942–1974,* 1974; Lee, *The United States Army in World War II, Special Studies: The*

Number of American women who have served in military conflicts:	
Civil War	400–1,000
Spanish-American War	1,500
First World War	33,000
Second World War	400,000
Korea	120,000
Vietnam (in theater)	7,000
Grenada (in theater)	170
Panama (in theater)	770
Gulf War (in theater)	41,000

Women as percentage of deployed force:	
Grenada	2%
Gulf War	8.6%
Somalia	11%
Bosnia	10.4%

—United States Department of Defense; United States Army, Defense Manpower Data Center.

Employment of Negro Troops, 1965; Moore, *To Serve My Country,* 1996; Putney, *When the Nation Was in Need,* 1992.

—Elizabeth Lutes Hillman

EDMONDS, SARAH EMMA EVELYN EDMONDSON (born December 1841, St. John, New Brunswick, Canada; died 5 September 1898), married surname Seelye, also known as Private Franklin Thompson, Flint Union Greys (later Company F, Second Michigan Infantry Regiment). Nurse, United States, American Civil War.

Sarah Emma Evelyn Edmondson was born in Canada to a father who wanted boys to help with the farming. His only son was epileptic; thus he put his five daughters to work in the fields during the day and the house at night. Sarah, the youngest girl, managed to attend grammar school, where she learned the basics; her desire for learning stayed with her throughout her life. When she was about fifteen, her father announced that she was to marry a much older man, and she ran away to Salisbury, Canada, where she spent two years as a dressmaker and changed her name to Edmonds. Then, for reasons still not clear, she disguised herself as a man, "Franklin Thompson," went to the United States, and became a successful Bible salesman. Her work took her to the Flint, Michigan, area in 1860, and a year later she volunteered for the Flint Union Greys, a company in the 2nd Michigan Infantry Regiment.

Edmonds enlisted on 17 April 1861 (some sources give 17 May), still in her assumed identity as Franklin Thompson; no physical examination was performed. She volunteered for duty as a (male) nurse and was present at several major battles of the Civil War, including Bull Run (First and Second), the Peninsula Campaign, and the Battle of Fredericksburg. She also served stints as postmaster and as a mail carrier, a job that entailed carrying heavy loads of mail over long distances, riding unaccompanied. But after two years of service, in April 1863, Edmonds suddenly deserted the army near Lebanon, Kentucky. The reason for her desertion is not known, but it was common for women serving in male disguise to desert if they feared that someone had discovered their true identity. Edmonds herself eventually claimed that she left because she suffered from a recurring fever, and her request for a leave of absence had been denied. If she had been hospitalized, her identity might have been discovered. Fellow soldiers later testified that "Thompson" had indeed taken ill with fever during the Peninsula Campaign. However, one source asserts that Edmonds's reason for desertion was to follow a lover who had left the service. Within weeks of her desertion Edmonds was again serving the army as a nurse, this time as her true self; such work was physically demanding and would seem to belie her claim of ill health. After the war she went back to New Brunswick, where she married Linus H. Seelye, bore and lost three children, and adopted two more.

Soon after leaving the service, Edmonds composed an autobiography, published first in 1864 and reissued in 1865. *Nurse and Spy in the Union Army* was written in the first person under her true name. The book quickly became a best-seller (more than 175,000 copies sold) and gave her national prominence. Edmonds wrote about her service in the field hospitals and her adventures behind enemy lines as a spy for the Union army, but made no mention of her masquerade as Thompson or her desertion. She explained that she wrote her story after "leaving the Army because of illness." She donated a large share of her royalties to the Sanitary Commission and the Christian Commission.

Undated portrait of Sarah Emma Evelyn Edmondson Edmonds. (Library of Congress)

Edmonds gained new celebrity as her claims were publicized in the press and Congress. In interviews she disclosed more details about her past. Unfortunately, she often changed the specifics. Her biographer, Sylvia Dannett, pieced much of her story together from firsthand interviews with relatives, letters, and official documents. The result confirmed Edmonds's service as a Union nurse but noted that Edmonds herself, when asked if her book could be considered authentic, had replied, "Not strictly so." Historian Betty Fladeland wondered how Edmonds could have kept her secret from all her male comrades. Her research discovered two Civil War diaries, written by close male colleagues of Edmonds, that make it clear that the writers were aware that "Frank" was female long before she left the army. Indeed, Edmonds had revealed her true identity to one of the diarists, her good friend Jerome John Robbins, in late 1861; he kept her secret except in the pages of his diary.

The most controversial claim in *Nurse and Spy,* Edmonds's alleged service as a Union spy, went unchallenged at the time and was the major reason for her subsequent popularity. Today, however, it is clear that her spying exploits were fiction. Although many writers have mentioned her espionage (some skeptically), not one, including her biographer, cited any primary source as corroboration other than Edmonds's own book. Furthermore, official Civil War records that should have revealed something of her work—for example, that she was personally selected to spy by General George McClellan and two other generals—make no mention of Thompson or the espionage he claims to have performed. Nor is there any anecdotal evidence to support Edmonds's claim of eleven spy missions, such as her claim to have entered Richmond in 1862 disguised as a Negro boy to collect details of Confederate fortifications, or that she later operated in the West posing as a Confederate soldier to determine enemy strength, or that dressed as a Southern civilian, she shot and wounded a Confederate captain to protect her cover. Finally, and most telling, neither she nor her former colleagues made any mention of her espionage services in connection with her pension application. She wrote a sequel to her book, but it was never published and the manuscript has been lost.

If anyone questioned the story of Sarah Edmonds's service as told in *Nurse and Spy*, nothing was said until 1882, when circumstances forced her to apply for an army pension and the reality of her army life was revealed. She had been reluctant to request a pension because she knew that the army had no record of a nurse named Sarah Edmonds. Thus she was forced to apply as Franklin Thompson, explaining that she had assumed that identity throughout her more than two years of service. As corroboration, she obtained affidavits from her startled but cooperative former comrades of all ranks after attending a reunion and convincing them that she was the young soldier they had known as Thompson. But these developments raised new problems since Thompson's record showed that he had deserted in 1863. Sarah countered that she had had to "leave unannounced" to get medical help that would have revealed her gender had she remained. The charge of desertion was removed from Franklin Thompson's record by Congress, and Edmonds was awarded the standard army pension of twelve dollars a month. Edmonds moved to Texas in her last years and became the only woman elected to the George B. McClellan Post No. 9 of the Grand Army of the Republic at Houston, Texas. When she died at age 57 on 5 September 1898, she was buried in the GAR cemetery in Houston with full military honors.

Although the motivations of Sarah Edmonds for her male disguise and for the spying stories remain a mystery, the quality of her army service, under extraordinarily difficult and unusual conditions, has never been denied. When Edmonds applied for her pension, testimony from her former comrades, summarized in a congressional report, was uniformly supportive, stating that "Franklin Thompson" was "never absent from duty, obeying all orders with intelligence and alacrity, his whole aim and desire to render zealous and efficient aid to the Union cause." Her story is important as the best-documented case of an American woman successfully performing military service in male disguise over an extended period of time.

Bibliography: Edmonds, *Nurse and Spy in the Union Army*, 1865 (reprinted as *Memoirs of a Soldier, Nurse, and Spy*, 1999); Fishel, *The Secret War for the Union*, 1996; Fladeland, "Alias Franklin Thompson," 1958, and "New Light on Sarah Emma Edmonds," 1963; Leonard, *All the Daring of the Soldier*, 1999.

—Hayden Peake

EGER, WOMEN IN SIEGE OF (1552). Castle Eger, Hungary.

In June 1552 the Ottoman sultan Suleiman initiated a multipronged campaign against the Habsburgs, seeking revenge for their successes in regaining influence in Transylvania, but was stopped by the small garrison of Castle Eger. At the time, it was defended by Captain Istvan Dobo and with a garrison consisting of some 1,920 men, plus about 90 women (40 of them ladies of the nobility). With Hungary at the time divided into three segments, there was no help to be had from any quarter. By 13 September Turkish forces under Ali Pasha, Ahmed Pasha, and Arslan Beg had encircled and laid siege to Castle Eger with their united force of 27,000 troops. In a show of defiance, when a Turkish messenger brought a written demand for surrender, Captain Dobo forced him to eat half the note, then burned the other half. Then Dobo placed a huge black coffin prominently on the castle's ramparts to indicate to the Turks that the garrison was committed to fight to the death. Before being encircled, the women of Eger distinguished themselves by their refusal to escape from the castle to safety and by their vow to assist in the defense of the city.

By 29 September, after heavy shelling, the Turks ran out of ammunition, but continued the siege using hand-to-hand combat, attempting to breach the castle walls. In describing this desperate phase of the siege, sources throughout the centuries have highlighted the names of women as having exhibited special valor: Margit Homonnay, Eufrozina Gyulafy, and Mrs. Job Paksy, as well as Katica Dobo, daughter of Captain Istvan Dobo. In addition to feeding the soldiers and caring for the wounded, the women helped defend the walls of the city, greeting the Turks who climbed onto ramparts with boiling water and burning faggots soaked with tar. Katica Dobo, in particular, is recorded to have led the women to the ramparts like a troop of Amazons, "her long brown hair streaming from under her helmet." Thus fighting on ruined castle walls side by side, the men and women repeatedly repulsed the continuous Turkish attacks. The siege failed in the face of such cohesive bravery. On 13 October 1552 the Turkish forces refused to obey the further orders of their commanders, and the campaign of revenge came to an end. The women of Eger made up only a small percentage of the defensive force; however, Hungarian folklore and popular heritage remember them in a collective sense as a lasting example of unity and collaboration in the face of desperation.

Bibliography: Farkas, Emod, *Magyarorszag nagyasszonyai* (Grand Ladies of Hungary) (Budapest; 1911).

—Antal Leisen

EGOROVA, ANNA ALEKSANDROVNA (born 23 September 1916, Tver' region, Russia), married surname Timofeeva. Senior lieutenant (ret.), pilot, Red Army Air Force, Soviet Union, Second World War.

One of the most famous women to fly in a male regiment was Anna

Timofeeva-Egorova, who flew the Il-2 "shturmovik" ground-attack aircraft. For several months she was a prisoner of war of the Germans. She became a **Hero of the Soviet Union**.

Egorova was an ambitious and energetic woman. As a teenager, she left her village to move to Moscow and work on the subway being constructed there, one of the most advanced projects of its time. At the same time, she attended night school and joined the flying club run by the Metro Construction Trade School. During the 1930s she worked as a fitter, caulker, and librarian. Despite the arrest of her brother during the purges, she managed to get accepted to the prestigious Kherson Flying School and was trained in navigation. She served as a flying-club instructor in 1940–1941.

Egorova started the war as a pilot in a communications squadron on the Southern Front. She was unhappy flying an unarmed biplane, especially after some hair-raising encounters with the enemy, and requested an assignment to a combat regiment. After persistent efforts she achieved a transfer in early 1943 to the 805th Ground Attack Aviation Regiment, which flew the Il-2 "shturmovik" ground-attack aircraft. In that unit she flew many combat missions in the North Caucasus, where the regiment suffered heavy casualties. Egorova's picture appeared on the front page of the military newspaper *Krasnaia Zvezda* on 1 July 1943 in recognition of her achievement in flying the Il-2. In early 1944 the regiment received the new two-seat model of the Il-2, and Egorova's armorer, Dusia Nazarkina, became her gunner. The regiment was transferred to the First-Belorussian Front, where it supported the huge summer offensives of 1944.

By late summer the 805th was based near Warsaw. On 20 August 1944, on her 277th mission, Egorova was shot down and captured by the Germans—possibly the only female aviator to become a prisoner of war (Nazarkina apparently died in the crash). Egorova suffered extensive injuries when she was shot down—broken arms, a broken leg, back and head injuries, and burns—and received no medical treatment from the Germans. She was in the camps for five months until she was released by a liberating Soviet tank unit on 31 January 1945. She then endured extensive questioning by the Soviet secret police, which suspected her, as it did all POWs, of having been a German collaborator. She was invalided into the reserves at the end of the war and married her former division commander, Viacheslav Timofeev. One of their two sons became an air force pilot. It was only in 1965 that Egorova finally received the Hero of the Soviet Union award; the delay was undoubtedly related to her brother's arrest in the 1930s and to her own capture by the Germans, which made her family politically suspect.

Another Il-2 pilot, HSU Tamara Konstantinova (a friend of Egorova's from before the war), flew with the 999th Ground Attack Aviation Regiment. Konstantinova was the only female Il-2 pilot on the Leningrad Front; her gunner was also female. Other women who flew the Il-2 included Lidiia Ivanovna Shulaikina (Baltic Fleet) and Mariia Tolstova.

After the war, an interviewer asked Egorova how she had learned to master the "flying tank," which not every man could fly. Egorova replied, "The Il-2, of course, is not a 'lady's' aircraft. But after all, I'm no princess, but a metal worker who helped build the Moscow subway system."

Bibliography: Cottam, *Women in War and Resistance*, 1998; Murmantseva, *Zhenshchiny v soldatskikh shineliakh*, 1971; Pennington, *Wings, Women, and War*, 2001; Timofeeva-Egorova, *Derzhis', sestrenka!*, 1983; Toropov, *Geroini*, 1969; Vershinin, *Chetvertaia Vozdushnaia*, 1975.

—Reina Pennington

EHYOPHSTA (born ca. 1826; died August 1915, Tongue River Reservation), also known as Yellow Haired Woman. Warrior, Cheyenne Nation, Indian Wars, United States.

Unlike most **Native American women** warriors, Ehyophsta entered battle to die and not for revenge. Her husband, Walking Bear, was killed by an accidental discharge of his own gun. Thereafter, her wish to follow her husband into death led the forty-year-old Ehyophsta into a series of fierce engagements against both white soldiers and enemy tribes.

Her first foray into combat after Walking Bear's death was with the army scouting party of Major George Forsyth at Beecher's Island in September 1868. Forsyth's command of fifty-one "Plainsmen" scouts was to find and report locations of Indian camps so that they could be forced onto reservations as part of the Southern Plains Campaign that year. The fight took place on the Arickaree Fork of the Republican River in Colorado. After having surrounded the scouting party on a small island, the Cheyenne Dog Society warriors, or Dog Soldiers, as they were more popularly known, kept the soldiers under siege for eight days. On a horse given her by her father, Ehyophsta joined in four mounted charges against the scouts. She was not injured in this battle, but her clothing was riddled with bullet holes. She did not count coup in this fight, although afterwards she was known and accepted as a woman warrior.

Her next battle was against the Shoshoni later in 1868, on Beaver Creek near the Big Horn Mountains in Montana (as remembered by Wooden Leg, in 1873, but most sources cite 1868). Plains tribes had gathered for the great autumn buffalo hunt, and this was also a time for horse raids. Both the Cheyenne and Shoshoni camps, near each other, had been out raiding when contact was made and skirmishes broke out. After one particular attack the Cheyenne chased and trapped a Shoshoni war party in a deep ravine. At the end of a four-day battle the Shoshoni were either dead or captured. One captured Shoshoni warrior was about to be questioned by the Dog Soldiers when Ehyophsta came up and said that she would do it. She then raised the warrior's arm and stabbed him twice in the armpit, causing his death. She then took his scalp. For this and two other coups counted during this fight, she was admitted to the Crazy Dog Soldier Warrior Society, the society of her husband. Descriptions of the other two coups have not been located.

Ehyophsta was interviewed twice by George Bird Grinnell, on 24 September 1908 and again on 31 July 1912, three years before her death. These accounts provide most of what we know about Ehyophsta. Father Peter J. Powell's voluminous works of Cheyenne history include a pictograph entitled "A Woman Warrior" from the Spotted Wolf–Yellow Nose Ledger (*People of the Sacred Mountain*, 1:134–135). Powell hypothesizes that this is a representation of Ehyophsta.

Bibliography: Grinnell, *The Fighting Cheyennes*, 1915, and *The Cheyenne Indians*, 1923; Niethammer, *Daughters of the Earth*, 1977; Powell, *People of the Sacred Mountain*, 1981.

—Rodney G. Thomas

ELEANOR OF AQUITAINE (born 1122? in Poitiers or Belin, near Bordeaux; died 1 April 1204 in Poitiers or Fontevrault). Duchess of Aquitaine, countess of Poitu, queen of France (1137–1152), duchess of Normandy, countess of Anjou, and queen of England (1152–1204).

Eleanor of Aquitaine was a woman of wealth whose power was based on her control of the important province of Aquitaine. Throughout her long life Eleanor used her power to support the military efforts of kings and would-be kings, raising armies in the attempt to sway political events—often successfully. She was a strong-willed and influential woman who sometimes flouted convention. Eleanor was directly involved in military conflict, beginning with her participation in the Second Crusade in 1147 and ending with her successful defense of a besieged castle in 1202.

Eleanor of Aquitaine and her daughter-in-law Isabella of Angoulême led into captivity by Eleanor's husband Henry II in 1173 after their rebellion. Twelfth–thirteenth-century fresco, Chapelle de Sainte Radegonde, Chinon, France. (The Art Archive/Dagli Orti)

The single event that undoubtedly had the greatest impact on Eleanor's life was her inheritance of the duchy of Aquitaine and county of Poitu. By the end of the tenth century Eleanor's ancestors acquired the ducal title confirming their governance over the region of Aquitaine. The ensuing dukes fortified their claim over Aquitaine, which led to the independent status that principality enjoyed within the French kingdom. Little is known of Eleanor's childhood or formal education. However, her lineage and the large role her female descendants played in Aquitaine's history attest to the strong probability that Eleanor received some form of education. Many of Eleanor's female relatives were active and influential figures who were generally characterized by a deep sense of piety, as is evident by the numerous noblewomen of Aquitaine who were linked to the founding of many great abbeys, including Bourgueil, Vendôme, and Fontevrault.

Eleanor's father, William X, died in 1137, and the laws in that region made her the primary heir of a vast estate—more land than the king himself controlled. Fifteen-year-old Eleanor was the richest marital prize in all of Christendom. She became a ward of the French king, Louis the Fat, who quickly married her to his son Louis VII. It is believed that the marriage was an unhappy one for Eleanor, whose lively spirit craved a more compatible husband than the pious Louis.

Eleanor's military involvement began with the Second Crusade. In 1146 she pledged thousands of her vassals to support the new crusade; moreover, she went to Jerusalem herself. This journey in 1147 is often described as merely accompanying her husband Louis VII, but Eleanor provided troops for the effort and took an escort of other noblewomen. Although it was not uncommon for women to participate in the **Crusades,** some observers regarded Eleanor's presence as setting a bad example to other women. (A papal bull regarding the Third Crusade forty years later would expressly forbid the participation of women.)

A large body of work exists describing a group of armed women dressed in masculine style who allegedly accompanied Eleanor on the Second Crusade—an army of Amazons among the crusaders. This account is largely derived from the medieval chronicler Nicetus Acominatus, who was not a firsthand observer. The original passage does not specifically refer to Eleanor, so there is no conclusive evidence supporting the claim that the queen and her entourage dressed as men during the crusade. Still, biographers continue to use this passage to dramatize Eleanor's role in the crusade. Both traditional and women's historians have enhanced or condemned Eleanor's role in the crusade. Most recently Petronelle Cook in her work *Queen Consorts of England* wrote that "[Eleanor and her entourage] designed their own uniforms (white tunics embroidered with red crowns and slit at the sides to show scarlet tights and long red leather boots)." There is no direct evidence for such a description, or even for Eleanor's physical appearance. However, it would not have been out of keeping with Eleanor's character to portray herself as an Amazon. In any event, even if Eleanor and her women dressed as Amazons, there is no indication that any of them actually fought.

The reaction of Eleanor's husband, Louis VII, to her involvement in the crusade is not known, though some speculate that he feared leaving her alone in Paris more than he disliked taking her on crusade. There were reportedly disputes between them over strategy in the Holy Land, with Eleanor taking the side of her uncle Raymond, the prince of Antioch, against Louis. Louis failed in his attempt to retake Jerusalem, and thereafter the marriage was increasingly acrimonious. During their fifteen-year marriage Eleanor gave birth to two daughters, but failed to provide Louis with the required male heir. This perhaps motivated Louis to finally agree to an annulment on the grounds of consanguinity in 1152. Eleanor retained her estates.

That same year the thirty-year-old Eleanor married Louis's vassal, twenty-year-old Henry Plantagenet, who in two years became king of England. Through conquest and marriage Henry ruled Normandy, Anjou, Aquitaine, and Poitu (later termed the Angevin Empire) in addition to England. During the first ten years of their marriage Eleanor gave birth to eight children, five boys and three girls, and it appears that for a time Eleanor and Henry were happy together. But a close examination of Eleanor's life reveals the repressive nature of both her marriages. Henry began to take mistresses, and Eleanor often lived apart from him, setting up residence in Poitiers.

The great rebellion of 1173 certainly indicated a rift between Henry and his wife, although Eleanor's involvement is shrouded in mystery. In 1170 Henry II crowned Henry, his eldest son, king of England. By coronating his son in his own lifetime, Henry declared his son's right to inherit the crown, endowing young Henry with the prestige of kingship without any real power. This caused great resentment and was ultimately the catalyst behind the rebellion of three of Henry's sons: Henry, aged eighteen, Richard, aged sixteen, and Geoffrey, aged fifteen. The youthful age of Henry's sons at the time of the rebellion has led many to speculate that Eleanor masterminded the rebellion, which pitted father against son and husband against wife. She may have encouraged her sons in the attempt to overthrow their father in the hope of improving her political position in the kingdom. However, the rebellion was quickly subdued by Henry. Eleanor, dressed as a man, attempted to evade Henry's army and join her sons in Paris, but she was captured and subsequently imprisoned for more than sixteen years for her role in the rebellion. This severe punishment indicates that her husband believed that Eleanor was involved, or even largely responsible. Henry's sons were only asked to pay a fine for their role.

When Henry II died in 1189, his eldest surviving son Richard the Lionheart succeeded to the throne and immediately released his mother from prison. Eleanor superseded Richard's wife at his coronation, taking the female place of honor opposite the newly crowned king. The aging queen energetically involved herself in public life, as her numerous charters at this time attest; Richard was away on crusade and as a hostage much of the time, and Eleanor was active in administering the kingdom. After Richard's death in 1199 Eleanor's political clout and military skill proved to be the determining factor that secured the succession of her youngest son John Lackland to the Angevin Empire. Eleanor at the age of seventy-seven organized and led a group of mercenaries who captured Anjou, while John claimed the Angevin treasure at Chinon. Eleanor laid waste the lands of those who supported John's rival claimant and nephew (and Eleanor's grandson), Arthur of Brittany. Eleanor's political and military influence was so great that Arthur attempted to take his grandmother prisoner. Arthur besieged Eleanor at the castle of Mirebeau, but she successfully defended the castle and sent a secret message to John at Le Mans informing him of the attack. John and his army arrived two days later after a forced march, just as Arthur secured the outer perimeter of the castle. Trapped by the castle walls on one side and John's troops on the other, Arthur was soon captured.

Eleanor died two years later at the age of eighty-two. Her body is buried at Fontevrault, where she lies entombed beside her husband, Henry II, and her son Richard the Lionheart. A woman of noble lineage, Eleanor ultimately lived to see her power and prestige confirmed through her own efforts.

Bibliography: Kelly, *Eleanor of Aquitaine and the Four Kings,* 1952; Kibler, *Eleanor of Aquitaine: Patron and Politician,* 1976; McNamara and Wemple, "The Power of Women through the Family in Medieval Europe, 500–1100," 1974; Owen, *Eleanor of Aquitaine,* 1993; Richardson, "The Letters and Charters of Eleanor of Aquitaine," 1974.

—Natalie Forget

ELEANORA OF ARBOREA (born ca. 1360s; died 1402/1404 in Oristano?, Sardinia). Daughter of Mariano IV, wife of Genoese warlord Brancaleone Doria, called Branca. Last of the giudici or "judges" of Sardinia, wrote an important law code, and ruled for two decades.

Eleanora of Arborea succeeded to the rulership of Arborea, one of four administrative regions of the island kingdom of Sardinia, after the assassination of her brother Ugone III in 1383. She quelled the rebellion that followed, reconquered territory held by her father, and kept Arborea in her family's hands. She is considered a national heroine for her efforts in securing the independence of Sardinia, which was in the fourteenth century essentially an Aragonese colony. Eleanora was also well known for her astute political negotiations and mediation that brought about the release of her husband from a long Spanish imprisonment. In the 1390s Eleanora revised and expanded a judicial code begun by her father, the famous *Carta de Logu*. Her version remained in force well into the nineteenth century. She ruled Sardinia for more than twenty years, leaving a legacy of peaceful independence and prosperity, in addition to an outstanding legal code.

Although her administrative and judicial activities are well documented, they have overshadowed her role as a military commander. Conservative sources such as P. Tola's *Dizionario degli uomini illustri della Sardegna* have tended to disregard as spurious local legends and written accounts of her military prowess. However, in her century and later it was not uncommon for women, and especially the female members of ruling dynasties, to participate in bellicose actions. It should also be remembered that the same historians who discount her possible heroism on the battlefield praise the formidable military expertise of her various male relatives. More research on this subject is needed.

Bibliography: Boulding, *The Underside of History,* 1976; Carta Raspi, *Storia della Sardegna,* 1987; Castellani, *Italian Women, Past and Present,* 1939; Day, Anatra, and Scaraffia, *La Sardegna medioevale e moderna,* 1984.

—Gloria Allaire

ELIZABETH I (born 1533; died 1603). Queen of England (1558–1603).

Born in 1533 to Henry VIII and Anne Boleyn, Elizabeth I ruled England and Ireland from 1558 to 1603. Contemporaries celebrated her reign as unusually peaceful, an image the queen promoted, in spite of England's clashes with Scotland and Ireland and its wars with France and Spain. For the most part these were interventions by Europe's most Protestant nation, aimed at maintaining an equilibrium of power or defending besieged Protestant allies, or they were retaliatory strikes meant to quash an already-aggressive foe. Embracing a philosophy of defense, Elizabeth craved détente, vastly preferring to defuse international threats with beguiling diplomacy—at which she was a virtuoso—whenever possible. Modern detractors of her government find in the queen's behavior in military crises an almost pathological fear of war. But in fairness, Elizabeth prevailed as queen for forty-five years in a toxic era in European history; she kept England secure through several attempted invasions and coups d'état; and she faced down her brother-in-law Philip II of Spain, the most feared warlord of the age.

Watercolor by Issac Oliver of Elizabeth I, 1588. (Dover Pictorial Archives)

Far more than Philip, whose bankruptcies were notorious for unleashing mutinous soldiers upon the world, Elizabeth was mindful of the costs of war. She inherited a crown still in debt, in part from the capricious border wars on Scotland by Henry VIII and her brother Edward VI. Throughout her reign Elizabeth struggled to contain military expenditures, which were enormous even in frugal hands. She raised funds by licensing privateers, civilian pirates who were authorized under her name to seize merchant ships of enemy nations; taxing her subjects, collecting rents on Crown lands, and, on a few occasions, extracting forced loans from wealthy families; borrowing from the City of London; and liquidating the Crown's real estate. Throughout the reign the shortage of funds was the queen's preeminent liability, made worse by infrastructures that pitted the Crown's interests against those of noble families. Other constraints included intelligence information of inconstant reliability; slow communications with field commanders; the lack of a standing army and only a modest navy; her sex; and some hawkish courtiers too eager to gain gold and glory in war. These seldom proved insurmountable to a woman determined to exercise her lawful sovereignty. While it was unthinkable that the queen would jeopardize her safety by making war alongside her commanders, as English kings had done for centuries, she was known to exercise close, continuous supervision of her forces from afar. To effect control she sometimes appealed to her commanders' chivalry, a game made easier by her unmarried status; but that status also made her rule problematic for men accustomed to patriarchy, and for the most part she prevailed by sheer will. On land commanders were micromanaged from London; at sea they had more liberty, yet in several important sea campaigns they allowed bravado and greed to overwhelm good sense, to their own and the queen's regret.

If Elizabeth suffered setbacks by her commanders' disobedience in the field, she demoralized them too with what seemed to be capricious changes in goals or strategies. Hot tempered, she was known for issuing rash orders to leaders of military expeditions, only to reverse herself in cooler moments. Contemporary commentators and her own advisors also reproached her for hedging decisions in crises when swift and certain military responses were needed, and for scrimping on materiel for her armies and navies to save expenses. Modern military historians often echo these recriminations, especially of the queen's legendary indecisiveness, but social historians and recent biographers often take a more balanced view, seeing Elizabeth's reluctance to make war as being less a failure of nerve than a deliberately conservative strategy, in keeping with her personal motto *semper eadem* (always the same).

From her half sister **Mary I**, whom she succeeded, Elizabeth inherited a war with France that had stung England in January 1558 when it lost Calais, an English territory since the fourteenth century. Elizabeth made peace with France early in 1559, which proved short-lived when later that year it became clear that French troops resident in Scotland were prepared to assist Mary, Queen of Scots, in her much-rumored plans to depose her cousin Elizabeth. Their struggle for power would aggravate the first half of Elizabeth's reign. Late in 1559 she made a clandestine alliance with Protestant nobles in Scotland, sending £3,000 to help them overthrow Mary. France threatened reinforcements, and in 1560 Elizabeth sent 8,000 soldiers but not enough supplies to support so large an army. Worse, she issued half-hearted rules of engagement, perhaps because as a monarch she was ambivalent about aiding any rebellion against any sovereign. The soldiers were prohibited from besieging

Edinburgh Castle, for example; instead, only outlying cities such as Leith could be attacked, and even there, poor planning squandered some 1,500 English lives. Once Elizabeth increased supplies, England prevailed. Yet historians attribute the victory less to Elizabeth's army or strategy than to the mushrooming civil wars in France, which kept its armies at home.

Victory notwithstanding, Elizabeth was bedeviled for years by the Stuart claim on England's throne. In 1567 she found herself in the peculiar position of vowing to send an army to Scotland again, now to defend Mary, under arrest as an accomplice to murder, even though Mary's intention of deposing Elizabeth had not been seriously retracted. Yet Elizabeth's threat of military action in defense of Mary never materialized. The year 1569 brought a fresh crisis connected to Mary's claim. A number of northern English earls with Roman Catholic sympathies rebelled, ostensibly against the state-run Anglican Church. As part of reverting to Catholicism, they intended to liberate Mary from an English prison and proclaim her queen of England. While the earls' ambitions may have constituted a serious breach of unity, Elizabeth overreacted. Her commander the earl of Sussex requested reinforcements of 500 men to stop the rebellion; the queen sent 10,000 at a cost of £30,000.

While Mary's advocates would continue to dog Elizabeth for some years more, no longer was their patron, Catholic France, very much a problem. For three decades French Catholics and Protestants would be locked in a bloody civil war. Not until the 1580s, however, would Elizabeth consider it enough of a threat to England's interests to intervene on behalf of Protestants there. Circumstances changed when it became likely that Spain would enter the conflict, representing the Catholic League. Elizabeth rightly feared such an alliance and countered by sending money and troops to the Protestant faction. In September 1589 she gave Henry IV, the new Protestant king of France, £20,000 and 4,000 men after his subjects rebelled en masse. She sent 3,000 men to Brittany in March 1591 and 4,000 to Normandy in June, and more again in 1592 and 1594, in spite of evidence suggesting that Henry was making poor use of them. From September 1589 the queen expended more than £145,000 helping royalists in France, only to have Henry IV convert to Catholicism in 1593.

As early as 1569 Elizabeth's most trusted advisor, William Cecil, Lord Burghley, had observed that Spain had displaced France as their greatest threat. This came about only in slight part because of a rare military provocation by Elizabeth when in December 1568 she seized Spanish merchant ships that had taken sanctuary in Southampton from storms at sea; they were bound for the Netherlands. Spain retaliated by declaring war on England; while this conflict ended in a draw in 1574, no significant damage having been inflicted by either side, it set the stage for the more consequential hostilities with Spain that broke out in 1587 and continued beyond Elizabeth's death in 1603.

That war began after Elizabeth sent soldiers to the Low Countries to aid Protestants in over throwing Spain's hegemony. Philip had had soldiers in the

Netherlands since 1566, but they had seemed to pose a threat to England only since the mid-1570s. In 1576 mutinous Spanish troops had sacked Antwerp and struck fear in London of its being their next target. At that time the Dutch had appealed to Elizabeth for help, but she had equivocated. In 1585 they renewed their petitions, and Elizabeth—now having intelligence that Philip was laying plans to attack England—responded with 5,000 foot soldiers and 1,000 cavalry. In return the queen demanded from the Dutch favorable terms on military loans and control of the towns of Flushing and Brill as collateral. In December 1585 she sent the earl of Leicester, a hawkish favorite at court, to command the English troops, who were already in disarray because of shortages of pay, food, and leadership. Almost instantly Leicester disobeyed his sovereign by accepting from the Dutch the title of governor general of the Netherlands. This act inflamed Philip's fear that she hoped to take the Low Countries for England. All evidence suggests, however, that Elizabeth intended to be no more than its staunch ally, motivated by economic and religious bonds with Protestants and by national security interests. The alliance was renewed in 1598 by a treaty that called for English soldiers to remain in Holland. Intervention there proved to be one of the most costly operations of Elizabeth's reign.

By the spring of 1587 England was braced for Spain's attack. The navy had been strengthened and provisions made for the better training of soldiers on land. Wisely the queen chose to attack first, an uncharacteristic decision that suggests the flexibility of her strategic thinking. She authorized Sir Francis Drake to attack the Spanish fleet moored at Cadiz. Fortunately the queen's second set of orders, restricting Drake to attacks only on Spanish ships on the high seas, did not reach him before he sailed, and his mission was a great success. After Drake had burned or captured some 37 ships, he then conquered Cape Vincent, from which he ambushed another 100 or so Spanish ships laden with war materials. The raid delayed the Armada's attack by a year.

Historians debate the degree to which England's victory over the Armada in the summer of 1588 was owed more to luck or skill. Luck brought storms to the English Channel, but skill and strategy were integral to victory. Early in 1588 the queen's advisors were divided as to strategy. Burghley, a moderate on military affairs, advocated a defensive posture, while Drake urged the queen to launch a second surprise attack against Spain's fleet before it sailed. Elizabeth rightly worried that Drake's proposal left England no naval barrier, yet it won approval. But when storms drove the English warships back into port in early June, Elizabeth reversed herself, ordering her navy to patrol England's southern coast instead. Her admirals were furious at having the mission rescinded, and the queen relented. On 8 July they sailed again, only to be forced back to port again by storms. This was fortuitous, as Spain's Armada had left its port on 12 July. It arrived on 19 July, 130 warships and 30,000 men strong. For a week or so the English navy kept up a running battle with the Armada as it moved up the English Channel to its anticipated rendezvous with reinforcements from Philip's army in the Low Countries. English naval vessels were operated in guerrilla fashion. Outgunned, outmanned, and outskilled by the Spanish, the English took advantage of the maneuverability of their smaller vessels, more mobile cannon, and proximity to supplies, brilliantly exploiting the winds by sacrificing small vessels, which were set afire to drift into the Armada's midst.

As a military leader, Elizabeth's own most noted appearance came on the heels of the Armada's defeat when in August 1588 she reviewed troops mustered at Tilbury. In an unprecedented display of her royal person in what might still have been a battle zone, Elizabeth appeared before some 12,500 soldiers as an armed Pallas Athena, carrying a truncheon and wearing a partial suit of armor. While costumed as a warrior, she delivered her most famous speech. She is alleged to have declared, "I know I have the body but of a weak and feeble woman, but I have the heart and stomach of a king, and of a king of England, too. . . . Rather than any dishonor shall grow by me, I myself will take up arms, I myself will be your general."

England's defense against the Armada drained the queen's coffers, so as soon as immediate danger had passed, she reduced the army and navy accordingly, demonstrating the financial if not strategic prudence that sometimes flabbergasted her commanders. By 17 August the Tilbury camp had been dismantled. On 1 August the queen's navy had 197 commissioned vessels; by 4 September only 34 remained on active duty. Even while reducing her forces, Elizabeth contemplated retaliation; in 1589 she attacked. Unlike Drake's preemptive strike in 1587, this one failed. Drake and Sir John Norris sailed in April 1589. They largely ignored Elizabeth's instructions to destroy Spain's remaining warships before filling their own coffers with loot pirated from Spanish merchant vessels. In their bumbling they lost not only thousands of pounds of war booty, but an estimated 3,000 to 11,000 English sailors to disease and storms before straggling home in June 1589.

The nation faced more "armada scares" in 1590, 1591, 1592, and 1594 as the war continued. In 1596 Spain attempted without success to reach Ireland, its fleet of nearly 100 ships again being cut in half by foul weather; in 1597 Sir Walter Raleigh and the earl of Essex again attacked Spain's coast with 120 English and Dutch ships and 6,000 men under Essex's command. Yet, as had happened in 1589, Essex ignored his instructions once at sea. Jettisoning the plan of attack approved by the queen, they went after Spanish merchant vessels carrying gold and silver from the Americas to the Azores Islands. The mission failed because they showed poor judgment. However, its failure also reflects poorly on Elizabeth's judgment. In the waning years of her reign she continued to bestow command upon select favorites such as Essex, even when their interests no longer paralleled her own. Indeed Essex, an erratic and hawkish sycophant, posed a particular danger that would only gradually become clear to the queen. In 1597 she sent him to Ireland to defeat the nationalist rebel Tyrone, who had been appealing to Philip II to help Ireland break free of English rule. (This was the latest uprising; others had occurred in 1560–1567, 1569–1573, and 1579–1583. Elizabeth saw Ireland as a nuisance, spending £930,000 in her reign to maintain an English presence on Irish soil, keeping it from France and Spain.) Essex, having led the failed 1597 raid on Spanish ports, campaigned vigorously for the assignment as a way to win back the queen's affection. In choosing him again, she put flattery before judgment. Before departing, he made a series of escalating demands for men and supplies, including an army of 16,000 men, which she satisfied at great cost. In Ireland Essex dallied, then engaged insignificant targets but skirted Tyrone, then impetuously returned to London, his job unfinished. Some evidence hints at his having corresponded with Tyrone before departing London for Ireland, perhaps to recruit him for an anticipated coup d'état against the queen. On 8 February 1601 Essex did march on the royal residence of Whitehall, and historians conclude that it was his weak resolve, not the strength of the queen's defenses in London, that preserved her from being overthrown. Essex was executed, and the queen herself died quietly in 1603, having earned the grudging, bitter respect of her enemies and deep affection of her subjects.

Elizabeth is not seen as a militant figure, yet she was a ruler who in many ways artfully directed the military affairs of her state. She ruled in an age when religious wars tore apart and bankrupted powerful states like France and Spain, yet Elizabeth made England rich and powerful. Although she avoided war, she was hardly a weak personality. Elizabeth was physically active; she danced, sang, rode, hunted, and had a famous temper; she could speak several languages and swear like a pirate. A half-century after her death the poet Anne Bradstreet would honor her as "our dread Virago."

Bibliography: Fraser, *The Warrior Queens*, 1989; MacCaffrey, *Elizabeth I: War and Politics, 1588–1603*, 1992; May, "Recent Studies in Elizabeth I," 1993; Neale, *Queen Elizabeth*, 1934; Somerset, *Elizabeth I*, 1991.

—Alan Shepard

"EMILIA PLATER" INDEPENDENT WOMEN'S BATTALION (formed 3 June 1943 in Sel'tse, near Moscow, Soviet Union; ordered to disband 25 May 1945, Warsaw, Poland). Women's infantry battalion, the eastern front, Second World War.

The Emilia Plater Independent Women's Battalion was the first cohesive, all-female Polish combat unit, founded in 1943 in response to the large influx of volunteers who had been exiled to the Soviet Union after the German occupation of Poland and invasion of Soviet territory. The battalion was intended to fight alongside Soviet troops, although, ironically, it was named after **Emilia Plater,** one of the leaders of the 1830–1831 Polish insurrection directed against tsarist Russia. Veterans of the battalion were to be called "Platerówki," often applied to all Polish military women who served on the eastern front. Its commanders and deputy commanders were men, while political/education officers were women: Halina Zawadzka, Irena Sztachelska, and Ludwika Bobrowska. Initially attached to the 1st "Tadeusz Kosciuszko" Polish Division, the battalion swore the oath alongside the division on 15 July 1943. It was directly subordinated to the 1st Polish Corps on 19 August 1943 and then to the First Polish Army on 17 July 1944. Unlike auxiliary female units in the West, the battalion was not subjected to special military regulations.

As of 18 August 1943 the battalion consisted of the command element, one fusiliers company, two infantry companies, one machine-gun company, a company of handheld antitank grenade launchers, and six platoons—mortar, reconnaissance, signals, medical, engineer, and logistics. At the end of 1943 one transport platoon was added. The battalion's permanent strength at the time was 691, including 48 officers, 163 NCOs, and 480 privates. Personnel strength fluctuated, however, as the battalion provided basic training to women subsequently assigned to other units or schools.

Some members, particularly teenagers, reportedly could not cope with the intensive training. The inability of some women to complete arduous training assignments was used as a pretext to gradually transform the battalion from a first-line combat unit to one assigned mainly sentry and military police duties, which the women carried out in an exemplary fashion. Yet the organizational structure was maintained, and members successfully took part in all major combat-training exercises alongside male soldiers. It seems that the battalion's changed status was due at least as much to the reluctance on the part of senior commanders to expose its women soldiers, "the future wives and mothers," to the very heavy losses suffered by Polish troops (one-third of the First Division was wiped out in the Battle of Lenino in Smolensk Region) as to the shortcomings of its soldiers. (Interestingly, Soviet women were actively involved in all types of combat assignments during this time.)

Only the Fusiliers Company took part in the Battle of Lenino, in auxiliary capacities: sentry and police duties, administration of first aid, and escorting German prisoners. The battalion as a whole remained at Sel'tse until early January 1944, at which time it was transferred to Smolensk Region. Two months later it was based in Ukraine. After several other moves, on 12 October 1944 the battalion was ordered to transfer to Praga, a suburb of Warsaw located on the eastern shore of the Vistula. In Praga the battalion became the sole military authority to guard military and civilian facilities, including private property. On 16 December 1944 all detached battalion subunits returned to the battalion. Finally, on 17 January 1945, upon the liberation of Warsaw, the first group of the battalion crossed the frozen Vistula, and on 21 March the battalion became directly subordinated to the Polish General Staff. The battalion's duties in Warsaw were the same as in Praga.

About 70 members of the battalion were killed. In May 1945 it had roughly 500 women members, representing a small percentage of the total number of

Polish women serving in the two Polish armies, with estimates ranging widely from 8,500 to 14,000, including some former members of the battalion who, like **Emilia Gierczak,** had been appointed platoon (or company) commanders in all-male units.

On 23 July 1945 the battalion was partly demobilized, and the discharged "Platerówki" were issued a complete uniform and, oddly, two towels. Some of them secured employment in military institutions, and about fifty became pioneer farmers in Platerówka, a village named after them in the new western territories. On 20 June 1993 a commemorative plaque dedicated to "Platerówki" (women soldiers of the First and Second Polish Armies) was unveiled. The ceremony took place in the Cathedral of the Polish Armed Forces in Warsaw, and the Holy Mass was celebrated by Chaplain General Major-General Sławoj Leszek Głódź.

Representative Participants

Bielawska-Pietkiewicz, Halina. Lieutenant-colonel (ret.). Former commander of the Officer Cadet Corps in the Infantry Officer School in Ryazan'. She was the third in her division to be promoted, and her decorations included the Cross of Valor.

Jabłońska, Helena. Second lieutenant (ret.). As private of the Fusiliers Company, she had been wounded in the Battle of Lenino. She was awarded Virtuti Militari, the highest Polish military decoration, and the Cross of Valor.

Krzywoń, Aniela. Private of the Fusiliers Company. She fell in the Battle of Lenino and was the only Polish woman soldier awarded the **Hero of the Soviet Union** medal.

Wołanin, Janina. Major (ret.). Former commander of a company of 82mm mortars in the 3rd Polish Division. She was wounded three times and was awarded Virtuti Militari and the Cross of Valor.

Bibliography: Chodziński, Wojciech, "Strzeżcie w sercu prysięgi," *Żołnierz Wolnosci* 20 October 1988: 1–2; Pawlowski, Edward; "Platerówki," *Wojsko Ludowe* June 1985: 93–95; Przeciszewski, Roman, "Ogólnopolski Zlot Platerówek," *Żołnierz Wolnosci* 8 September 1986: 1; Rychlewski, Czesław, "Szły znad Oki," *Żołnierz Polski* 11 (17 March 1985): 14–15, 18–19; Stasiński, Adam, "Platerówki," *Polska Zbrojna* 18–20 June 1993: 4, and "Uhonorowanie Platerówek," *Polska Zbrojna* 21 June 1993: 1.

—Kazimiera J. Cottam

EMMA FITZOSBERN (died ca. 1098), also known as Emma Gauder, Emma of Hereford, Emma of Norfolk. Countess of Norfolk.

Emma FitzOsbern was the daughter of Adliza of Tosny and William FitzOsbern. Her marriage in 1075 to Ralph Gauder, earl of Norfolk, sealed an alliance between her new husband, her brother Roger (earl of Hereford), and Waltheof of Northumbria against William the Conqueror. Their plot against William was betrayed almost immediately to Archbishop Lanfranc by Waltheof, who claimed to have been coerced into agreement. Moving ahead despite the loss of the element of surprise, both Roger and Ralph were checked by forces loyal to the king. Ralph retreated to his castle of Norwich under pursuit, but was later able to escape to Denmark to seek an invasion fleet. In his absence the defense of the besieged castle was left to a garrison under Emma's command. She endured for three months, then surrendered the castle after negotiating a settlement that allowed her and her retainers to leave Norwich unmolested. They were given forty days to leave England, and although most of them lost their lands, they avoided the penalties of death or mutilation. The countess went to Brittany, where she rejoined her husband after his unsuccessful naval invasion. In 1096 Emma accompanied her husband and her youngest son on the First Crusade, where they participated in the siege of Nicaea. Both Emma and Ralph died in the Holy Land around 1098.

Bibliography: *Dictionary of National Biography;* Douglas, *William the Conqueror,* 1964; Echols and Williams, *An Annotated Index of Medieval Women,* 1992.

—Annette Parks

EMMA, QUEEN OF FRANCE (died 935), also known as Emma Capet, Emma of Burgundy, Emma of Neustria.

Emma, daughter of Duke Robert (Capet) I of Paris, married Ralph of Burgundy, whom she helped elevate to the throne of France in 923 with the help of her brother, Hugh the Great. Emma defended her husband's right to the throne against the Carolingian claimant, Charles III "The Simple," and her brother-in-law, Herbert II of Vermandois. In 927 she defended the important town of Laon against a siege by Herbert II; after being forced to give up the town, she returned to besiege the town with her own forces before finally yielding to Herbert II at the request of her husband. She captured Avalon in 931 and in 933 led the siege of Château Thierry against Herbert II with King Ralph's army; the castle was reportedly surrendered to Emma and not to her husband. She died two years later, in 935, without an heir.

Bibliography: Flodoard of Rheims, *Les Annales de Flodoard,* 1905; Glaber, *Historiarum libri quinti,* 1886; Riché, *The Carolingians,* 1993; Stafford, *Queens, Concubines, and Dowagers,* 1983.

—Christopher Corley

ENGLE, REGULA (born 5 March 1761, Zurich; died 25 June 1853, Zurich). Soldier, French revolutionary and Napoleonic armies, 1792–1815.

Known best through her memoirs, which were published in 1821 and later titled *L'amazone de Napoléon: Mémoires de Regula Engle,* Engle followed her common-law husband, Colonel Florian Engle, into the service of France. During two decades of service, primarily on campaign with Napoleon Bonaparte, Engle bore twenty-one children, of whom ten died. Widowed after Waterloo, she turned to literary pursuits in order to survive.

Although French military archives contain no record of Florian or Regula Engle, it was not uncommon during the Napoleonic Wars for women to accompany their officer husbands on campaign. According to Engle, such was her capacity. Unlike most wives, however, she followed her husband onto the battlefield rather than remaining in the army train. She allegedly participated in campaigns as well known as the Egyptian Campaign, the Battle of Marengo in the Italian Campaign, and Waterloo, after having followed Emperor Napoleon I to Elba during his first exile. She was fifty-four years old at Waterloo. During the years of "bivouacs, voyages, battles, joys and immense suffering" (according to her memoirs), Engle's varied activities and her large family earned her the epithet of "Mother Courage" that is widely used by biographers.

Bibliography: Barnum, Emily Kenne, "The Thrilling Adventures of a Swiss Amazon: Regula Engel-Egli," *Swiss Monthly,* 1930–1931; Engle, *L'amazone de Napoléon,* 1985.

—S.P. Conner

ERAUSO, CATALINA DE (born 1592 in San Sebastián, Spain; died 1642?, Mexico), also known as Erausú, Erauzú, La Monja Alférez, the Nun Ensign, the Lieutenant Nun; masculine aliases include Alonso Díaz Ramírez de Guzmán, Antonio de Erauso. Ensign, Spanish army, Chile.

Catalina de Erauso, hiding her identity as a woman, fought for the Spanish Crown on the Chilean frontier against the Araucanians. For her bravery she was awarded the rank of ensign. Despite the disclosure of her true sex, Erauso was granted a pension by the king and papal sanction to continue living as a man.

Although placed in a convent at age five by parents who hoped for a chaste and cloistered future for their daughter, Erauso defied her destiny by escaping when she was fifteen. For three days she hid, during which time she began her transgendering process. Fashioning breeches from her habit and affecting a young man's comportment, she discarded by the side of the road both her tresses and her femininity and began her travels throughout Spain. Like many Iberian youths of the era, Erauso saw the immense opportunities for adventure and wealth in the New World and made her way to South America. Moreover, she realized that the quickest route to fortune and honor was through battle, and at that time the greatest action was to be found on the Chilean frontier. Upon arriving in Concepción, Chile, under the name of "Alonso Díaz Ramírez de Guzmán," she fell in with a regiment of fifty sol-

Erauso, Catalina de

Catalina de Erauso dressed as a man. (Perry-Castañeda Library)

diers who ironically were commanded by her older brother Miguel de Erauso, whom she had never met.

Not only did Erauso successfully rise to the challenge of warfare, she excelled at it. She proved herself on one occasion when a group of Araucanians, greatly outnumbering the Spanish, ambushed the regiment. As defeat for the Spaniards seemed imminent, the inspired "Alonso," crying "Santiago!" (the traditional call of Spanish offensive), rallied the surviving troops and managed to hold off the native soldiers. For her bravery her commander awarded her the rank of ensign.

Erauso played perfectly the role of a young man with battlefield experience, alternating between courting the innocent maidens of the Spanish Empire and defending her manly honor throughout many a duel. Her proclivity to violence, however, ultimately led to the disclosure of her greatest secret. With a trail of murdered victims behind her, including the tragic accidental slaying of her brother, Erauso was called before the bishop of Huamanga, in front of whom she confessed the truth.

It was the incredibility of her story, coupled with the fact that she was both a nun (albeit unprofessed) and a virgin, that saved her. Not only was she not punished, but the king of Spain awarded her a generous pension, while Pope Urban VIII in 1627 gave her official sanction to spend the rest of her life dressed as a man. With pension in hand and in men's dress, "Antonio de Erauso" returned to the New World, this time to Mexico, where she spent the rest of her life as a muleteer.

A reputed memoir by Erauso was not published until 1829, although her story was widely known in folklore and tradition. (The book was reissued in 1996 as *Lieutenant Nun: Memoir of a Basque Transvestite in the New World*.) The memoir is written like a novel, and there is some dispute about its true authorship. However, there is no debate as to Erauso's existence and military experience. Official religious records confirm her baptism on 10 February 1592 as well as her father's payment to the convent for his daughter's internment. Erauso's petition to the Crown for a military pension still exists, as do affidavits by fellow officers to assist her in securing the allowance. Lastly, during her lifetime, several Erauso-inspired broadsides were penned. Her portrait can be seen in the Museo del Ejercito Militar in Madrid.

The life of Erauso inspired at least one imitator. C.R. Boxer describes the case of Dona Maria Ursula de Abreu e Lencastre, born in Brazil, who sought to escape an arranged marriage by assuming a male identity as "Baltasar de Couto Cardoso," enlisting as a marine on a warship and sailing for Lisbon and from there to India in 1699. She served some fourteen years without being discovered. Her identity was finally revealed when she was seriously wounded in the course of rescuing her captain, Afonso Teixeira Arrais de Melo e Mendonça, from an enemy attack. She later married her captain and bore a son named João. She and her husband received a small grant of land where they apparently lived out their lives.

Bibliography: Boxer, *Women in Iberian Expansion Overseas*, 1975; Erauso, *Lieutenant Nun*, 1996; León, *Aventuras de la Monja Alférez*, 1973; Merrim, "Catalina de Erauso: From Anomaly to Icon," 1994; Salas, *Soldaderas*, 1990; Vallbona, *Vida i sucesos de la Monja Alférez*, 1992.

—Alexa Samuels

ERKETÜ QATUN (born ca. 1551; died 1612, Mongolia). Erketü Qatun is a title (powerful lady) rather than a name and was probably given to her later in her life. Her original name is not known. Also known as *San niangzi* (literally, the Third Lady, because she was the third wife, and the second concubine, of Altan-qan, a powerful Mongolian leader), *Noyanču jünggin* (the empress living in the inner quarters). Mistress of Loyalty and Obedience. Mongol regent of the noble Borjigin clan during China's Ming dynasty (1368–1644).

Erketü Qatun was wife to four Mongol chieftains of the Borjigin clan, commanded military power, and was instrumental in keeping peace with the Ming court from the 1570s to her death in 1612. She was the granddaughter of Altan-qan (Anda in Chinese, 1508–1582), a Mongolian prince of the Borjigin clan, by his daughter. In 1570 she was to marry a prince of Ordos (in today's Suiyuan Province), possibly a nephew of Altan-qan, but struck by her beauty, Altan-qan decided to keep her for himself (marrying his own granddaughter). To appease the Ordos prince, Altan-qan compensated him with the intended second wife of his grandson Baγa-ači. Baγa-ači was furious, so he went over to the Chinese settlement under the Ming government, which historically had been in conflict with the Mongols. He took with him his first wife and more than a dozen subordinates, offering themselves to Fang Fengshi, the grand coordinator of Datong.

Fang Fengshi received the emperor's permission to try to bring peace to the region. Negotiations resulted in the opening of trade between the Mongols and the Chinese. Baγa-ači was returned to Altan-qan, to his grandfather's great joy, and, in 1571 the Ming emperor granted Altan-qan the title prince of obedience and righteousness—a title that would be much contested for the next several decades. Erketü Qatun had three sons by Altan-qan; of the three, Budaširi was to distinguish himself and play an important role later.

Erketü Qatun

In 1582 Altan-qan died, leaving Bayising and his personal troops in the command of Baγa-ači, his favored grandson. As his favored concubine, Erketü Qatun held the official seal of prince of obedience and righteousness, used for communication with the Ming officials—paying tributes, holding horse fairs (to exchange goods with ethnic Chinese), receiving annual gifts, commanding other chieftains, and so on. Altan-qan's almost unchallenged power was thus divided. Nevertheless, before she remarried and another enfeoffment was made, Erketü Qatun alone, in possession of the princely seal, wielded the power of the inheritance.

Given Erketü Qatun's prestige with the Ming court and her command over a good portion of the Mongol nobility, Altan-qan's eldest son Sengge (Huang-taiji in Chinese) soon realized that he needed a marriage alliance with Erketü Qatun in order to improve his own standing with both the Mongols and the Chinese. Since in Mongol and general central Asian custom a son could inherit his deceased father's concubine, such a marriage was not extraordinary. In 1582 Erketü Qatun and Sengge were married, and in April 1583 Sengge was appointed the new prince of obedience and righteousness.

To Erketü Qatun, the marriage was as much one of convenience as it was to Sengge, who basically pursued a dissipated life of women and pleasure. That Erketü Qatun was of strong character is shown in two ways: first, before the marriage she forced Sengge to renounce all his previous wives; second, after the marriage she took advantage of Sengge's neglect of duty by taking control of his best troops and stationing herself in the Western Patrol, far from Sengge's headquarters at the Eastern Patrol.

A few months later, in August 1583, Baγa-ači died in an accident, leaving his command of Bayising and Altan-qan's troops to his widow Baγa-beyiji, since their sons were very young. Erketü Qatun hoped that Baγa-beyiji would marry her son Budaširi to consolidate another branch of power; however, in 1584, to her great chagrin, Baγa-beyiji married Sengge's son Čürüke (Chelike in Chinese), his father's rival. For a while, tension was high, and all sides placed their troops at strategic locations and made small raids and counter-raids: Budaširi moved south of Bayising, Erketü Qatun readied her troops for the imminent situation, and Čürüke sent his troops into the mountains north of Bayising.

Some time between 1585 and 1586 Sengge passed away. This second marriage had provided Erketü Qatun with much more military power. She hid the princely seal, hoping to benefit her own son Budaširi later. Meanwhile, Čürüke was finding it hard to improve his standing with the Chinese without Erketü Qatun's backing. The Chinese also pressured him to make a marriage alliance with Erketü Qatun, maintaining that the princely title would go to someone else if he did not. Since Erketü Qatun had by then more than 10,000 well-trained troops, the marriage alliance would indeed bring Čürüke much power and prestige with both Chinese and Mongols. He therefore divorced all his wives, including Baγa-beyiji, the widow of "the favored grandson," and married Erketü Qatun in 1586. Some time later Baγa-beyiji married Erketü Qatun's son Budaširi, just as she had hoped, and the tensions between the two factions finally dissolved through marriages.

In 1587, well respected by Chinese garrisons at the border as well as among the Mongol forces, Erketü Qatun was given the title of Mistress of Loyalty and Obedience by the Ming court, both as an acknowledgment of her military power and as appreciation for the continued peace along the borders she had guaranteed. For the next twenty years the area between Xuanda and Gansu was at peace.

In 1606 Čürüke fell very ill, and the princely seal was once again in Erketü Qatun's hands. After he died in 1607, contenders for his title of Prince of Obedience and Righteousness included his grandson Bušuγtu (Pushitu in Chinese) and Erketü Qatun's own grandson Sodnam (Sunang in Chinese). Sodnam was the most legitimate heir, as the direct grandson of Altan-qan, partly to forestall her from marrying Bušuγtu, who held an inferior position in the clan (as the grandson of Altan-qan's grandson Čürüke).

Erketü Qatun herself, now about sixty years old, resisted another marriage for as long as she could. However, the Chinese were not willing to leave the title of Prince of Obedience and Righteousness unfilled indefinitely, since it was an important channel of communication and a guarantee for border peace; they seemed to believe that anyone who married Erketü Qatun would inevitably be influenced by her pro-China stance. It might have been logical for her to ally with Sodnam, but she could not marry her own grandson.

A marriage between Erketü Qatun and Bušuγtu would provide some stability for the Chinese, even though he was a minor figure in the Mongol scene. The marriage took place on 21 June 1611, but still did not end all the difficulties. Erketü Qatun apparently did not entirely trust her newest husband and did not give him the seal of the Prince of Obedience and Righteousness, although at one point she did show him the gold, silver, brocades, furs, and gems that she had accumulated and stored in the Buddhist temple Hongci si (Vast Kindness Temple) through her life with three generations of Princes of Obedience and Righteousness. The sight must have impressed him greatly, for he turned to ask the Chinese for increased subsidies and a new title, an indication that he did not receive the old seal upon his marriage with Erketü Qatun.

Amid all these uncertainties, Erketü Qatun suddenly fell ill and died on 24 July 1612 (some sources say 1613, but 1612 is probably more accurate). The Ming court sent military representatives to attend the ceremony at the governor-general's headquarters, and the emperor sent seven altars of offerings and gifts of silk and cloth to various relatives of the dead. Thus ended Erketü Qatun's life as a Mongol regent respected by both the Mongols and the Ming.

Probably the most significant military role Erketü Qatun played was commanding substantial military power to keep the peace. For the forty years of her married life she had under her command the best troops among the Mongol nobilities. Although she had clashed with other chieftains, there were no records of any major battles fought by her in either Chinese or Mongol records. Yet she was revered by both sides, and the border between the Ming Chinese and Mongols enjoyed the longest peace in history. Her power therefore goes well with an informal semantic interpretation of the Chinese character for military strength (*wu*): to deter with weapons.

Bibliography: Qu Jiusi, *Wanli wukung lu (zu ben)* (Taipei: ca. 1979); Serruys, "Two Remarkable Women in Mongolia," 1975; Zhang, *Ming shi*, 1974.

—Sherry J. Mou

ERMENGARDE OF NARBONNE (ca. 1120s–ca. 1194). Viscountess of Narbonne.

Ermengarde was still a child when she inherited Narbonne from her father Aymeri II, which resulted in attempts by both Count Alphonse-Jourdain of Toulouse and his son Raymond to take the viscounty in order to "protect" her and her lands. Once she reached adulthood, Ermengarde took an active role in protecting her inheritance. By 1143 she led troops in battle against her feudal overlords, the counts of Toulouse, and consolidated her power over her own vassals, who had unsuccessfully revolted against her. It is clear from contemporary charters that Ermengarde ruled her own lands and that her husbands—Ermengarde married at least twice and possibly three times—did not play a part in their governance. She is described as distinguishing herself by the wisdom of her government. In 1192 she retired in favor of her nephew.

A strong royalist, she supported King Louis VII of France in the king's campaigns in southern France as a counterbalance to the powerful counts of Toulouse. Ermengarde personally led troops in battle in 1172–1173 when the nobility of Aquitaine revolted against the harsh rule of the duke of Aquitaine, King Henry II of England, the husband of **Eleanor of Aquitaine**.

Bibliography: Echols and Williams, *An Annotated Index of Medieval Women*, 1992; *Gesta Henrici* (Stubbs ed.); Labarge, *A Small Sound of the Trumpet*, 1986; Vaissete, *Histoire générale de Languedoc*, 1889.

—J.M.B. Porter

ERMESSEND OF CARCASSONE (ca. 980s–ca. 1050s). Countess of Barcelona. Coruler and regent of County of Barcelona, Catalonia, Spain.

Ermessend, as befitted an eleventh-century countess, was responsible for defense and maintenance of the peace in the County of Barcelona; she also fought for her son's claim to the county and was later engaged in a civil war with her grandson Count Ramon Berenguer I.

After Ermessend's marriage to Count Ramon Borrell of Barcelona, the new couple worked together to strengthen their hold over the county, to recover control over the castles of the marcher lords, and to extend their political and military influence north into Languedoc (where Ermessend's family held power) and south into Muslim territory. When Ramon Borrell died in 1018, Ermessend assumed two successive coregencies with her brother, Bishop Pierre of Gerona: the first was for her son Beruenger Ramon I, and the second, after the latter's death in 1035, was for her grandson Ramon Berenguer I. Throughout this period Ermessend, facing the opposition of recalcitrant viscounts and bishops in addition to external enemies, exercised considerable political and military power, presiding over courts, intervening in disputes, calling on the military services of allies, and even owning castles in her own right or receiving the homage of castellans. Interestingly, one of these castellans, **Gidinild,** was also female—she was a noblewoman who had captured Cervera in 1026 and had constructed a fortress there. However, when Ermessend's grandson Ramon Berenguer I reached the age of majority, the two came into conflict. Their struggle for power lasted from the 1040s until 1057, when her claims were relinquished for a payment of 1,000 ounces of gold. Ermessend's career is an excellent example of the great power often wielded by noblewomen in Spain and southern France after the late ninth century, when the increasing importance of hereditary succession and family ties to the construction and maintenance of military and political authority enabled many women to exercise such authority in their own right.

Bibliography: Echols and Williams, *An Annotated Index of Medieval Women,* 1992; Lewis, *The Development of Southern French and Catalan Society, 718–1050,* 1965; Shahar, *The Fourth Estate,* forthcoming; Shneidman, *The Rise of the Aragonese-Catalan Empire, 1200–1350,* 1970.

—David Hay

ERXLEBEN, HEATHER. Private, 3rd Battalion, Princess Patricia's Canadian Light Infantry, Canada.

Heather Erxleben was the Canadian Forces' (CF) first female infantry soldier, 1989. In 1979 Servicewomen in Non-traditional Environments and Roles (SWINTER) was introduced by the CF to determine the impact on operational effectiveness of employing servicewomen in sea, land, and air near-combat units and at isolated units. In 1987, in response to the mandate of the Canadian Charter of Rights and Freedoms to expand service opportunities for women, the CF created Combat-Related Employment for Women (CREW). CREW trials became the first stage in the total integration of women into the CF in every capacity, with the exceptions of submarine duty and Roman Catholic chaplain. The CF had already eliminated restrictions on mixed-gender employment in the air force. Private Erxleben participated in international exercises as an infantry soldier. Canada's nonparticipation in combat situations has meant that Erxleben and other qualified women have not had actual combat experience.

> *Women serve in the armed forces of thirteen of NATO's sixteen member nations. (Luxembourg has no women in uniform, and Iceland has no armed forces.) Highlights of each nation's national status report (1998 figures) included the following:*
>
> Belgium: About 3,120 women make up 7.18 percent of the Belgian armed forces. In February, for the first time, nine women began serving on a minehunter, thereby making up 19 percent of the vessel's crew.
>
> Canada: About 6,700 women make up 10.8 percent of the regular force; another 5,800 or so make up 18.6 percent of the reserve. Women continue to move into senior leadership positions, and this year the air force appointed its first woman squadron commander. A woman also became a search and rescue technician, the first to serve in one of the toughest, most demanding occupations in the Canadian Forces.
>
> Denmark: About 870 women make up 5 percent of Denmark's regular military personnel (excluding conscripts).
>
> France: About 10,600 women make up 7.5 percent of the force. French officials are considering opening all posts to women except those involving direct and prolonged contact with hostile forces; flying carrier-borne aircraft; and service aboard submarines, in marine or commando units, and in some gendarmerie posts.
>
> Germany: About 3,800 women make up 24 percent of the German military's medical service. Another 37 women serve in military bands. These are the only two branches where women are allowed to serve.
>
> Greece: The 717 women in uniform make up about 3.75 percent of the Greek armed forces.
>
> Italy: A law passed in 1999 finally mandated the admission of women to the ranks of the Italian armed forces and military academies. Large numbers of Italian women reportedly applied, but as of 2001, they still comprised only a fraction of a percent of total military personnel.
>
> The Netherlands: About 4,000 women make up about 7.2 percent of the Dutch armed
>
> *(continued)*

Bibliography: Director of Women Personnel, *Women in the Canadian Forces,* 1986; Capt Marsha Dorge, "CREW," in *Sentinel,* 25(3) 1989, 8–9; Thomas, "Women in the Military," 1978.

—Patricia Bowley

ETCHEBÉHÈRE, MIKA (fl. 1930s, Argentina, Spain). Commander of a military unit, Spanish Civil War.

Mika Etchebéhère, an Argentinian, was married to Hypolito Etchebéhère, an Argentinian of French Basque extraction. Mika acquired French citizenship through her husband. Both were part of the anti-Stalinist Communist Opposition and, before falling out with Trotsky, were part of the group that founded the Fourth International. They were in Germany in 1932 and 1933, where they saw firsthand the devastating impact of Nazism on German Communists. In Spain at the start of the **Spanish Civil War** Hypo became commander of a Partido Obrero de Unificación Marxista (POUM) unit, while Mika carried out noncombatant administrative work in the unit. After Hypo was killed on 16 August 1936, Mika remained with the battalion and eventually came to lead a combat unit. She showed decisiveness with her unit at the terrible siege of the Siguenza Cathedral in which 500 soldiers and local townspeople were surrounded and subjected to prolonged bombardment. Subsequently she was placed in command of a unit in active combat in the trenches at the Moncloa front in Madrid.

Her experiences are particularly interesting because it was unusual for a woman to lead male soldiers. She insisted that under her command *milicianas* were to be treated as "militiawomen, not domestics," and that in the enterprise of winning the war men and women were equal. Etchebéhère has noted that it was necessary to retrain the men in her command before they accepted her regimen. The fact that several militiawomen were keen to transfer into her unit suggests that the environment she created was not typical (see Spanish Civil War, women in).

Etchebéhère wrote about her time in Spain in a substantial volume of recollections, *Ma guerre d'Espagne à moi*, which reveals a highly intelligent woman devoted to her husband's memory and to the welfare of the soldiers she led. Her great strength of character and administrative skills drew on her strong common sense and what might be called her "womanly virtues." She was at pains to ensure that her troops were warmly clothed, and when they were immured in the trenches in Madrid, she negotiated with the supreme command so that groups of five men at a time could have breaks away from the front, usually to go to the brothel, in order to overcome the ennui and wretchedness that overwhelmed men confined in stinking trenches. Above all, she ensured that her unit was properly fed. She instituted hot-food patrols and employed ingenious methods such as the cook using a wheelbarrow to trundle huge cauldrons of steaming food through the trench up to the line. Her real stroke of genius, however, was to acquire several hundred small thermos flasks in which hot coffee could be delivered individually to soldiers isolated at their posts. When it was impossible under fire to provide them with a proper meal, a thermos of coffee would lift the men's spirits. Having originally been opposed to alcohol on "revolutionary principles" and having been "shocked" at the volume of alcohol consumed at the front, she came to accept its virtue in keeping up morale. At the end of the day Etchebéhère would go through the trenches, a flagon in each hand, dispensing a shot of cognac to the men as they bunked down for the night. While these efforts to keep her troops clothed, fed, and in good morale might be described by some as oddly "feminine," it is a truism that an army marches on its stomach; a male commander who made similar efforts would undoubtedly be viewed as a wise leader.

Etchebéhère attributed some of her success to the fact that she was careful never to indicate any sexual interest in any of the men and immediately rebuffed any overtures they made toward her. As she

forces. This summer, the first woman will complete the Air Force Advanced Staff Officer Course and, for the first time, the navy will promote a woman to the rank of captain.

Portugal: About 2,200 women make up about 5 percent of Portugal's armed forces. Women have served on Portuguese navy ships since 1993, but they are prohibited from serving in some combatant specialties.

Spain: About 2,400 women make up 2.3 percent of Spain's armed forces. Women officers and NCOs can apply for any post, but women troops and sailors are prohibited from tactical and operational postings in legion units, special operations and paratroop units, submarines, marine landing forces, and small ships without appropriate accommodations.

Turkey: Women are only admitted in the Turkish armed forces as officers, not as NCOs or enlisted personnel. About 680 women officers now serve in the Turkish army, navy, air force, and gendarmerie. Sixty women attended military academies last year; 274 are attending this year.

United Kingdom: About 2,890 women make up 7.8 percent of the Royal Navy, with 745 at sea aboard 50 ships. About 73 percent of all navy posts are open to women. The British army's 7,432 women comprise 6.7 percent of the force; as of April 1, women can serve in 70 percent of the army's slots—up from 47 percent. About 5,000 women make up 8.9 percent of the Royal Air Force. About 20 are pilots and 26 are navigators. Women remain excluded from the Royal Marine Commandos, the Infantry, Royal Armoured Corps, and the Royal Air Force Regiment.

United States: About 200,000 women make up 14 percent of the active duty force; 88.2 percent of the military's 1.1 million jobs are open to women. About 225,000 women serve in the reserve components and comprise 15.5 percent of their strength.

—Adapted from an article by Linda D. Kozaryn, American Forces Press Service, reported by DefenseLink news on the Web at http://www.defenselink.mil/news/Jun1998/n06171998_98061.72.htm.

said, her soldiers saw her "neither as a man or a women" but as a sort of "hybrid" who was "wrapped in her inapproachability and her legend." She thought that it helped that she was not Spanish; her foreign origins made it easier for the men to accept her untraditional role. She had acquired a light carbine rifle, but in her descriptions of battles it appears that more frequently than using the gun herself, she passed it to the fourteen-year-old boy who was a volunteer in the militia and her faithful and courageous runner. Overall, it is probable that although she engaged in combat and took her turn on watch, it was Etchebéhère's administrative skills rather than her prowess in battle that bonded the men to her.

Removed from her command as part of the disarming of the independent political militias, Etchebéhère remained in Spain until 1938, afterwards settling permanently in France. Perhaps by dint of her common sense, she managed to survive the swathe of assassinations that Stalin's agents cut through her political friends and associates.

Bibliography: Etchebéhère, *Ma guerre d'Espagne à moi*, 1976; Low, *Red Spanish Notebook*, 1937.

—Judith Keene

ETHERIDGE, ANNA (born 3 May 1844, Detroit, Michigan; died 1913), born Lorinda Anna Blair, also known as Annie. Daughter of the Regiment, 2nd, 3rd, and 5th Michigan Volunteers, Union army. United States, American Civil War.

Anna Etheridge is probably the most famous of army women in the **American Civil War;** unlike many other women, she was famous in her own time and was widely regarded as a heroine. In 1860 she married James Etheridge, and when he enlisted in the 2nd Michigan Volunteers in 1861, she signed on as a Daughter of the Regiment. Although her husband apparently deserted soon after, "Annie" stayed with the army, transferring to the 3rd Michigan Volunteers, where she knew many of the soldiers. For three years she served as the 3rd's "daughter," a position that entailed various responsibilities ranging from cook, water carrier, and nurse to color bearer and a member of the regiment who marched and drilled with the men. When the 3rd Michigan disbanded three years later, Annie transferred to the 5th Michigan Volunteers along with many other veterans of the Third.

Etheridge's activities are well documented in letters and other records. She accompanied her regiment into battle, tending to the wounded and assisting with their transport to the rear. She is noted for rallying the regiment at the Battle of Chancellorsville (1–3 May 1863), riding along the line to encourage the men to stay and fight; one man who met her there wrote that "she is always on hand and ready to bear the same privations as the men; when danger threatens, she never cringes." She is remembered for her ability to convince retreating Michigan Volunteers to turn and hold their position at Spotsylvania (6 May 1864). Etheridge served with the Army of the Potomac in 28 engagements, including the Battles of Fredericksburg, Chancellorsville, Gettysburg, the Wilderness, and Petersburg. Her "uniform" was a green riding habit, and she was said to have worn a pair of matched .45 pistols over her skirt. She was wounded only once during this time, though her skirts were frequently marked with bullet holes. If she ever used her pistols, it was not mentioned in the available sources. However, her bravery under fire and patient endurance of field conditions were frequently remarked by those who observed her.

Etheridge mustered out in July 1865 when her regiment demobilized. She thereafter worked as a clerk in the Pension Office and in 1870 married Charles E. Hooks. In 1886 a senator helped her gain a pension for her wartime service. When she died in 1913, she was buried in Arlington National Cemetery. For her service in the Federal army, Anna Etheridge was awarded the Kearny Cross, given to enlisted men of the 1st Division, 3rd Army Corps, who distinguished themselves in battle.

Undated portrait of Anna Etheridge. (Library of Congress)

Bibliography: Dannett, *Noble Women of the North*, 1959; Hall, *Patriots in Disguise*, 1993; Moore, *Women of the War*, 1866.

—Susannah U. Bruce

EURYDICE II (died 317 BCE), also known as Adea. Macedonian queen and general.

Eurydice-Adea received a military education from her mother, **Cynane**. She grew up in a family of military innovators who had achieved the unification of Macedon, the conquest of Greece under Philip II, and the conquest of the Persian Empire under Alexander. She probably began to play a role in the military side of power politics after her marriage in 320 to the Macedonian king Philip Arrhidaeus. In July of that year Eurydice appealed directly to the Macedonian army in an attempt to influence negotiations among the successors of Alexander. Although this failed, she obtained an enormous amount of power when her husband was named king.

In 317 Eurydice declared her support for the dynast Cassander, an enemy of Alexander's mother **Olympias**. As a result, Olympias, whose power was based in the neighboring kingdom of Epirus, marched on Macedon with an army. Eurydice hoped to decide the matter in a single battle and led her forces against Olympias in what one Greek historian called "the first war between women." Dressed in helmet, breastplate, and greaves, Eurydice met Olympias at Euia but never got the opportunity to fight. Her troops defected en masse when they saw the mother of Alexander, and handed over Eurydice and her husband to Olympias, who then took control of Macedon. She imprisoned Eurydice and Arrhidaeus in a small room, killed him, and forced her to commit suicide. Eurydice was unrepentant to the end, insisting that she was queen of Macedonia.

The so-called Tomb of Philip in the royal Macedonian necropolis in Verginia contains a room with the grave of a woman surrounded by military grave goods—gold-embroidered greaves, a silver quiver, and arrowheads. Some scholars have hypothesized that this woman is Eurydice.

Bibliography: Arrian, *History of the Successors of Alexander* 1.33; Athenaeus, 13.560; Diodorus Siculus, *Historical Library* 18.39.2, 19.11.2–9; Macurdy, *Hellenistic Queens*, 1975.

—Alexander Ingle

F

FANY CORPS (FIRST AID NURSING YEOMANRY CORPS). Founded 1907, United Kingdom. Volunteer military support services, including nursing, driving, and espionage.

The uniquely British First Aid Nursing Yeomanry (FANY) Corps was formed at Holborn, near London, in 1907, by retired cavalry sergeant-major Edward C. Baker. His intent was to free men for battle by substituting trained women to perform certain essential noncombat duties. To get started, Baker advertised in the British national papers for women volunteers who would pay their own way. By 1911, when the corps was taken over by Miss Grace Ashley-Smith, the volunteers paid to join the corps, provided their own uniforms, funded their own training and equipment, including horses, held military ranks, and stood ready in case of need. In addition to first aid and home nursing, FANY members were required to pass a course in "horsemanship, veterinary work, signaling and camp cookery." By 1918, after nearly five years in the field, their duties had evolved from picking up wounded and transporting them in horse-drawn hospital wagons to driving motorized ambulances and soup kitchens and operating hospital canteens. Although bombarded and gassed, they had disproved "all the old arguments against women serving at the front." In 1937 the corps's name was officially changed to Women's Transport Service (FANY). This unconventional designation described the newly assigned primary function while preserving the name by which it was fondly and widely known; it is called FANY to this day.

In 1938, after a battle for control with the army, the corps was split in two. One element, designated the Women's Transport Service (FANY), was assigned to the army to provide transport drivers. This was the unit that Princess

Elizabeth joined in 1945. The original FANY headquarters element became known informally as the Free FANY, or simply the Corps. It continued training in Britain and overseas. When war broke out, it provided radio operators, nurses, ambulance drivers, and support to "the Org." "The Org" was the FANY term for the secret Special Operations Executive (SOE) formed in July 1940 to support resistance movements in occupied areas. The relationship began when SOE Colonel (later General) Colin Gubbins, then the Director of Operations and Training, was contacted by his FANY friend, Miss Phyllis Bingham. He was looking for women to do secret unspecified work and wondered whether the FANYs could meet the need. Free FANYs were soon packing parachutes, operating radios, and assisting in agent training. Seventy-three were themselves trained as agents, and thirty-nine of these were dropped into France. Several led resistance groups that sabotaged enemy facilities. Many were captured and escaped to survive the war. Thirteen FANY agents died in Nazi concentration camps. After service in all theaters of operations, more than 100 FANYs were decorated at the war's end.

Members of the First Aid Nursing Yeomanry (FANY) running to their ambulances during a "stand to" in the Southern Command during World War II. (Hulton/Archive by Getty Images)

The FANYs exist today and continue to follow their motto *Arduis Invictus* (in difficulties unconquered). Based in London, they remain on call to the police major-incident unit and stand ready to meet the wartime need should it arise once again.

Representative Participants

Her Royal Highness the Princess Alice, countess of Althone, honorary commandant in chief, Women's Transport Service (FANY). Princess Alice provided royal patronage for the corps from 1933 until 1981. She was succeeded by HRH Princess Anne.

Rudellat, Yvonne, FANY/SOE (born 11 January 1897, Maisons-Laffitte, France; died 23/24 April 1945, Bergen-Belsen Prison Camp, Germany). M.B.E. (honorary). Born and educated near Paris, petite Yvonne Cerneau met and married Alex Rudellat during a trip to London in 1920. They settled in England and had a daughter in 1922. A grandmother by the Second World War,

FANY Corps (First Aid Nursing Yeomanry Corps)

she was recruited by FANY/SOE and became the first woman agent to enter France. Code-named "Jacqueline," Rudellat joined the PROSPER network in 1942, where she served as a courier, handled agent receptions, carried out railroad sabotage, and eventually, with coworker Pierre Culioli, set up and commanded the ADOLPHE subnetwork. In June 1943 "Jacqueline" was severely wounded in the head while trying to run a roadblock. Barely alive, she refused to talk and was sent first to Ravensbrück and then to Bergen-Belsen, where she died of typhus less than six weeks before it was liberated.

Szabo, Violette, ensign, FANY/SOE (born 26 June 1921, British Hospital, Paris; died January 1945, Ravensbrück Prison Camp, Germany). George Cross (posthumous), Croix de Guerre (posthumous). Violette Bushell, the daughter of a British "lorry driver" and a French mother, was nineteen when she married French Foreign Legion officer Etienne Szabo. She joined the Auxiliary Territorial Service (**ATS**) in September 1941 and served in an artillery battery until the birth of her daughter, Tania, in June 1942. A few months later her husband was killed at El Alamein. Before Tania was six weeks old, Szabo joined FANY/SOE, bent on completing her husband's work. Code-named "Louise," Szabo became one of the best marksmen in SOE. Trained as an agent, she made two missions to France, where she served with the Maquis. Arrested twice by the Gestapo, she escaped both times. The third time, however, she was surrounded by Gestapo agents in a farmhouse. Using a Sten gun, she exchanged shots, killing several, until her ammunition was exhausted and she was arrested. When she refused to talk, she was sent to the Ravensbrück concentration camp and shot in February 1945.

Violette Szabo, ca. 1945. (Hulton/Archive by Getty Images)

Witherington, Pearl, captain, FANY/SOE (born 1914, France; died 198?). M.B.E., French Medal of Resistance, Croix de Guerre, Legion of Honor. At the outbreak of the Second World War Witherington was working for the British air attaché in Paris. She returned to Britain, and when she was refused an application for FANY/SOE, she reported to the headquarters on her own and was accepted. On 23 September 1943 Witherington, code-named "Marie," was dropped by parachute into France, where she served as a courier and member of the sabotage team in the WRESTLER circuit. In May 1944, when the Gestapo arrested her superior, she assumed command of the 3,500-man circuit. Under her extraordinary leadership WRESTLER conducted more than 800 railway interdictions and sabotage missions while killing more than 1,000 soldiers and taking more than 20,000 prisoners. Witherington was nominated for a Military Cross, for which women were ineligible. She received instead the lesser civil M.B.E., which she returned with the observation that "she had done nothing civil." The Air Ministry then offered an M.B.E. (Military), which she accepted.

Bibliography: Foot, *SOE in France,* 1966; King, *"Jacqueline,"* 1989; Popham, *FANY,* 1984; Rossiter, *Women in the Resistance,* 1986; Thomas, "Women in the Military," 1978; Ward, *F.A.N.Y. Invicta,* 1955.

—Hayden Peake

FEDUTENKO, NADEZHDA NIKIFOROVNA (born 30 September 1915, Rakitnoe, Belgorod Region, Russia; died 28 January 1978, Kiev, Ukraine, Soviet Union), also known as Nadia. Guards major (ret.), Red Army Air Force. Squadron commander, Soviet Union, Second World War.

Nadezhda Fedutenko accumulated several thousand flying hours prior to enlisting in the women's 587th Bomber Aviation Regiment (later redesignated the **125th "M.M. Raskova"** *Borisovsky* **Guards Bomber Aviation Regiment**), which flew twin-engine, medium-range Pe-2 dive bombers and was one of the three women's regiments to emerge from the **122nd Aviation Group** formed by **Marina Raskova**. Due to her outstanding airmanship and leadership skills, Fedutenko was immediately appointed squadron commander and quickly gained the respect and affection of her subordinates.

Ever since an aircraft landed near her home when she was a small child, Fedutenko dreamed of becoming a pilot. In her teens, as she was learning the turner's trade, she was active in an aircraft-modeling club in the evening. In 1935 she qualified both as a pilot and as a parachuting instructor upon graduating from the Tambov School for Civil Air Fleet Pilots.

For about six years before the war Fedutenko flew passengers and cargo by day and night, mastering several types of aircraft. Serving at the front from the beginning of the war with the Kiev Special Civil Aviation Group on the Southwestern Front, she flew the R-5, usually at a low level and without an escort, delivering ammunition, equipment, food, and medications; evacuating the wounded from enemy-held areas; transporting members of the General Staff; and on reconnaissance missions.

After joining the 125th, she took part in that regiment's first operational mission on 28 January 1943, over Stalingrad; the two squadron commanders and their crews flew twice that day in formation with an experienced men's squadron. She led a flight into battle on her second sortie. Her squadron flew in excess of 500 missions (1943–1945) in support of Soviet ground troops on the southern, Don, North Caucasus, Western, 3rd Belorussian, and 1st Baltic Fronts, breaking through antiaircraft fire and repelling fierce enemy fighter attacks.

Fedutenko frequently acted as deputy divisional leader on operational missions. For example, on 2 September 1943 her divisional leader was shot down and Fedutenko replaced him. With her navigator Zubkova, she led fifty-four aircraft to a target, a fortified area near El'nia, enabling friendly troops in this sector to go over to the offensive.

In 1946 Fedutenko was demobilized and transferred to the reserves. In addition to the **Hero of the Soviet Union** awarded to her on 18 August 1945, she received the Order of Lenin, the **Order of the Red Banner** (twice), the Order of Patriotic War I Class, and many medals.

Bibliography: Anishchenkov and Shurinov, *Tret'ia vozdushnaia,* 1984; Bashkirov and Semenkevich, "Geroini Sovetskogo neba," 1969; Cottam, *Women in War and Resistance,* 1998; Ol'khovskaia, "Otvazhnaia eskadril'ia," 1971; Shamiakin, *Navechno v serdtse narodnom,* 1984; Toropov, *Geroini,* 1969.

—Kazimiera J. Cottam

FEITOSA, JOVITA ALVES (born 8 March 1848, Brejo Seco, Ceará, Brazil; died 1867?). Second sergeant, Second Voluntários da Pátria Corps of Piauí, Brazil, War of the Triple Alliance (Paraguayan War) (1864–1870).

After the Paraguayan invasion of Brazil in late 1864, a wave of patriotic indignation swept Brazil. Among the thousands who volunteered to defend their country was Jovita Alves Feitosa, who at the age of seventeen headed for Teresina, the provincial capital of Piauí. Cutting her hair, she enlisted in July 1865 in a volunteer battalion. Pierced ears quickly gave her away as a woman, and the police investigated her.

Overnight she became a local celebrity. A newspaper in Teresina published the transcript of her police interrogation and her portrait; provincial authorities eventually permitted her to remain in the battalion. She embarked with the rest of the provincial contingent for the national capital of Rio de Janeiro. Her fame increased in every port where the troop transport stopped en route. In Salvador, Bahia, she was feted by the population and provincial authorities, and a newspaper reporter declared that her patriotic example had encouraged hundreds of men to volunteer.

Jovita Alves Feitosa, 1865. (Perry-Castañeda Library)

Despite her celebrity, Feitosa's military career came to an abrupt end on 16 September 1865 in Rio de Janeiro when the War Ministry ruled that because women could not serve in any branch of the armed forces, she should be discharged and returned home. Instead of returning home, however, Feitosa remained in Rio de Janeiro, committing suicide over a failed love affair two years later. While Feitosa did not see combat (unlike the many **Paraguayan women** who were caught up in the later defense of their country), her example, like that of **Maria Quitéria,** reveals the extent to which wars opened up possibilities for women to break out of traditional roles.

Bibliography: "Carta Particular," Bahia, 5 September 1865, *Jornal do Commercio* (Rio de Janeiro) 10 September 1865; Lamédica, Gabriela Mirande, "La fotografia en los tiempos de guerra: La historia de la Sargento Jovita Alves Feitosa durante la Guerra del Paraguay," *Memoria del V Congreso de Historia de la Fotografia* (Buenos Aires: Comite Ejecutivo Pernamente, 1997), 57–59; Um Fluminense, *Traços biographicos da heroina brasileira Jovita Alves Feitosa, ex-sargento do 2.o Corpo de Voluntários do Piauhy, natural do Ceará* (Rio de Janeiro: Typ. Imparcial de Brito & Irmão, 1865).

—Hendrik Kraay

FIGUEUR, THÉRÈSE (born 1774, Talmay, France; died 1861), also known as Madame Sans-Gêne (meaning "carefree" or "without worries") because of her intrepidness. Dragoon from 1793 until 1815 in the French revolutionary and Napoleonic armies.

Stories vary about why Thérèse Figueur joined the military. Some say she was abandoned by a lover, causing her to renounce her sex and join the military; others claim she was trained by an uncle in the skills of a soldier. Figueur did, in fact, don a male disguise and join the Légion des Allobroges in 1793 (later the 15th Regiment of Dragoons), where she continued to serve until 1795. During that time she met Napoleon Bonaparte at Toulon and served in the Army of the Pyrenees as well.

In spite of the law of 30 December 1793 that proscribed women from service as soldiers (see **French Revolution and Napoleonic Era, women soldiers in**), Figueur remained in the military. Such a position was uncommon, but not impossible, if officers and peers allowed her to continue her service in disguise. From 1795 to 1799 she served in the Army of Italy, where she was wounded four times, had horses shot out from under her twice, and was made a prisoner of war. During examination of the injuries she had received prior to her being captured, her sex was detected by the enemy. She was returned to the French, who pensioned her from the military in January 1800. Her pension, based on valorous service and combat-related injuries, amounted to 200 francs per year.

Returning to the military later in 1800, again in disguise, Figueur served in the 9th Dragoons in Austria and Prussia. In 1809 she campaigned in Spain, where three years later she was captured by guerrillas and made a prisoner of war until she was freed after Napoleon's abdication in 1815. In 1831, after repeated petitioning, Figueur was granted an additional pension of 100 francs based on her former husband's service record. By that time she had become the widow of Clément-Joseph-Melchior Sutter, a soldier like herself, who had died in 1823.

In 1839, feeling the long-term results of her injuries and without sufficient resources to subsist, Figueur petitioned for admission to the Hospice des ménages in Paris. Her requests were granted because of her inability to care for herself and because she was denied admission to Les Invalides, the national veterans' hospital (a denial based solely on account of her sex). For the first several years of her confinement at the Hospice, Figueur dictated her lengthy memoirs, which were published in 1842. Among the reasons for publishing her memoirs was Figueur's desire to claim her military nickname of "Sans-Gêne," which had, in other circles, been given to the former laundress and wife of Napoleonic general François-Joseph Lefebvre, whose outspokenness was widely known.

Although Figueur was never granted the Legion of Honor, which she coveted, her distinguished service was attested by five Napoleonic generals: Lannes, Augereaux, Noguez, Lemoine, and Laborde. She was finally granted the Medaille de Sainte-Hélène for her service, which, in Augereaux's words, "had been above and beyond the courage expected of her sex."

Bibliography: Cère, "Mme Sans-Gêne," 1893, and *Mme Sans-Gêne*, 1894; Dossier XR48, "Figueur," Service historique de l'Armée, Vincennes, France; Leduc, *Les campagnes de Mlle Thérèse Figueur*, 1842.

—S.P. Conner

FOMICHEVA-LEVASHOVA, KLAVDIIA IAKOVLEVNA (born 25 December 1917, Moscow; died 5 October 1958, Moscow, USSR), also known as Klava. Guards lieutenant-colonel, Red Army Air Force. Squadron commander, Soviet Union, Second World War.

Klavdiia Fomicheva was considered both a valued colleague and a talented flying instructor, with outstanding leadership skills. Calm, composed, quick to react, patient, and persistent, she quickly gained everyone's respect. While training in **Marina Raskova**'s **122nd Aviation Group,** she had wanted to become a fighter pilot, but Raskova persuaded her that she had the necessary qualifications to become a competent bomber-group commander. Initially flight and deputy squadron commander of the 587th Bomber Aviation Regiment (redesignated the **125th "M.M. Raskova"** *Borisovsky* **Guards Bomber Aviation Regiment**), Fomicheva replaced Evgenia Timofeeva as the second squadron commander when the latter became the deputy of the regimental commander V.V. Markov.

After graduating from a primary school, Fomicheva became a bookkeeper apprentice at the Moscow Regional Office of the State Bank. In 1935 she enrolled in a gliding course provided by her employer. As one of the best glider pilots in her group, she was given the opportunity to learn to fly aircraft in the evening. In 1938 she became a full-time instructor at the Reutovsky Flying Club of the Moscow Region.

On her first two combat missions in January 1943, near Stalingrad, the most crucial sector of the Soviet-German front at the time, Fomicheva flew with an experienced male squadron. Afterward she operated independently. Mindful of the basic tasks of a bomber-group leader, such as skillful evasive actions, appropriate target approach in formation, flying absolutely straight on a bomb run, and clearing targets correctly, she assisted her squadron commander Timofeeva in thoroughly training their subordinates. Consequently, not a single aircrew of their squadron was lost in the fierce fighting over the Kuban' area of North Caucasus. During the famous battle of 2 June 1943 Fomicheva's squadron, abandoned by its fighter escort, engaged eight enemy fighters and shot down four. Therefore, the squadron's performance was held up by the air force commander in chief as an example for all of Soviet bomber aviation to follow.

During the stubborn fighting near Smolensk, at which time the squadron virtually doubled its effectiveness by doubling its rated bomb load, Fomicheva continued improving her subordinates' performance. However, she was shot down on 17 September 1943 near El'nia, at which time her gunner was killed. Seriously injured, Fomicheva was put out of action for several months. Shortly after her return, early in 1944, she assumed command of her squadron. She was shot down again on 23 June 1944 at the beginning of the Belorussian offensive. Again, her gunner was killed. However, Fomicheva and her new navigator Galina Dzhunkovskaia were operational by 28 August 1944. She especially distinguished herself, both as to leadership qualities and outstanding airmanship, in the difficult, stubborn fighting in the Baltic area and East Prussia. Altogether Fomicheva flew fifty-five operational missions and was credited with shooting down eleven enemy aircraft in group combat.

Awarded the **Hero of the Soviet Union** on 18 August 1945, Fomicheva was also presented with the Order of Lenin, the **Order of the Red Banner** (twice), the Order of the Red Star, and many medals. She took part in the Victory Parade on Moscow's Red Square on 24 June 1945. Afterwards she was employed for four years as a flying instructor at the Air Force Academy. Grounded due to her deteriorating health, she taught air tactics at the Military Aviation School for Pilots in Borisoglebsk. She died prematurely from an illness related to her wartime injuries and was buried at Moscow's Novodevich'e Cemetery. Here a tiny aircraft soars above the white marble obelisk. Underneath there is an epitaph, an excerpt from M. Gorky's "The Falcon Song": "Though you have died, you will live forever as an inspiration to those who are bold and strong in spirit!"

Bibliography: Cottam, *Women in War and Resistance,* 1998; Markova, "Youth under Fire," 1983; Shamiakin, *Navechno v serdtse narodnom,* 1984; Shkadov, *Geroi Sovetskogo Soiuza,* 1988; Sinitsyn, "Put' muzhestva i otvagi," 1983; Toropov, *Geroini,* 1969.

—Kazimiera J. Cottam

FOOTE, EVELYN PATRICIA (born 19 May 1930, Durham, North Carolina, United States). The second woman (after General **Mary Clarke**) to command a United States Army post. Retired army brigadier general.

Brigadier General "Pat" Foote has been a strong supporter of opening all military roles to women, including combat roles, and has spoken out on issues ranging from supporting the deployment of mothers during the Persian Gulf War to educating the public about the important role of servicewomen. She served during a time of great change for American military women. When Foote was commissioned in 1960, the highest rank women could attain was lieutenant colonel. As she once said, "The Army has come a million miles in opening the opportunities for women."

Foote earned a B.A. in 1953 from Wake Forest University and an M.S. degree in government and public affairs from Shippensburg State University in 1977. She entered the army in 1960, advancing through the ranks to brigadier general in 1986. Foote served in a variety of positions throughout her army career, including platoon officer, **WAC**, at Fort McClellan, Alabama; public affairs officer during a tour of Vietnam in 1967; personnel management officer; faculty member at the United States Army War College; and from 1983 to 1985, commander of the 42nd Military Police Group in Mannheim, Germany. In October 1988 she became the commanding general of Fort Belvoir, Virginia, the first woman to command that army post. At Fort Belvoir General Foote supervised the transition of the post from the Army Engineering School to a Military District of Washington post. When she retired from the army on 1 September 1989, she was the only woman in command of an army post.

Foote was recalled to active duty in November 1996 to serve as the second-ranking member of a panel convened by Army Secretary Togo D. West, Jr., to review policies on sexual harassment and make recommendations for ending it in the army. She retired again in October 1997. She spoke to the House National Security Military Personnel Subcommittee, making a strong statement in support of gender-integrated training: "I do not agree . . . that same sex platoons should be organized within gender integrated units. Nei-

---

Statement of Brigadier General Evelyn "Pat" Foote, United States Army (retired), 17 March 1998, before the House National Security Military Personnel Subcommittee concerning the report of the Federal Advisory Committee on Gender Integrated Training and Related Issues:

My experience as a basic training battalion commander reaffirmed the unquestionable logic of training men and women together who subsequently will serve together in combat support and combat service support units worldwide. By integrating men and women within platoons down to the squad level, these soldiers learned the value of teamwork early and learned to appreciate the skills each soldier contributed to the successful accomplishment of group tasks. A natural byproduct of the pride soldiers feel when they get the job done is mutual respect of one another regardless of gender. Cohesive squads are the essential building blocks of cohesive platoons. . . .

When the decision was made at the Department of the Army to end gender integrated basic combat training in 1980 or 1981, I was both shocked and dismayed. My experience and the experience of numerous fellow training battalion commanders had been very favorable. We believed that we were producing better trained soldiers who had a far greater appreciation for the value of each soldier, both male and female, brought to his or her unit. . . . Many valuable training years were lost before gender integrated training was reintroduced into the Army in 1994. . . .

Gender integrated training, unconstrained by artificial barriers, enhances military readiness and produces soldiers who are highly motivated team members, equally trained and ready to serve the nation as they have sworn to do.

—http://www.house.gov/hasc/testimony/105thcongress/3-17-98foote.htm.

ther do I agree that separate barracks for male and female recruits are essential. Both these Kassebaum-Baker recommendations seem to infer sexual harassment is a byproduct of basic training. My experience as the commander of a basic training battalion provided no evidence that the incidents of sexual harassment with which I dealt had any cause and effect connection with the fact that male and female recruits trained together at the platoon level and shared separate, secure living space in the same barracks."

In March 1998 Foote spoke at the Pentagon during the Department of the Army's observance of Women's History Month, stating, "In 1960 when I was commissioned in the Women's Army Corps, my job was described to me not in terms of its opportunities, but by its limitations. There were many jobs I could not perform. I could not command men, I could never aspire above the rank of lieutenant colonel and an enlisted woman could not aspire above the rank of E-7 or sergeant first class. I have been able to ride on the razor blade of incredible social change, and seen what is impossible today become possible tomorrow. . . . I just happened to be the right person at the right place and I stayed the course." General Foote's awards include the Distinguished Service Medal, Legion of Merit with oakleaf cluster, and the Bronze Star.

Bibliography: Army News Service 23 March 1998; Foote, Evelyn P., "War Is No Time to Make Changes," *Washington Post* 19 February 1991: 17A; Holm, *Women in the Military*, 1992.

—Vicki L. Friedl

Female vanguard marching on Versailles. (Citizen's army and women march on Versailles to protest against King Louis XVI.) French engraving of October 5, 1789. (The Art Archive/Musée Carnavalet Paris/Dagli Orti)

FRENCH REVOLUTION AND NAPOLEONIC ERA, WOMEN SOLDIERS IN (1792–1815). France and French military campaigns throughout Europe.

The French revolutionary era (1789–1799) was characterized by dramatic change in the government of France as the Old Regime monarchy of Louis XVI was swept away and replaced by the First French Republic in 1792. In the course of governmental change, the French privilege system was dismantled, law codes and the courts were altered significantly, civil war broke out, international war was declared, and gender roles were debated and ultimately defined with a rigor that had never before characterized French law. Prior to the restrictions against women that were legislated beginning in

1793, women formed political clubs, petitioned to participate fully as French citizens, marched in street demonstrations, entered the military in both supply and combat roles, and were recognized and honored as women soldiers by the French government in the *Annales du civisme*.

The French had long debated women's roles; in fact, from the fifteenth century on, the vitriolic literary debate called the *querelle des femmes* (debate over the woman question) raged in France. Women's actual roles, however, throughout prerevolutionary French history were less defined by sex than by privilege or lack of privilege. Women were defined by their relationship to a male member of the family, by their relationship to the Roman Catholic Church (e.g., as a nun or abbess), or by regional legal variations concerning property holding or inheritance (e.g., the legal status of *femme couverte* or *femme sole*). In the eighteenth century, which was also the century of the Marquis de Sade, Casanova, Defoe's Moll Flanders, "molly clubs" (homosexual social clubs) in Paris and London, and what some contemporaries charged was a more liberal and/or "libertine" environment, there was a serious and fundamental redefinition of gender roles. According to historians like Thomas Laqueur, there was a certain sexual ambivalence, or a gender blurring, that defined the century.

Women soldiers, albeit most often in disguise, served in the royal armies of Europe, often going to substantial trouble to feign maleness (using a tube or narrow funnel to appear to be male when urinating). On campaign, for example, at the time of the French and Indian War and its European counterpart (1754–1763), women were so ubiquitous in supply and in the army train that as the battlefront shifted, they were frequently engaged in hostile action as well. The heralded and widely reported case of the Chevalier d'Eon de Beaumont, who declared himself to have been a woman soldier in the service of the king (although a later autopsy disproved his allegations), further confirms the shifting gender boundaries of the eighteenth century even at the highest levels of society. It was in this context that women of the French revolutionary era adopted certain roles, including militancy and the military, that were later disparaged and then denied to them.

The roles of common women in the French revolutionary era can be placed

---

"Common wisdom" about women's military roles:

Warfare is . . . the one human activity from which women, with the most insignificant exceptions, have always and everywhere stood apart. . . . Women . . . do not fight . . . and they never, in any military sense, fight men.

—John Keegan, *A History of Warfare* (New York: Knopf, 1993), 76.

Women have never taken a major part in combat—in any culture, in any country, in any period of history.

—Martin Van Creveld, "Women of Valor: Why Israel Doesn't Send Women into Combat," *Policy Review* (Fall 1991): 65–67.

Almost universally war is an all-male enterprise.

—Lionel Tiger, *Men in Groups* (London: Nelson, 1969), 80.

Women's unsuitability for combat is made apparent by the fact that they have never engaged in it.

—Jeff M. Tuten, "The Argument against Female Combatants," *Female Soldiers—Combatants or Noncombatants? Historical and Contemporary Perspectives*, ed. Nancy Loring Goldman (Westport, CT: Greenwood Press, 1982), 239.

in a number of categories: (1) unprivileged women who remained engaged in traditional labor, for example, secondhand dealers, workers in service employment, textile workers, farm laborers, "fishwives," and "market women," (2) women whose traditional labor was supplemented by prepolitical and political activity, for example, women who marched to Versailles in 1789 to demand that the king return to Paris to deal with the nation's economic scarcity, (3) women who formed political associations and petitioned for citizenship rights, including the right to form "battalions of amazons," and (4) women who actually served in the military, either in supply or as soldiers in combat situations. Most studies of women during the French Revolution in recent years have dealt with the first three categories rather than with women soldiers. In the premodern world of the eighteenth century armed actions in the streets (forms of militancy) were often conflated with military combat.

During the first several years of the French revolutionary decade a number of women worked on negotiating the boundaries of citizenship. These are women who could be placed in categories (2) and (3). **Théroigne de Méricourt,** for example, a talented peasant girl from Austrian Luxembourg, arrived in Paris at the moment of the first revolts of spring 1789. By October, instead of being a spectator to the events of the French Revolution, she had become a participant and leader in the Women's March to Versailles. Widely reported by witnesses, Méricourt dressed as an "amazon" in a green riding habit, plumed hat, and occasional weapon at her side, recruited women to the cause of the revolution in a short-lived political club that was housed at her dwelling, frequented the galleries of the revolutionary clubs and public assemblies, demanded that women be allowed to participate in any manner of revolutionary activities, and died insane in the women's prison in Paris in 1817, having been placed there in 1794 by her brother. The widely known image of Théroigne de Méricourt mounted on a cannon in the Women's March of 1789 is among the reasons for her frequent inclusion in works on women soldiers, although "armed militant" would be a better category for her actions.

Similar women included Olympe de Gouges, Claire (Rose) Lacombe, and Pauline Léon. Olympe de Gouges, who was born Marie Gouze, was the daughter of a trinket peddler from Montauban, France. A "grub-street" literary figure in Paris, Gouges is best known for her 1791 political pamphlet *Les droits de la femme et de la citoyenne* [The rights of woman and of the citizen]. The pamphlet, radical in its intent, proposed prenuptial agreements for women and demanded that if capital punishment was to be inflicted upon women, then at least they deserved the right to participate in creating the laws that disciplined and punished them. In her pamphlet she had said, "A woman has the right to mount the scaffold; she must also have the right to mount the tribune." A quixotic champion of the rights of women and a militant in her actions, fundamentally Gouges was a playwright, constitutional monarchist, and journalist.

Like Gouges, Lacombe and Léon were clubwomen, organizers, and militants who fomented for women's equal treatment and who demanded that women have the right to fight for their country. Founders of the *Société des citoyennes républicaines révolutionnaires* (May 1793), Lacombe and Léon had already participated in the significant events of the French Revolution prior to establishing this first women's political association. Léon, in particular, had organized street demonstrations as early as 1789, had organized an assault on a Parisian journalist's home in 1791, and later that year had taken a petition signed by 300 women to the National Assembly to allow women to form militias, to train, and to march to the front should war be declared. Frequently armed in street demonstrations, Lacombe and Léon steered their women's group in 1793 into the camp of the Jacobins who controlled the government of France. Later that year they discovered that the Jacobins had a different agenda when it came to women.

At the same time that some French women were continuing their working-class lives while others were petitioning and demonstrating in the streets of Paris (sometimes violently), other women had already joined the armies of France in a number of capacities. In the southwest and west of France a number of women took part in the civil war that had been prompted by the nationalization of Catholic Church lands and the requirement that clergy become servants of the state. Among the best-known women of the French counterrevolution were **Rénee Bordereau** and **Marie-Louise Victorine de La Rochejaquelein,** whose combat is well documented. Other women who served in the counterrevolutionary contest and whose exploits are known from the memoirs of other combatants include Jeanne Robin, Madame Bulkeley, Madame Regrenil, Madame de Fief, and the chevalier "Adam."

The international conflict that began with France's declaration of war against Austria in April 1792 brought other women into the French revolutionary armies. Although documentation is limited concerning their experiences, the records of at least seventy-eight women soldiers can be found in

Olympe de Gouges giving her patriotic remarks to Louis XVI, September 1788. Engraving of Frussolte. (Perry-Castañeda Library)

various locations, including dossiers at the French Army Archives (Vincennes, France), the *Annales du civisme* (an official French revolutionary listing of heroic deeds and awards), memoirs, and local and regional records. The extent of the international war provided an opportunity for French women to engage themselves in the contest. As women like Claire Lacombe offered the French Assembly in Paris the "strength of her arms and her courage to massacre the tyrants," the government relaxed laws concerning enlistment, basic training, and minimum height and age. Women did not need to disguise themselves; the French National Assembly, when questioned concerning the enrollment of women in the military, noted that no law precluded women's service. The volunteers of 1792 were typically young, rural, and smaller than former recruits, and they came from nearly everywhere, serving reasonably short tours of duty.

According to existing records, the first twenty-six women who joined the French revolutionary armies during the revolutionary era served, in general, about a year to eighteen months, came from families with few resources, and were as likely to be married as not. Their positions in the military varied as greatly as their personal circumstances: seven served in the infantry, three in heavy artillery, three were riflemen, and one was a grenadier. Additionally, one served in transport, one as a trumpeter, one in military police, three as aides-de-camp, and one as a sailor. The remaining women who joined the French revolutionary armies after 1792 followed a similar pattern, being predominantly provincial by birth and engaging in a variety of military occupations, while being slightly older than the first recruits and more likely to be married.

Among the various motivations for military service, according to army dossiers, were the following: Geneviève Prothais joined the military when men "ignored the call to arms"; **Sophie Julien** said that she had been "born with male traits" and the military served her needs; **Marie Angélique Duchemin** found military service more satisfying than civilian service; Claudine Rouget fled from her home on the day of her marriage to an "old man"; and scores of others cited family and economic incentives as primary. The military during the French revolutionary period provided security of rations, potentially stable employment, and support for one's family without the necessity of cross-dressing to hide one's sex. Among the other women who served during the early years of the Revolution were **Rose Barreau, Rose Bouillon, Reine Chapuy, Marie Charpentier, Thérèse Figueur, Virginie Ghesquière, Marie-Barbe Parent, Madeleine Petitjean, Anne Quatsault,** and **Marie-Henriette Xaintrailles.**

The environment for women's service to France had been reasonably open, if not heralded, in 1792 and until the fall of 1793. Then the Jacobins came to dominate French politics and took a very different position. Police informants reported that women joined the military only to receive incentive pay before they deserted, that they trafficked in the sale of military uniforms, that they were a serious "contagion" that emasculated military men, and that they were particularly unsuited for positions of responsibility. These "reports" were consistent with the Jacobins' reactionary position against equal rights and citizenship for women. To the Jacobins (the party of Maximilien Robespierre and the Committee of Public Safety), the army undergirded the state. As the new nation was viewed, only a "virile" state could prosper, one in which active citizens were clearly defined as masculine.

In April 1793 the first significant piece of revolutionary legislation proscribing women's roles was debated and passed; it designated women who traveled with the armies as either "useful" or "useless." All women, including the wives of officers, women in supply beyond the acceptable limit of four per battalion, prostitutes, and women soldiers, were given passes and a small sum of money to send them home. Even after the legislation was ratified, however, evidence shows that women soldiers serving on the front distant from Paris remained reasonably unaffected by the new directives. By the fall of 1793, with the widely reported executions of the queen, salon women, and women like Olympe de Gouges "who wanted to be men," the environment for women soldiers turned notably hostile. In the Convention in Paris Pierre-Gaspard Chaumette, a member of the municipal government of Paris assailed women's visible roles in revolutionary activities. In particular, Chaumette noted, "Should women lead our armies? If there was a Joan of Arc, it was only because there was a Charles VII; if the destiny of France was placed in the hands of a woman, it was because the king did not have the head of a man and because his subjects were less than nothing" (*Le Moniteur*, 17 November 1793). A final decree of 30 December 1793 demanded that all "useless" women be excluded from the army and from military encampments. Stringent measures were designed to require appropriate adherence to the provisions of the law: Generals, for example, could be court-martialed for failure to expel women soldiers.

In an environment in which Jacobin support for hearth, home, and children left little to be debated, most women in the military were summarily drummed out of the armies. Women remained in the military only if they served in disguise or if their regiments chose to retain them while being fully aware of the possibility of disciplinary action. According to extant military dossiers, more than 80 percent of the discharges of women in 1793 were on account of their sex. Often, regardless of valorous service or intercession by officers and male compatriots, women had no recourse to allow them continued service in the military. The Jacobin repression of women, however, was only the first step in legislative action. By 1795 the Directory government of France restricted the collective action of individuals and women's employment opportunities. By the time of Napoleon's regime women were further relegated to the "private sphere." Not only did Napoleon demand that women be severely restricted to the control of their husbands, but he oversaw the creation of the Code Napoléon, which not only denied them any significant employment opportunities (including the military) but consigned them to second-class citizenship, severely restricted divorce rights, limited guardianship of their children, restricted ownership of property, and required dependence on a male in circumstances of litigation.

While revolutionary legislation was successful in expelling most women from the French revolutionary armies, a few women managed to remain with their companies, including Thérèse Figueur, who served extensively in the Napoleonic contest in Spain and Portugal and published her memoirs during her later years in a convalescent home, Angélique Duchemin, who served in seven campaigns, Marie Xaintrailles, who left the military during the Napoleonic contest and who reminded government bureaucrats that she had gone to war as a soldier, not as a woman, and **Marie-Jeanne Schellinck**. One additional Napoleonic soldier, **Regula Engle,** is known only by her memoirs, which were published in 1821.

Regardless of the presence of a few women in the military, according to army sources, Napoleon was insistent that absolute order had to be maintained and that intractable rules had to be observed in the professional army that he was constructing. Among those rules was the exclusion of women from the military service. While generals' wives might follow their husbands in the army train and while authorized women could still serve in supply, there was no room for women soldiers. The Code Napoléon carefully defined a "private sphere" for women, a sphere in which no transgressions from the "natural" roles of women would be allowed. After the Napoleonic era ended, in which the coveted civilian and military medal of the Legion of Honor was created, thirty-six years passed before the first woman soldier of the armies of the French Revolution and Napoleonic period received the coveted award. Angélique Duchemin, soldier and prisoner of war, received her cross of the Legion of Honor not from Napoleon I, but from his nephew Napoleon III during the Second Empire of France.

Bibliography: Bertaud, *La vie quotidienne,* 1985; Brice, *La femme et les armées,* n.d.; Cère, *Mme Sans-Gêne,* 1894; Dekker and van de Pol, "Republican Heroines," 1989, and *The Tradition of Female Transvestism in Early Modern Europe,* 1989; France, Service historique de l'Armée de Terre (Vincennes, France), cartons XR48 and XR49 (Dossiers "Alibert" to "Xaintrailles"); Hacker, "Women and Military Institutions," 1981; Kates, *Monsieur d'Eon Is a Woman,* 1995, and "The Transgendered World of the Chevalier/Chevaliere d'Eon," 1995; Laqueur, *Making Sex,* 1990; Tranchant and Ladimir, *Les femmes militaires de la France,* 1866; Yalom, *Blood Sisters,* 1993.

—S.P. Conner

G

GAO, first name unknown (ca. 695, China). District Mistress of Loyalty-Displaying. Military leader during the Tang dynasty (618–907).

When Mochuo, the khan of Tujue, attacked Feihu District (now Laiyuan in Hebei Province), Gao, whose husband Gu Xuanying was the district magistrate, led the successful defense. After a considerable time the siege was unresolved, and Mochuo left. Wu Zetian (r. 684–704) sent a decree to honor Gao, stating: "A while back Mochuo attacked the district, and everyone was worried that the district would be captured by the enemies. Under such a situation, even with men guarding the place firmly, it might not hold out. Yet this woman, with loyalty in mind, not afraid of flying arrows, brought peace back to an endangered place, and won people's gratitude. If she is not commended and promoted, how can we encourage others to do the same? The wife of Gu Xuanying is given the honorary title of District Mistress of Loyalty-Displaying."

Interestingly, Wu Zetian reigned as empress after her husband, Emperor Gaozong (650–683) died in 683. In 690 she changed the dynasty title to Zhou, styling herself "Emperor Holy Almighty," and became the only female emperor (as opposed to empress) in Chinese history. (For a similar case, see the entry on the female pharaoh **Hatchepsut**.) Thus Gao was honored for her unusual military achievement by a woman of unusual political achievement.

Bibliography: Liu Xu, *Jiu Tang shu*, 1975; Ou-yang Xiu, *Xin Tang shu*, 1975.

—Sherry J. Mou

GARCÍA MORALES DE CARRERA, PETRONA (born 1817, Mataquescuintla, Guatemala; died 17 August 1857), also known as Toña. Revolutionary fighter, Guatemalan Revolution, 1837.

Petrona García Morales played a prominent role in the Guatemalan Revolution of 1837 as the wife and comrade in arms of the caudillo Rafael Carrera (1814–1865). The parish priest in Mataquescuintla, Francisco Aqueche, had befriended Carrera, an itinerant veteran of the Central American Civil War of 1827–1829, and arranged his marriage with "Toña" in early 1836. The daughter of a respectable rancher in this poor rural district, Petrona raised Carrera's status in the community. Known for being as devout and prudent as she was beautiful, Toña García was also haughty, strong-willed, and jealous. When Carrera assumed leadership of the successful peasant uprising against the liberal government of Mariano Gálvez in 1837, Petrona accompanied her husband into battle. A French observer, Alfred de Valois, related that Madame Carrera matched her husband in courage and daring, able to handle pistol or lance effectively. He said that she cut off the noses and ears of her husband's mistresses and boasted of having mutilated many of Rafael's lovers. The U.S. chargé d'affaires in Guatemala, Charles DeWitt, reported Toña's brutality toward other women in a different context. In a letter of 24 October 1838 he related that after one battle she had "amputated the ears and breasts of several soldiers' wives who had been taken prisoner with their husbands, and thus mutilated sent them hither."

These atrocities occurred in a revolution characterized by barbarity. When government soldiers entered Mataquescuintla in June 1837, they raped Petrona and destroyed the houses of Carrera and of Petrona's family. In 1838 the Central American federal president, Francisco Morazán, took the town, ordered her father shot, and then commanded that his head be cut off, fried in a bucket of oil, and displayed on the corner of his house as a warning to all rebels, despite the fact that the elder García had not participated in the revolt.

When her husband became chief of state of Guatemala (1844–1848 and 1851–1865), Petrona was an important asset to his political career as first lady and hostess at frequent dinners and celebrations. She bore him five children and astutely managed Carrera's assets as they amassed considerable wealth. Petrona became a victim of the cholera epidemic of 1857, which Carrera's troops had brought back from Nicaragua after defeating the filibuster William Walker; she was buried in a crypt in the Cathedral of Guatemala.

Bibliography: Carrera, *Memorias*, 1979; Granados, *Memorias*, 1877–1893; Woodward, *Rafael Carrera*, 1993.

—Ralph Lee Woodward, Jr.

GHESQUIÈRE, VIRGINIE (born ? near Lille in northern France; died 1855 in France), also known as *joli sergeant* (pretty sergeant). *Voltigeur* (skirmisher) who served in disguise in the Napoleonic armies, 1806–1812.

In 1806 Virginie Ghesquière assumed a male disguise and duped enlistment officers into believing that she was her brother (whose name remains unknown), whose health was too fragile to allow him to serve in the military in spite of his conscription. Allegedly Ghesquière's likeness to her brother was uncanny. She served in the 27th Light Cavalry, where she remained undetected as a female until she was injured in 1812. According to various stories, she saved her captain at Wagram during the Napoleonic Wars and fought valorously under the command of General Andoche Junot in Spain in 1808. During the Peninsular War she moved forward to save her wounded colonel during a particularly bloody engagement. In the fray two Spanish officers attempted to stop her, but she managed to take one of them prisoner. Reportedly she was jokingly nicknamed the *joli sergent* (pretty sergeant) because of her facial features; even though her male disguise was accepted, she was seen as a "pretty boy."

Other records and contemporary stories noted her somewhat cavalier attitude toward the military. Allegedly Ghesquière had deserted the French army briefly in 1806. She had then been granted amnesty to return to her post, only to desert again in 1810. She then received an additional amnesty that allowed her to remain in the military until her discharge in 1812, when her disguise was detected and her commander discharged her according to French law.

Although some sources note that she received the Legion of Honor, the highest French decoration, such was not the case. Her closest claim to fame was her popularization in a song written by a man named Cadot. The song was then printed in the *Journal de l'Empire* in October 1812. The verse, in its refrain, granted "the star of honor" to "Virginie, in spite of her modesty but in the name of the Emperor."

Bibliography: Brice, *La femme et les armées,* n.d.; Cère, *Mme Sans-Gêne,* 1894; La Bédollière, *Beautés des victoires,* 1839; Tranchant and Ladimir, *Les femmes militaires de la France,* 1866.

—S.P. Conner

GIDINILD (fl. 1026), also known as Guidinild, Guidinhild. Castellan of Cervera, Spain.

While very little is known about Gidinild herself, it is nevertheless certain that she played a direct role in the capture and subsequent defense of the town of Cervera. Having taken the town, Gidinild built a tower or castle there and was confirmed as its castellan by **Ermessend,** countess of Barcelona, in 1026. Gidinild's career provides a good indication of the prominent military and political roles that some women were able to fulfill in eleventh-century Spain and southern France.

Bibliography: Bonnassie, *La Catalogne,* 1975–1976; McLaughlin, "The Woman Warrior," 1990; Shahar, *The Fourth Estate,* forthcoming.

—David Hay

GIERCZAK, EMILIA (born 1925, Poland; died 17 March 1945, Kolobrzeg, Pomerania, Poland), also known as Elka. Second lieutenant, Polish army formed in the Soviet Union. Platoon commander, 10th Infantry Regiment, Poland, Second World War.

Polish patriotism was an important element in Emilia Gierczak's upbringing. It was assumed that Gierczak's parents (Józef and Leontyna) had named her "Emilia" because the name had been made famous by the legendary **Emilia Plater**. Like other Polish women officers, Gierczak was trained at the Infantry Officers' School in Riazan', Soviet Union. Her ambition was to study medicine after the war had ended.

At first Gierczak served in the famous **"Emilia Plater" Independent Women's Battalion,** initially subordinated to the Polish "Tadeusz Kosciuszko" 1st Division created in the Soviet Union in 1943. She was among battalion soldiers trained as line commanders of male companies and platoons in the various branches of service and front-line units of the new Polish army. Women were trained for these roles due to the drastic shortage of Polish male officers (in part caused by the massacres of Polish officers at Katyn and in Soviet POW camps at Kozel'sk, Starobel'sk, and Ostashkov), as well as the unsuitability for officer training of many of the Polish army's male recruits and the availability of qualified female volunteers.

Gierczak distinguished herself during the fighting on the Pomeranian Rampart and fell in combat while leading an assault group. An eyewitness account of her death on 17 March 1945 states that Gierczak "was charged with a difficult task—to capture a stoutly defended building. She could have waited until it got dark to do this, but she didn't want anyone to think that she was afraid. . . . She was never to learn that the mission was accomplished, for she was hit in the forehead as she moved up to embolden her men, who had earlier gone to ground, by stirring them to action by her personal example. So the devil himself took possession of the platoon! Henceforward, no obstacle could deter the men from avenging the death of their woman-commander!"

Gierczak's memorabilia, including her last letter to her mother, are on display in the Military Museum in Kolobrzeg, a seaside resort and port on the Baltic formerly known by its German name Kolberg. A street in Kolobrzeg and a number of schools and Polish scouting organizations have been named after her. However, the otherwise comprehensive *Ludowe Wojsko Polskie* (Polish People's Army) published in Warsaw in 1973 does not even mention her.

Bibliography: Cottam, K. Jean, "Veterans of Polish Women's Combat Battalion Hold a Reunion," *Minerva* 4 (1986): 1–7; articles from *Żołnierz Polski* (Polish soldier): Jagielski, B., "Bylo to w Kolobrzegu . . . ," 10 (11 March 1979): 14–15; Kulinski, Rajmund, "Ponad historia legenda," 10 (11 March 1979): 10; Rychlewski, Czeslaw, "Szy z nad Oki," 11 (17 March 1985): 14–15, 18–19; Unger, Piotr M., "Ludowe Wojsko Polskie 1943–1945," 4 (27 January 1980): 22.

—Kazimiera J. Cottam

GIRALDA DE LAURAC (born second half of the twelfth century, Laurac, France; died 3 May 1211, Lavaur, France), also known as Geralda, Guirande, Giraude de Lavaur, Dame de Lavaur. Defended the fortified town of Lavaur during the Albigensian Crusade, early thirteenth century.

Giralda de Laurac belonged to the noble Laurac family, a family supporting the Cathar (Albigensian) sect, a religion that the papacy and the king of France considered heretical, but that had a large following in Languedoc, the southern region of France. Her mother and three of her sisters were "perfects," particularly dedicated celibates. Giralda became lady of Lavaur through her marriage. As persecution of Cathars became common, numerous perfects, both men and women, came to take refuge at Lavaur, a heavily fortified town.

In 1209 Pope Innocent III organized a bloody crusade against the Albigensians in southern France, lasting some twenty years. During the spring of 1211 Giralda acted as castellan and shared custody of the town of Lavaur with her brother Aimery, defending against attacks by crusaders from the north led by Simon de Montfort (see **Alice de Montmorency**). Though Giralda did hold an unusually strong position as *domina,* none of the sources report that she actually participated in combat. As castellan, however, she would have played a large role in organizing Lavaur's defense. The defenders of Lavaur consisted of 80 knights, a garrison of undetermined size, and 400 inhabitants of the town. The town was besieged for several weeks between March and May 1211. The crusaders were unable to get close enough to assault the walls because the inhabitants of Lavaur had a tunnel dug into the dry fosse surrounding the town and could dig it out at night as the crusaders filled it in during the day.

After a frustrating siege and skillful defense Lavaur was successfully stormed by the crusaders. In accordance with both the current conventions of war and the fact that the inhabitants were considered heretics or friends to heretics, the crusaders executed many of the defenders after the siege, including Giralda. The crusaders tossed her down a well and threw stones down upon her until she was crushed to death. It is tempting to see Giralda de Laurac as a martyr in the cause of freedom of religion or as an example of how medieval men treated women who transcended gender roles, especially military ones. It should be noted that 400 others of both genders were burned in the aftermath of the siege, and Giralda's brother Aimery, plus the 80 knights of the garrison, were also executed, killed by the sword after their surrender.

Consequently, there is nothing particularly noteworthy about Giralda's execution other than the unusual manner of her death. The circumstances of her death were striking to her contemporaries; the story was told in the Provençal language by William of Tudela and in Latin by Peter of Les Vaux-de-Cernay and William of Puylaurens. William of Tudela writes that Giralda "was taken, shouting, weeping, screaming, and thrown sideways into a pit; she was there buried under stones to the great agitation of the crowd." The sources do not list the specific reason why the crusading army chose that particular method for her execution, other than that perhaps because she was a noblewoman it was thought that she should not be burned or cut to pieces. Chroniclers partisan to Simon de Montfort accused Giralda of being pregnant by her brother or even by her son. But for William of Tudela her death "was a pity and a crime" because she was renowned as a generous and gracious woman. An eighteenth-century fresco in the cathedral of Lavaur commemorates the execution of Giralda and her brother.

Bibliography: Marks, *Pilgrims, Heretics, and Lovers,* 1975; Sumption, *The Albigensian Crusade,* 1978; Vaux-de-Cernay, *Histoire albigeoise,* 1951; Wakefield, *Heresy, Crusade, and Inquisition in Southern France,* 1974; William of Puylaurens, *Chronique, 1203–1275,* 1976; William of Tudela and Anonymous, *The Song of the Cathar Wars,* 1996.

—Laurence W. Marvin and Marie-Thérèse Lalaguë-Guilhemsans

GRIZODUBOVA, VALENTINA STEPANOVNA (born 31 January 1910, Khar'kov, Ukraine; died 1 May 1993, Moscow, Russia). Colonel (ret.), Red Army Air Force. Regimental commander, Soviet Union, Second World War.

Valentina Grizodubova was the daughter of S.V. Grizodubov (1884–1965), a self-taught pioneer aircraft designer, and inherited his enthusiasm for flying. At fourteen she flew a glider solo. She graduated from the Penza Flying Club (1929), Khar'kov Flying School, and Advanced Flying School in Tula (1933). She owed her admission to flying training to Sergo Ordzhonikidze, an old Georgian Bolshevik and Stalin's ally. As an instructor pilot, Grizodubova trained eighty-six male pilots before the war and flew many types of aircraft. As a test pilot, she set seven world records, including an altitude record of 3,267 meters on 15 October 1937 in a seaplane, three speed records, and one distance record on 24 October 1937, when she covered about 1,500 kilometers from Moscow to Akhtiubinsk, Kazakhstan, with **Marina Raskova** as her navigator.

Grizodubova's most famous achievement was when she piloted an ANT-37 she named *Rodina* (Homeland) on 24–25 September 1938 nonstop from Moscow to the Far East. With copilot Polina Osipenko and military navigator Marina Raskova she flew a distance of 6,450 kilometers, setting a new women's distance world record. However, the flight was plagued by difficulties: communications failed early in the flight, and the crew ran out of gas short of their destination. Unable to reach the planned landing field at Komsomolsk-na-Amur, Grizodubova was forced to belly-land the *Rodina* in the middle of remote and swampy taiga country. It took ten days for rescuers to locate them. For their achievement, the aircrew were awarded the **Hero of the Soviet Union,** becoming the first Soviet women to be thus honored. In 1988, in a press interview, Grizodubova was candid in discussing the very negative aspects of her famous transcontinental flight. She said that just before the takeoff, the aircrew's ground communications expert was arrested. Their regular radio, the emergency set, and the intercom all failed in flight. She attributed her shortage of fuel to the negligence of the mechanics who apparently did not refill the tanks after extensively testing the engines on the ground. Thousands of local inhabitants searched for the missing aircrew, and sixteen airmen lost their lives when two aircraft collided during the rescue operation.

During the Second World War Grizodubova avoided service in Raskova's "women's regiments" like the **125th Guards Bomber Aviation Regiment**. Instead, she was appointed commander of the all-male 101st Long-Range Aviation Regiment, the only Soviet woman to command an all-male aviation unit. Soon after assuming command in May 1942, Grizodubova flew a successful group mission to demonstrate that its Li-2 aircraft (a modified DC-3 with a crew of six) were suitable for use as night bombers. In June 1942, ordered to fly supplies to blockaded Leningrad, Grizodubova again carried out her mission successfully. She led her regiment into battle herself. At times she flew as copilot to monitor her pilots' performance, and she was noted for flying more than male regimental commanders did. She did not brook incompetence, even in senior officers; due to her intervention, a troublesome Soviet general, her superior, was removed from his post. Though strict, she was caring and compassionate toward her subordinates.

In September 1942 Grizodubova's regiment was placed at the disposal of the central headquarters of the partisan movement. Overcoming dense enemy flak and engaging enemy fighters, her aircrews flew more than 1,850 supply missions to partisan areas and on the way back evacuated wounded partisans and homeless children. In 1943 Grizodubova resisted her superiors' orders to decrease these flights. Hence grateful leaders of the Ukrainian and Belorussian partisans supported the application of her division to grant Guards status to her regiment. Eight of Grizodubova's subordinates earned the Hero of the Soviet Union, and on 17 May 1944 her regiment was awarded the honorific of *Krasnosel'sky* for its participation in the lifting of the blockade of Leningrad. This unit was awarded the **Order of the Red Banner** on 30 August and was subsequently redesignated the 31st *Krasnosel'sky* Guards Bomber Aviation Regiment.

When Grizodubova was recalled to Moscow in June 1944 to take up a senior post in civil aviation, she had about 200 operational missions to her personal credit. Overall, Grizodubova spent 18,000 hours in the air. In her postwar post she assisted the future astronaut Savitskaia in becoming a test pilot.

Grizodubova was awarded the Order of Lenin (twice), the Order of the October Revolution, the Order of Patriotic War I Class (twice), and many medals. In September 1973, on the occasion of the thirty-fifth anniversary of the *Rodina* flight, an obelisk in the form of a wing made of granite was erected in the Khabarovsk Territory in honor of Grizodubova's famous aircrew. In 1978 she personally visited the area, where a film titled *Wings* was made about her for Soviet Central Television. Grizodubova served on the Executive of the Soviet War Veterans Committee and was a member of the Veterans' Council of Long-Range Aviation and chairperson of her regiment's Veterans' Council.

> *Actions of Russian women during the Russian Civil War:*
>
> Women took naturally to irregular warfare, an activity unknown to them since the days of 1812 when the Baba-Bogatyr Vasilia (the peasant wife of a village elder) and her associates pitch-forked, scythed, and burned alive the stragglers from Napoleon's army. According to the memoirs and accounts of the Civil War, the rough, comradely ways of partisan life were congenial to women. They paid dearly for it when caught: the captured women of a Yakutsk group were frozen alive into icy statues. . . .
>
> Captured nurses were often treated with special brutality by Whites. Near Petrograd in 1919, three nurses were hanged in bandages from the beams of their field hospital with their Komsomol pins stuck through their tongues . . .
>
> The most prominent of the full-time women "commissars" . . . was Zemlyachka. . . . Now in her forties, the only vestige of her bourgeois origins was the pince-nez that she wore in grotesque contrast to her short hair, boots, pants, and leather coat. Merciless to her enemies, Zemlyachka would unblinkingly order the shooting of one of her female comrades caught at treason or would preside . . . er the execution of thousands of White officers and other members of the "bourgeoisie" after the conquest of the Crimea. . . . A number of women worked in the Cheka during the Civil War. . . . [It was reported that] a certain Grebennikova, "cigarette in mouth, enormous revolver at the belt," forced Red Guards to violate women and children before executing them. The anti-Bolshevik (and anti-Semitic) English journalist, Hodgson, reports the story of "a young Jewish girl who was known as Rose" who shot her victims joint by joint or boiled the skin off their arms (the well-known "glove trick").
>
> —Richard Stites, *The Women's Liberation Movement in Russia; Feminism, Nihilism and Bolshevism, 1860–1930* (Princeton, NJ: Princeton University Press, 1978), 318, 319, 321.

Bibliography: Bashkirov and Semenkevich, "Geroini Sovetskogo neba," 1969; Brontman and Khvat, *The Heroic Flight of the Rodina*, 1938; Cottam, *Women in War and Resistance*, 1998; Verkhozin, "Polkom komanduet zhenshchina," 1966, and "Komandir polka," 1969.

—Kazimiera J. Cottam

GWYNNE-VAUGHAN, DAME HELEN (born 21 January 1879, London; died 26 August 1967, Sussexdown, England), born Helen Fraser. Chief Controller of WAAC in France, February 1917–August 1918; commandant of WRAF, August 1918–December 1919; chairman, Women's Legion, 1933–1936; president, Emergency Service, October 1936–February 1939; director, ATS, July 1939–June 1941. C.B.E. (mil) 1918, D.B.E. 1919 (mil).

A botanist by training and profession, Helen Gwynne-Vaughan served in both world wars as one of Britain's senior female officers. In February 1917, after the death of her husband, Gwynne-Vaughan was selected to head the overseas section of the Women's Auxiliary Army Corps (**WAAC**) in February 1917. Initially facing prejudice, Gwynne-Vaughan fought unsuccessfully for status equal to that of male soldiers and officers and bitterly resented the camp-follower status of the corps. When the Women's Royal Air Force (WRAF) needed a new head in August 1918, Gwynne-Vaughan was selected because of the success of the WAAC in France.

During the 1920s Gwynne-Vaughan returned to her prewar academic life, but kept an interest in the military, serving in both the WRAF's and the WAAC/QMAAC's old comrades associations and in the revived **Women's Legion**. After the disbanding of the Women's Legion, Gwynne-Vaughan led and gained War Office assistance for its renamed officer-training section, Emergency Service. Gwynne-Vaughan played a crucial role in the discussions leading to the creation of the Auxiliary Territorial Service (**ATS**) in September 1938, merging Emergency Service with it. She was appointed director of the ATS in July 1939 despite being overage. Morale and public relations problems originating in the hasty formation of the ATS led to her replacement in July 1941.

Bibliography: Gwynne-Vaughan, *Service with the Army*, 1942; Izzard, *A Heroine in Her Time*, 1969.

—C.H.N. Hull

H

HANCOCK, JOY BRIGHT (born 1898, New Jersey; died 20 August 1986). Captain, United States Navy (retired), First and Second World Wars.

Joy Bright Hancock first enlisted as a naval yeoman (F) in the Women's Naval Reserve during the First World War. After the war she served as a civilian employee in the Bureau of Aeronautics, Department of the Navy. Hancock was a pilot herself and was twice married (and widowed) to naval aviators. She wrote her first book, *Airplanes in Action,* in 1938. At the outbreak of the Second World War she successfully lobbied for the reestablishment of the Women's Naval Reserve. She was commissioned as a lieutenant, USN(R), in October 1942. The highest-ranking woman at the Bureau of Aeronautics, she was the **WAVES** (Women Accepted for Voluntary Emergency Service) representative to both that bureau and to the deputy chief of naval operations (air), as well as the liaison between the bureau and Lieutenant Commander **Mildred McAfee,** the Women's Reserve director. McAfee and her advisors at the Bureau of Personnel had come from the educational and professional worlds; Hancock's naval expertise proved invaluable to them. Hancock has been described as "the only WAVE leader with a clear idea of how the navy operated." Hancock advocated gender-integrated specialist training at aviation schools for WAVES members

Joy Bright Hancock as a young woman in uniform. (Library of Congress)

and the opening of jobs in aviation mechanics to women. The decision that permitted 3,000 women to serve as aviation machinist's mates was a major breakthrough for women in the armed services. She was also instrumental in opening up overseas service to WAVES members.

After the Second World War Hancock remained on active duty, taking over as director of the Women's Reserve in 1947 and leading the effort to maintain the Women's Reserve as a peacetime organization. The Women's Armed Services Integration Act, passed on 12 June 1948, codified the WNR in law. Hancock left active duty in May 1953 upon reaching the statutory retirement age. In 1971 Hancock moved to a United States Navy retirement home in McLean, Virginia, where she maintained an active interest in women in the navy until her death in 1986. She continually advocated integrated training and service for navy women. Hancock's memoirs, published by the Naval Institute Press in 1972, are one of the most useful sources to date on the history of the WAVES.

Bibliography: Alsmeyer, *The Way of the WAVES,* 1981, *Old WAVES Tales* 1982, and "A Preliminary Survey," 1983; Hancock, *Lady in the Navy,* 1972; Weatherford, *American Women and World War II,* 1990.

—Patrick J. O'Connor

HARLEY, LADY BRILLIANA (born ca. 1598, Brill, Netherlands; died 31 October 1643, Brampton Bryan Castle, Herefordshire, England). Defender, English Civil War.

Lady Brilliana Harley, wife of Sir Robert Harley, defended Brampton Bryan Castle, Herefordshire, against royalist forces in 1643 during the English Civil War. She saw the war as a religious struggle between godly Puritans in Parliament and pro-Catholics led by King Charles I. Since the Harleys were the only important Puritan gentry in Herefordshire, their home became a center of opposition to the king and his religious policies.

When the Civil War began in 1642, Lady Brilliana asked her husband to allow her to join him in London (where he attended Parliament), but accepted the necessity of staying at Brampton Bryan as God's will. She gathered men and weapons because she feared a royalist siege, as both she and her husband were charged with treasonous activities. On 26 July 1643, 700 royalist soldiers surrounded the castle, with about 50 soldiers and 50 civilians inside. Lady Brilliana refused to surrender the castle. She used many delaying tactics in her

Undated painting of Lady Brilliana Harley. (Perry-Castañeda Library)

negotiations; as a woman, she claimed that she could not surrender the castle without knowing her husband's intentions. During the nearly seven-week siege Lady Brilliana witnessed the spoliation of her property and endured the offensive curses of royalists. Buoyed by her religious faith, she believed that God would eventually deliver her, and indeed the siege was lifted on 9 September when the royalists temporarily left. Emboldened, she then initiated a successful attack on nearby royalists. Lady Brilliana, however, was caught in the castle when royalists renewed their siege in October, and she died on 31 October from a lung ailment. Without her leadership, the castle fell in 1644; it was burned to the ground with her remains inside.

Bibliography: *Dictionary of National Biography;* Eales, *Puritans and Roundheads,* 1990; Fraser, *The Weaker Vessel,* 1984; Lewis, *Letters of the Lady Brilliana Harley,* 1854; Vann, "Women in Preindustrial Capitalism," 1977.

—Robert W. Gee

HART, NANCY (born ca. 1735; died 1830, Frog Island, Kentucky), also known as "Ann" and "Aunt" Nancy Hart. Patriot scout who captured and killed Loyalist troops in her home, United States, Revolutionary War.

Nancy Hart took part in the vicious partisan fighting between Loyalists and patriots in Georgia during the Revolutionary War. Unlike most women and children, who were evacuated from areas of heavy conflict in Georgia, Hart remained to protect her home in Wilkes County while her husband, Benjamin, served with the state militia. Possessed of great courage and proficiency with a rifle, Hart served as a spy and scout against the Loyalist forces on several occasions. Hart is most famous for holding five (or six) Loyalist men at bay with a rifle after they demanded food in her home. She shot and killed two of the men and held the others until patriot forces arrived and executed them outside her cabin. While Nancy Hart became a particularly fierce symbol of female assertiveness during the Revolutionary War, she exemplifies the martial resolve shown by many women with political convictions who were caught up in a war with no clearly demarcated "home" front.

Hart's family hailed from Pennsylvania, and it has often been asserted, with only dubious documentation, that Hart was the niece of General Daniel Morgan and the cousin of Daniel Boone. Hart and her husband, Benjamin, had lived on the southern frontiers of North Carolina and South Carolina before they settled in Georgia shortly before the Revolutionary War. After the war Benjamin Hart died, and Nancy Hart eventually moved to Henderson County, Kentucky.

During the Revolutionary War Nancy Hart combated Loyalists both around her home and occasionally as a scout or spy. Hart reportedly took several Loyalist prisoners: one at her home while she was making soap and one while she was gathering information for patriot forces across the Savannah River in South Carolina. Other exploits attributed to Hart include spying in a British camp while dressed as a man and defending an unspecified fort from Loyalist attack. The specific details of these stories cannot be verified, but they testify to Hart's involvement in all aspects of partisan warfare. Hart not only protected her home, but seized opportunities to fight on behalf of the local patriots.

Hart's most famous exploit, the capture of the five Loyalists in her home, appeared in print in 1825, and the story was popularized by Elizabeth Ellet in her 1848 book *The Women of the American Revolution*. Ellet claimed to have personally verified the facts of the story on a trip to Georgia. A Loyalist party was chasing a patriot whom they suspected Hart of helping to escape when they stopped at Hart's cabin and demanded that she feed them. While the men were eating, Hart seized their weapons, trained her rifle on them, and, as Ellet put it, claimed "she would kill the first man who approached her." After one man tried to move, Hart shot him. Hart sent her daughter to get help as she detained the other four, one of whom she also had to shoot before the militia arrived to hang the survivors. She also served on occasion as a scout and spy, according to oral traditions in Georgia first recorded by George White in 1854.

During the Civil War a group of women who named themselves the **"Nancy Harts"** formed a Confederate militia company in La Grange, Georgia. Commemorations of Nancy Hart's Revolutionary War heroism include Hart County, Georgia, which was named after her, and a monument erected to her memory by the Daughters of the American Revolution in Henderson, Kentucky.

Bibliography: Coulter, "Nancy Hart," 1955; Ellet, *The Women of the American Revolution*, 1848–1850; White, George, *Historical Collections of Georgia* (New York: 1854).

—Sarah J. Purcell

HATCHEPSUT (born ?, ruled ca. 1504–1482 BCE; died ca. 1482 BCE, Egypt). Pharaoh of Egypt.

Hatchepsut ruled Egypt as a female pharaoh for more than twenty peaceful years. Evidence is now growing that her military prowess may have been seriously underestimated due to preconceived ideas of feminine pacifism. While either unwilling or unable to expand Egypt's sphere of influence in the Near East, Hatchepsut was certainly prepared to fight to maintain the borders of her country.

From the time of her coronation onwards Hatchepsut wished to be recognized as a conventional pharaoh. She therefore abandoned conventional female dress and had herself depicted with the clothing, actions, and body of a male king. Tradition dictated that the pharaoh should be a mighty warrior capable of leading troops against Egypt's long-standing enemies to the south and the northeast. For a long time it was assumed that a woman would be an instinctive pacifist and therefore incapable of conforming to this stereotype.

In consequence, the images of Hatchepsut the warrior were dismissed as mere propaganda. Hatchepsut's reign was long believed to be completely free from military activity, and it was assumed that Egypt's international position must have been seriously weakened by her lengthy rule.

It is now recognized that Hatchepsut's mortuary temple at Deir el-Bahri, Thebes, provides us with clear evidence for military activity. Here an engraved scene shows the god Dedwen leading a series of captive Nubian towns toward the victorious Hatchepsut, each town being represented by a name topped by an African head. Elsewhere Hatchepsut is portrayed as a sphinx trampling Egypt's foes, and there is a damaged description of a Nubian campaign in which Hatchepsut apparently emulated her father's military practices: "As was done by her father . . . a slaughter was made among them, the number of dead being unknown, their hands were cut off."

As the evidence from the temple is a mixture of official pronouncements and conventional scenes, it may be argued that these battles have been "borrowed" from an earlier pharaoh. However, the graffito of a man named Ti confirms that there was fighting in the south: "I saw him [i.e., Hatchepsut] overthrowing the Nubian nomads, their chiefs being brought to him as prisoners." Supporting evidence is provided by an eyewitness named Djehuty who saw Hatchepsut on the field of battle, gathering the spoils of war.

Limestone statue of Queen Hatchepsut. (Dover Pictorial Archive)

There is less evidence for campaigning to the north, although the Deir el-Bahri temple does hint at some action. It is the military career of Tuthmosis III, Hatchepsut's successor, that confirms Hatchepsut's firm control over her northeastern territories. When Tuthmosis became ruler of Egypt, the client states in Syria and Palestine, seeing Hatchepsut's death as a weakening of Egypt's position, rebelled. As Tuthmosis had inherited a disciplined army in a state of readiness, he was able to launch an immediate and successful counterattack. Tuthmosis went on to push back the boundaries of the Egyptian Empire until Egypt became the dominant force in the Mediterranean world.

The varying interpretations of the extent of military activity under Hatchepsut, the female pharaoh, are revealing of the ways in which history is affected by gender bias. Depictions of military activity that would not be questioned for a male pharaoh were ignored or assumed to be misattributed simply because Hatchepsut was female.

Bibliography: Robins, *Women in Ancient Egypt*, 1993; Tyldesley, *Hatchepsut*, 1996.

—Joyce Tyldesley

HELLSTROM, SHEILA (born ?, Lunenburg, Nova Scotia). Brigadier general, Canadian Forces.

Sheila Hellstrom was the first female general in the Canadian Forces (CF), being promoted to brigadier general in January 1987. She was subsequently joined by two other general officers—Brigadier General **Wendy Clay** and Commandre Laraine Orthlieb—as the highest-ranking women ever in the CF. Hellstrom, who retired in 1990 after thirty-six years in the CF, established her military career path before the Servicewomen in Non-Traditional Environments and Roles (SWINTER) and Combat-Related Employment for Women (CREW) programs were implemented to integrate Canadian women into combat- and non-combat-related positions. Hellstrom enlisted in the Royal Canadian Air Force (RCAF) approximately two years after the RCAF was authorized to accept women volunteers in March 1951. She became a colonel and then director of women personnel in 1983. Concurrent with her promotion to brigadier general in 1987, she became director-general for personnel, dealing with careers for officers in all three services of the Canadian armed forces. In 1989, as Canada's highest-ranking female officer, she was appointed chair of the Committee on Women in the NATO Forces. Throughout her long career Hellstrom saw herself as a role model for women as valuable members of the CF and as proof that the CF would be stronger with women personnel.

Bibliography: Butt, Major Robert, "A Long Way from Lunenburg," *Sentinel*, 25(3) (1989): 10; "Hellstrom Becomes Canada's First Woman General," in *Minvera*, 5(1) (1987): 46–47; Thomas, "Women in the Military," 1978.

—Patricia Bowley

HENRIETTA MARIA (born 26 November 1609, Paris, France; died 21 August 1669, Colombes, France). Queen consort of Charles I (king of Great Britain and Ireland, 1625–1649). Led royalist troops in Yorkshire campaign, English Civil War, February–July 1643.

Even before hostilities opened between King Charles I and the "Long Parliament" in 1642, Henrietta Maria used her contacts on the Continent to procure money, arms, and troops. In February the queen left for Holland to purchase arms and supplies, returning to England the next year. When Henrietta Maria landed at Bridlington on 22 February 1643, she was discovered by a parliamentary squadron the next morning, and her ships and lodgings were shelled; the queen herself was forced to take refuge in a ditch. The enemy failed to capture her or the stores she had brought in. In April Henrietta Maria set out with "my regiment of infantry and cavalry." Together with General Goring, she urged taking Leeds by storm, but was overruled. Henrietta Maria also proposed sending 3,000 troops to Lancashire "to clear out that country, which I hope will be done in ten or twelve days."

The queen wrote the king from Newark on 27 June that "I carry with me three thousand foot, thirty companies of horse and dragoons, six pieces of cannon, and two mortars," describing herself as "her she-majesty generalissima, and extremely diligent, with 150 waggons to govern, in case of battle." On 14 July 1643 she entered Oxford with Charles. As the military situation deteriorated, Henrietta Maria was forced to flee to France in July 1644, where she died twenty-five years later. She was not a battlefield figure, but Henrietta Maria was militant in the attempt to retain her throne and was at least in part responsible for the organization of royalist military efforts.

Bibliography: Bone, *Henrietta Maria*, 1972; *Dictionary of National Biography*; Gardiner, *History of the Great Civil War*, 1904; Green, *Letters of Queen Henrietta Maria*, 1975; King, *Women of the Renaissance*, 1991.

—Curtis F. Morgan

Henrietta Maria, Queen Consort of England, 1901. Florence MacKubin. (Library of Congress)

HERO OF THE SOVIET UNION, WOMEN RECIPIENTS, 1938–present, Soviet Union and Russia. Awarded primarily for proficiency in combat during the Second World War.

The title Hero of the Soviet Union (HSU) was introduced on 16 April 1934; on 1 August 1939 the Hero of the Soviet Union Medal was added (renamed Gold Star Medal on 16 October 1939). The award was conferred on those who distinguished themselves by their heroic deed or deeds. The first Soviet women to receive the HSU were **Marina Raskova,** Polina Osipenko, and **Valentina Grizodubova,** who were recognized for their pioneer nonstop transcontinental flight to the Far East in September 1938.

By 5 May 1990, ninety-five women had been granted the HSU, including Aniela Krzywon, who served with the Polish **"Emilia Plater" Independent Women's Battalion** during the Second World War and was killed in the Battle of Lenino. Much-belated HSU awards were granted to three Soviet women combatants from the Second World War on 5 May 1990 by Mikhail Gorbachev: Senior Lieutenants **Lidiia Litviak** and Ekaterina Zelenko (posthumously) and Chief Petty Officer Ekaterina Mikhailova-Demina. Two Soviet female cosmo-

Hero of the Soviet Union, Women Recipients

Notaliio Meklin posing in front of a Po-2 after being awarded the Hero of the Soviet Union. (Perry-Castañeda Library)

nauts were given the HSU in the postwar years: Valentina Tereshkova, the first woman in space, and Svetlana Savitskaia, the only woman to be awarded the HSU twice.

Soviet women HSUs can be divided into two major groups: thirty-five were airwomen, and twenty-nine were partisans or resistance activists from the Second World War. Twelve women HSUs who served in the ground forces and all three of the navy awardees were medical personnel who also were Second World War combatants. Given that women comprised 8 percent of Soviet military personnel in 1943, the number of Soviet women who received the HSU for their service during the Second World War appears disproportionately small when compared with the total of 11,500 HSUs earned by Second World War combatants. It would seem that senior commanders were not overly generous when awarding the HSU to women soldiers. Many women who served in the Red Army were primarily involved in so-called 'defensive' operations, such as engineering preparation of terrain, the building of fortifications, mine clearing, or air defense. Consequently, though women were exposed to great danger, they were less likely to receive the HSU, which was reserved for combatants who performed aggressive, offensive, and spectacular acts of heroism.

A substantial number of Soviet women who earned their HSU during the Second World War received the award only in the 1960s. This delay was often due to the recipients' alleged "political unreliability," a charge made if the individual had been a prisoner of war or had family members who had been arrested during Stalin's reign of terror. During the 1960s, however, such individuals were generally rehabilitated and belatedly awarded the HSU. In addition, awards were delayed for decades for individuals like Litviak who died in combat, but whose bodies were not identified until much later.

Representative Recipients

Chechneva, Marina Pavlovna (1922–1984), Guards major (ret.), Red Army Air Force. Squadron commander, Soviet Union, Second World War. Chechneva served in the **46th Taman'sky Guards Bomber Aviation Regiment,** flew 810 combat sorties with about 1,000 flying hours to her credit, and published several books and many articles about her unit. She was awarded the HSU on 15 August 1946.

Dzhunkovskaia-Markova, Galina Ivanovna (1922–1985), Guards major (ret.), Red Army Air Force. Squadron navigator, Soviet Union, Second World War. Dzhunkovskaia-Markova served in the **125th "M.M. Raskova"** *Borisovsky* **Guards Bomber Aviation Regiment,** flew sixty-nine operational missions, and published several books and articles about her regiment. She was one of the five airwomen of the 125th to be awarded the HSU on 18 August 1945.

Kolesova, Elena Fedorovna (1920–1942), private, Red Army. Partisan leader and saboteur, Soviet Union, Second World War. Kolesova served in the famous Commando Unit No. 9903, which contributed to the German defeat in the Battle of Moscow. Of five women partisans who received the HSU, two served in this unit, including the legendary Zoia Kosmodem'ianskaia. Kolesova was airdropped into occupied Belorussia on 1 May 1942 and was credited with instructing hundreds of partisans in mine laying. She was awarded the HSU on 21 November 1944.

Mikhailova-Demina, Ekaterina Illarionovna (born 1925), chief petty officer (ret.), Red Navy. Marine and medic, Soviet Union, Second World War. Mikhailova-Demina served in the 369th Independent Naval Infantry Battalion that advanced from the Azov Sea through the Balkans to Vienna. She took part in the capture of Belgorod-Dniestrovsky and the storming of the Yugoslavian fortress of Ilok. She was recommended three times for the HSU by her immediate superiors before she was finally awarded the HSU by Gorbachev on 5 May 1990.

Sosnina, Nina Ivanovna (1923–1943), private, Red Army. Leader of Malin Underground, Ukraine, Soviet Union, Second World War. With intelligence obtained by her brother Valentin and others, Sosnina planned and organized resistance based on missions assigned to her by the nearby "Kutuzov" Partisan Detachment. Along with her surgeon father, she died in a fire set by enemy soldiers during the performance of an operation on a wounded partisan. Censured by Soviet authorities for associating with "enemies of the people," she was rehabilitated and granted the HSU posthumously on 8 May 1965.

Soviet women on strength and endurance:

[Medic A. M. Strelkova]: "I don't know how to explain this. We carried men who were twice or three times our own weight. On top of that, we carried their weapons, and the men themselves were wearing greatcoats and boots. We would hoist a man weighing 80 kilograms on our backs and carry him. Then we would throw the man off and go to get another one. . . . And we did this five or six times during an attack."

[Partisan Vera Safronovna]: "Under extremely difficult conditions, however, women, these fragile, highly emotional creatures, proved stronger and capable of greater endurance than men. We marched for 30 or 40 kilometers; horses dropped dead, men fell, but women kept on walking and—can you imagine it?—singing."

[Partisan Raisa Khasenevich was sent to take a precious German typewriter to another partisan group]: "The men took nothing more than a rifle with them, but I carried a rifle, a typewriter, and Elochka [her two-year-old daughter]."

—Svetlana Alexiyevich, *War's Unwomanly Face,* trans. Keith Hammond and Lyudmila Lezhneva (Moscow: Progress Publishers, 1988), 56, 58, 217.

Zubkova, Antonina (Tonia) Leont'evna (1920–1950), Guards captain, Red Army Air Force. Squadron navigator, Soviet Union, Second World War. Zubkova served in the 125th "M.M. Raskova" *Borisovsky* Guards Bomber Aviation Regiment and was one of the five airwomen of this unit to be awarded the HSU on 18 August 1945. Her pilot **Nadezhda Fedutenko** was frequently appointed divisional deputy leader on bombing missions, and Zubkova led fifty-four aircraft onto the target on 2 September 1943. A postwar professor of mathematics at the "N.E. Zhukovsky" Air Force Engineering Academy, tragically Zubkova fell under a train and was killed in 1950 at the age of thirty.
Bibliography: Chechneva, *Nebo ostaetsia nashim,* 1976; Cherniaeva, *Docheri Rossii,* 1975; Cottam, *Women in War and Resistance,* 1998; Lebedev, "Znak osobogo otlichiia," 1979.

—Kazimiera J. Cottam

HERRERA, PETRA (fl. early twentieth century), also known as "Pedro." Soldier, 1910 Mexican Revolution.

During the 1910 Mexican Revolution Petra Herrera wanted recognition for her considerable talents as a soldier and the right to remain in the army in active service. Like many other female soldiers, Herrera had to disguise herself as a man in order to be a line soldier and become eligible for battlefield promotions. Disguised as "Pedro Herrera," she and her soldiers blew up bridges and infiltrated the camps of the enemy "Federal" forces, causing havoc among them. In May 1911 Herrera led Maderista soldiers in the storming of Torreon, Coahuila, and the routing of Federal forces. Her heroism was acknowledged in a ballad, "Corrido del Combate del 15 de Mayo en Torreon."

Once she established her reputation as a good soldier, Herrera let everyone know that she was a woman by wearing braids and soldiering under her real name. Many rival factions are said to have sought her services. In 1914 she was cited as a "captain and leading a force of 200 men." At the end of hostilities, about 1917, Herrera requested not only to be recognized as a general but also to remain in active service. Although the women's army she commanded of 300–400 female soldiers was disbanded, Herrera was made a colonel and was recruited by the Mexican government to be a spy in Jiménez, Chihuahua, operating as a bartender. But Herrera was feared as a potential leader of rebellion, so she was murdered in Jiménez, possibly by government henchmen.
Bibliography: Herrera-Sobek, *The Mexican Corrido,* 1990; Salas, *Soldaderas,* 1990, and "Soldaderas: New Questions, New Sources," 1995.

—Elizabeth Salas

HIND BINT'UTBA (fl. seventh century). Participant in the Battle of Yarmūk.
Hind bint'Utba was a formidable woman who left her mark on Arab his-

toriography, which represents her as fierce and prone to exact terrible vengeance upon her enemies. Mohammed's uncle, Hamza, had killed her father at the Battle of Badr (624 CE). Hind participated at the Battle of Uhud that same year, where she avenged herself upon Hamza; tradition credits her with eating part of his liver. At the Battle of **Yarmūk** in mid-August 636, near the Yarmūk River in Syria, she helped rally the Arabs at a critical moment. The forces of the Byzantine emperor Heraclius entered the Arab encampment; Hind, along with other women who had been left at the camp, not only exhorted the fleeing Arabs to fight but took up arms herself. Balādhurī reports that she urged the Arabs to circumcise with their swords all uncircumcised prisoners. Whether or not the hapless Byzantines suffered this fate is unknown. With Abū Sufyān, her third husband, Hind was the mother of Mu'āwiya, the founder of the Umayyad dynasty.

Bibliography: al-Balādhurī, *Futūh al-Buldān*, ed. de Goeje, p. 135; Donner, Fred, *The Early Islamic Conquests* (Princeton: 1981); *The Encyclopaedia of Islam* New Edition, Leiden: Brill, 1979.

—C.A. Hoffman

HOBBY, OVETA CULP (born 19 January 1905, Killeen, Texas; died 16 August 1995). Colonel (ret.), director, WAAC/WAC, 1942–1945, United States, Second World War.

In 1942 the thirty-seven-year-old Oveta Culp Hobby took over the newly formed Women's Army Auxiliary Corps (WAAC) with the support of General George C. Marshall and Secretary of War Henry Stimson. She saw the WAAC through its transition in September 1943 from WAAC to **WAC** (Women's Army Corps), when the organization shifted from auxiliary to full military status.

The accomplished Hobby brought prestige as well as talent to her task. A graduate of the University of Texas Law School, Hobby was named parliamentarian of the Texas legislature in 1925, became editor of the *Houston Post* in 1938, and was also the wife of a former governor of Texas and mother of two children before ever donning a military uniform. She worked tirelessly during the war to attain recognition and benefits for her troops and became the first American woman to garner numerous military and civilian honors, including the Distinguished Service Medal. Hobby resigned from the WAC in 1945, but remained a political figure active in civic and business affairs after the war.

Undated portrait of Oveta Culp Hobby. (Perry-Castañeda Library)

American women as percentage of army leadership, 1997:		
Rank	1987	1997
Sergeant first class through sergeant major	3%	11%
Second lieutenant through captain	15%	16%
Major through colonel	6%	12%
Generals*	1%	2%

*Four of 408 in 1987, 5 of 313 in 1997.

—United States Army, Defense Manpower Data Center.

President Eisenhower named her the head of the Federal Security Agency (later called the Department of Health, Education, and Welfare) in 1953.

Bibliography: Meyer, *Creating G.I. Jane,* 1996; Treadwell, *The Women's Army Corps,* 1954.

—Elizabeth Lutes Hillman

HODGERS, JENNIE (born 25 December 1844, Belfast, Ireland; died 10 October 1915, East Moline State Hospital, Illinois), also known as Albert D.J. Cashier. Soldier, United States, American Civil War.

Jennie Hodgers served in male disguise as Albert D.J. Cashier in the Union army for three years in the **American Civil War**—the longest verifiable service by a female soldier in that war. In fact, one of the unique things about Hodgers is that her service in male disguise is extremely well documented.

Little is known about Hodgers's early years. As a young girl in Ireland, she apparently worked as a farmer and shepherd and wore male clothing while performing her chores. Hodgers took on male identity when she boarded a U.S.-bound ship as a stowaway. She ended up working in Illinois and the outbreak of the Civil War was already living under the identity of "Albert D.J. Cashier." She never explained her reasons for assuming male identity, but some sources indicate that she had an uncle in the United States who was able to get a job for his "nephew" in a shoe factory. Undoubtedly it was easier and safer for a runaway like Hodgers to earn a living as a young boy rather than as a girl.

In 1862, when Hodgers was seventeen, she joined the 95th Illinois Volunteer Infantry. Five feet tall, "Albert Cashier" was one of the shorter soldiers to be accepted into the army; lack of height, however, did not keep her from being a good soldier. During her service the 95th Illinois was an active regiment, part of Ulysses Grant's Army of Tennessee. Former comrade C.W. Ives regarded Cashier as a good soldier and recalled several incidents in which she gained distinction, such as the time she climbed "a tall tree to attach the Union flag to a limb after it had been shot down by the enemy." Cashier's captain considered the soldier—who participated in forty battles without being wounded—fearless and dependable. Not one of her comrades seems to have suspected that "Cashier" was anything but "a brave little soldier," although a bit of a loner who kept to himself.

Like many other Civil War soldiers, Hodgers contracted chronic diarrhea, which plagued her from mid-1863 on, though she managed to avoid hospitalization, which might have exposed her identity. Albert Cashier remained with the 95th Illinois until "he" was honorably discharged in August 1865.

After the war Hodgers continued to live as Albert Cashier for many decades. By 1869 she had settled down to work as a custodian and "handyman" in Saunemin, Illinois. "Cashier" lived a solitary life, though she seems to have been accepted as almost a member of one local family for whom she worked. There is no evidence that she was ever sexually active or romantically involved with another person.

In 1890 she applied for a government pension, but it was denied when she refused to submit to a physical examination. In 1911, when she was sixty-seven years old, Hodgers had an accident and seriously injured her leg. In the course of treatment her sex was discovered by locals, but they kept the secret. Because she was unable to continue working, her only recourse was to apply for a government pension and nursing care in the Illinois Soldiers and Sailors Home. The aid of Ira M. Lish, a state senator, was enlisted, but everyone still kept Cashier's secret while helping the illiterate veteran gain admittance to the home.

Clipping from an old newspaper about Jennie Hodgers. (Library of Congress)

However, Cashier's identity did not remain secret long in the nursing home. Once revealed, she attracted a great deal of attention. The state of Illinois had her declared insane, at least in part because of her long masquerade in male identity, and transferred her to an asylum. She was forced thereafter to wear women's clothing, which she hated; she even pinned her skirt into the shape of pants. She died less than two years after her transfer to the asylum.

In the meantime, investigations by the military pension bureau confirmed that "Albert D.J. Cashier" had indeed served in the 95th Illinois. Her former comrades were uniformly surprised to hear of her true identity, but each corroborated to the investigators that she had performed all the duties of a soldier and in interviews then and for years to come continued to refer to Hodgers as "he" even after learning that "he" was a woman.

Hodgers had been a member of the Grand Army of the Republic, and upon her death in 1915, she was buried by it in her uniform with full military honors. Her tombstone bore the inscription "Albert D.J. Cashier, 95th Ill. Inf."

Bibliography: Clausius, "The Little Soldier," 1958; Davis, "Private Albert Cashier," 1989; Larson, "Bonny Yank and Ginny Reb," 1990; Leonard, *All the Daring of the Soldier*, 1999; National Archives, Washington, DC, RG 15, Records of the Veterans Administration, Albert D.J. Cashier Pension File, and RG 94, Military Service Records and Medical Records of Albert Cashier; Seeley, *American Women and the U.S. Armed Forces*, 1992.

—Gayle Veronica Fischer

HŌJŌ MASAKO (born 1157; died 1225), also known as "the nun-shogun." Wife of the first shogun; regent after his death. Japan.

Although she never saw a battlefield, Hōjō Masako was perhaps the most important woman in Japanese military history. Indeed, her celebrity as a figure of martial legend and folktale rivals those of her contemporary, **Tomoe Gozen,** and the mythical third-century empress-regent, Jingu Kogo; but unlike Jingu, Masako was real, and unlike Tomoe, she built her fame not as a warrior, but as a governor of warriors. For this wife of Japan's first national warrior chieftain and mother of two others was one of the principal architects of the Kamakura shogunate, a regime that dominated her country's military men and military affairs for more than a century and a half.

Masako began life in 1157 as the second child and the eldest daughter of a theretofore inconsequential warrior family in Izu Province (south of present-day Tokyo). In 1160 her father, Tokimasa, became the guardian to Minamoto Yoritomo, the thirteen-year-old son of a defeated warrior noble in the capital. Legends abound concerning the relationship that developed between the young exile and his guardian's daughter, but reliable information is scanty. According to one account, Yoritomo's initial interest was not in Masako, but in a younger sister of hers; according to another, a love-struck Masako was forced to elope with Yoritomo in the middle of the night because her father vehemently objected to the marriage.

In the event, Tokimasa could scarcely have chosen a better son-in-law, for two decades after his exile Yoritomo adeptly parleyed his pedigree, the localized ambitions of provincial warriors, and a series of upheavals within the imperial court into a new-fledged form of nationwide power and authority. He established a new institution, a sort of government within the imperial government, headquartered in Kamakura (southwest of present-day Tokyo) and charged with overseeing eastern Japan and the warrior class.

When Yoritomo died in 1199 and was succeeded by his son Yoriie, a power struggle developed rapidly between Yoriie's father-in-law, Hiki Yoshikazu, and his maternal relatives, led by his mother Masako and his grandfather Tokimasa. In 1203 the Hōjō attacked and eliminated the Hiki and then deposed Yoriie, replacing him with his younger brother, Sanetomo. Because Sanetomo was only eleven years old at the time, Tokimasa assumed real control of the Kamakura regime. A scant two years later Masako and her brother Yoshitoki deposed their father and then set about reorganizing the Kamakura power structure, redefining both the regime's role in national governance and their own place within it.

Masako, who had taken the tonsure shortly after the death of her husband, rapidly came to be known as "the nun-shogun"; Yoshitoki assumed the position of regent (*shikken*) over the still-adolescent Sanetomo. Over the next decade and a half the pair targeted and destroyed, one by one, every warrior house in Japan that appeared to be a serious competitor for control of the shogunate, often by cleverly inciting potential rivals into rebellion.

The assassination of Sanetomo in 1219 gave Masako and Yoshitoki the excuse to first declare martial law and seize dictatorial powers over the shogunate and then to arrange for the first of a six-generation series of infant court nobles and imperial princes serving as figurehead shoguns. Their most noteworthy achievement during this period was their victory in 1221 over an army raised by the former emperor, Go-Toba, and sent out to destroy the Kamakura regime. The shogunate's triumph in this conflict resulted in an expansion of its vassal network in western Japan and the appointment of shogunal deputies to govern the imperial capital city of Kyoto.

When Yoshitoki died in 1224—some say that he was poisoned by his wife, who hoped to shift control of the shogunate to her own family—it was the sixty-seven-year-old Masako who foiled the plot and ensured the succession of Yoshitoki's son Yasutoki as regent. Thus this remarkable woman remained a political player to be reckoned with almost to the moment of her death a year later.

Bibliography: Benton, "Hōjō Masako: The Dowager Shōgun," 1991; Butler, "Woman of Power behind the Kamakura Bakufu: Hōjō Masako," 1978; Nagahara Keiji, *Chūsei Seiritsuki No Shakai To Shisō* (Yoshikawa Kōbunkan, 1977); Sotomura Hisae, *Hōjō Masako* (Jinbutsu Nihon No Joseishi vol. 3; Shugeisha, 1975); Tabata Yasuko, *Nihon Chūsei No Josei* (Yoshikawa Kōbunkan, 1987); Watanabe Tamotsu, *Hōjō Masako* (Yoshikawa Kōbunkan, 1961).

—Karl Friday

HOLKAR, AHILYABHAI (born 1725; died 1795, India), also known as Ahilyabai, Ahalya Bai Holkar. Rani of Indore. Military and political leader, Maratha Confederacy, eighteenth century, India.

Ahilyabhai Holkar was born in 1725 at Chowndi, near Aurangabad, where her father, Manakoji Shinde, served as the village *patel*, or accountant. At the age of eight she was brought to the attention of Malhar Rao Holkar, the newly appointed Maratha commander of Malwa, who sought her as a suitable bride for his wayward son, Khande Rao.

Malhar Rao, like most Maratha military leaders of his generation, was a peasant soldier who had risen through the merit-oriented, semifeudal hierarchy that formed the basis of the Maratha Confederacy. Since the mid-seventeenth century the Marathas, a loose alliance of Hindu peasant-warrior communities, had fought a series of successful wars against the Mughal Empire and had come to dominate an area from Rajasthan in the north to Mysore in the south, and their armies had made extensive inroads against

Hyderabad and Malwa, as well as long-distance cavalry raids throughout the subcontinent. Under the leadership of the Poona Durbar, or the council of peshwas, the Marathas supported their troops by dominating trade and pilgrimage routes, plundering neighboring Mughal provinces, and efficiently administering their core territories. Malhar Rao's position was an extremely important one within the confederation, as his fief in Malwa formed the front line of attack and defense against the Mughal emperor of Delhi and his Rajput allies in central India, as well as an important trade route between western and northern India that followed the Narmada valley. His command of the region was made hereditary by the peshwa in 1734, shortly after Ahilyabhai was brought to live at his headquarters in Indore.

Malhar Rao trained Ahilyabhai from childhood to serve as his administrative assistant, continuing an education that had begun under her own family's direction. Ahilyabhai's early interest in languages and literature was combined with considerable skill in managing finances and intense devotion to the Shaivite sect of Hinduism. Under Malhar Rao's tutelage, she also acquired detailed knowledge of the practical aspects of warfare and statecraft, and the administration of the Holkars' province was largely left in her care. Malhar Rao, a better soldier than administrator, thus concentrated his efforts on military affairs.

Under Ahilyabhai's influence, Khande Rao began to take his duties as a prince seriously, and by all accounts their marriage was a happy one, producing a son, Malerao, in 1745, and a daughter three years later. However, Khande Rao was killed at Mahar Rao's side at the siege of Kumbheri in 1754, and it was only with considerable difficulty that the Holkars could persuade Ahilyabhai not to commit suttee on his funeral pyre according to custom. She chose instead to dedicate herself to state service and soon took her husband's place as Malhar Rao's chief lieutenant. She took part in her father-in-law's campaigns and commanded the Maratha artillery at the decisive Battle of Panipat in 1761. This engagement, fought north of Delhi, dealt a heavy blow to the Maratha Confederacy, as Abdali Shah and his Afghan forces shattered their armies and killed their most capable leaders. Malhar Rao and Ahilyabhai survived the disaster through a timely retreat and devoted their energies to consolidating their position in Malwa. Malhar Rao died in 1766, leaving Ahilyabhai in an extremely uncertain political position.

The Poona Durbar confirmed the succession of Ahilyabhai's son, Malerao, but his reign was cut short by the onset of insanity within eight months. Upon his death on 27 March 1767, enemies of the Holkar dynasty plotted their removal from command of Malwa, as Ahilyabhai's daughter, Muktabhai, was not eligible, under Hindu law, to succeed to the throne. Peshwa Raghoba and Malhar Rao's minister, Chandrachud, at once attempted to oust Ahilyabhai from Indore. She did not, however, collapse as quickly as the conspirators had hoped; she drafted a strong reply to the peshwas, promising loyalty if she were allowed to keep the Holkars' kingdom, but a disastrous war should any attempt be made to seize it from her. While Raghoba mustered troops, Ahilyabhai won a diplomatic victory by securing offers of support from Peshwa Madhav Rao and several other Maratha commanders. Her own army, under an able soldier, Tukoji Rao, was joined by considerable reinforcements under Bhonsle, the gaekwad of Baroda, and the raja of Mohan; they confronted Raghoba's 50,000 men at Ujjain a short time later, and the conspirators hesitated. Ahilyabhai issued a warning: "You seem to be under the misconception that I am just a frail woman. How frail I really am you will know best on the battlefield.... You will moreover find yourself faced with a battalion of women. If I am defeated, none will laugh at me, but if you are routed, you will not be able thereafter to show your face to anyone."

Raghoba pondered this message, as well as his intelligence reports of the excellent organization and training of Ahilyabhai's forces; he chose to save himself from disgrace by pretending that he had merely come to Indore to offer condolences upon the death of her son. Ahilyabhai permitted him to visit her, but demanded that his army remain on the border. The other members of the Poona Durbar quickly recognized Ahilyabhai's rule in Malwa, and Tukoji Rao was appointed commander of their forces in the region.

Ahilyabhai immediately turned her attentions to Rajasthan, where Malhar Rao had become deeply entangled in the conflicts among numerous petty states. The Chandravats, staunch enemies of the Holkars, challenged Ahilyabhai's governorship of Udaipur and Rampura, taking advantage of the absence of Tukoji Rao and her best troops, who were campaigning with the main Maratha army in northern India. For the time being, Ahilyabhai conciliated the rebels with a revenue-free grant of thirty-one villages, but the Chandravats raised a new insurgency in 1771, again striking while Ahilyabhai's forces were away on service. Under Sharif Rao, Ahilyabhai's reserves suffered a defeat at the hands of the Rajputs near Palsuda. A series of inconclusive campaigns ensued that only served to consolidate and increase the military strength of the Rajput states on the border of Indore; two major defeats were dealt to the Maratha forces, together with the loss of frontier territory and several towns in 1787. Ahilyabhai proceeded to address this emergency by mobilizing all of the resources at the disposal of her government; a larger army was raised, and an efficient bureaucracy ensured that it entered the field fully equipped and well supplied, unlike most Indian armies of the period. Despite heavy resistance, the Holkar forces, under Ahilyabhai's personal command, drove the Chandravats back to the fort of Amad, where a crushing defeat was inflicted, and the principal rebel commander, Saubhag Singh, was captured and executed.

The remaining years of Ahilyabhai's life were filled with personal disasters. Her daughter Muktabhai's son, Nathu, died of sickness during the last campaign against the Chandravats, and her son-in-law fell ill and died in 1791. Despite Ahilyabhai's pleas, her daughter commited *sati* upon her husband's funeral pyre. For a time it seemed that Ahilyabhai might crumble under the weight of these catastrophes, but she managed to rally her spirits and resume leadership of the state. The most significant reform of the last years of her reign was the beginning of the modernization of Tukoji Rao's army under the direction of a French mercenary named Dudrenec. Four battalions of 400 men each were thus equipped, organized, and trained according to the contemporary European model in 1792. Ahilyabhai's interest in military affairs was waning, however, and adequate resources were not available to support the new troops properly; their pay fell fourteen months in arrears by mid-1793. Politically, both the army and the state began to be compromised by Tukoji Rao's military rivalry with Mahadji Sindhia, the Maratha regent who controlled the Mughal emperor.

Moreover, Ahilyabhai's advancing age and the burden of her work and private losses were beginning to take their toll, and she herself died quietly on 13 August 1795 at the age of seventy. Following her death, Indore State, together with the rest of central India, passed into a period of internal political strife and civil war from which it would never fully recover. Ahilyabhai's army, still under Tukoji Rao's command, was largely destroyed in the general struggle for hegemony among the Maratha chiefs, and her rich treasury, together with the wealthy city of Indore, was plundered by Yashwant Rao's forces in 1800.

Ahilyabhai Holkar stands apart as one of the most able rulers of the Maratha Confederacy. Under her rule, Indore became a prosperous and powerful state, and Ahilyabhai's military effectiveness was largely based upon the strength of the local economy. Her earliest administrative reforms were the introduction of a mint and a standardized system of currency that facilitated trade; numerous temples were constructed throughout the province, the fairs of which became important markets, generating considerable revenue. Ahilyabhai generally followed a conservative economic policy, generating finances through tolls and licenses rather than heavy taxes upon agricultural production; considerable sums were devoted by the state to extending cultivation, securing trade routes, and developing charitable and educational institutions.

Today Ahilyabhai's statue may be found in front of her modest palace in the center of Indore, which has become a major industrial and commercial center. An annual festival in her honor continues to be held at the Ganapati temple in the village of Sawer, some twenty-nine kilometers north of Indore.

Bibliography: Burway and Ganesh, *Life of Subhedar Malhar Rao Holkar*, 1930; Kamath and Kher, *Devi Ahalyabai Holkar*, 1995; Nagrale, *Peshwa Maratha Relations and Malhar Rao Holkar*, 1989; Sarma, *Ahilyabai*, 1969.

—James W. Hoover

HOLM, JEANNE MARJORIE (born 23 June 1921, Portland, Oregon, United States). Director, Women in the Air Force (WAF) (1965–1973); first woman officer to attend Air Command and Staff College (1952); first woman general in the United States Air Force (1971); first woman major general in the armed forces (1973); author of *Women in the Military: An Unfinished Revolution*.

Major General Jeanne Holm was one of the most influential of the women's service directors. As Women in the Air Force (WAF) director, she took advantage of the great social changes occurring in the United States to make positive changes for air force women.

Holm enlisted in the Women's Army Auxiliary Corps (WAAC, later **WAC**) during the Second World War and graduated from Officer Candidate School in January 1943. After the war she left the service to attend Lewis and Clark College in Portland, Oregon, but in 1948 she returned to active duty in the newly formed air force, saying, "I was between semesters, had nothing to do anyway and was flat broke." As an air force officer, Holm served in a variety of staff positions, both in the United States and overseas. She was promoted to major in 1951 and completed the Air Command and Staff College the next year, the first woman to attend. In 1965 Holm was assigned to the Pentagon and in November was selected as director of Women in the Air Force and promoted to colonel.

Official portrait of Jeanna Margorie Holm. (Library of Congress)

As director, Holm enacted changes that positively affected all women in the armed forces. One of her first actions was to use the power of the Defense Advisory Committee on Women in the Services (DACOWITS) to assess the utilization of women in the military, including their potential for greater employment, recruitment, and retention. An interservice working group was established that proposed changes that led to the signing of Public Law 9-130 on 8 November 1967. PL 9-130 allowed women to be promoted to general and flag grades, opened promotions to colonel/captain, equalized retirement rules for men and women, and removed ceilings, including the 2 percent limit on the regular force strength of women in each service.

In 1968 Holm began her "New WAF" program, designed to enact important changes for the WAF through laws and regulations, including a new legal status, expanded strength, better utilization, improved promotion potential, higher morale, and improved retention. Under her guidance, air force women achieved a greater equality with men; all noncombat career fields were opened, rank restrictions were removed, and women could be assigned to bases worldwide, attend all professional schools, receive full active-duty and retirement benefits, and remain on active duty while raising a family. Holm faced resistance to every proposed change, but she persisted. She later said, "I went into this office in November of 1965 and at a crucial moment. I could

Positions and occupations open to active duty women in the United States, 1999:		
Branch of service	Percentage of occupations	Percentage of positions
Air Force	99	99
Army	91	67
Navy	96	91
Marine Corps	93	62
Coast Guard	100	100

—United States Army, Defense Manpower Data Center.

never have functioned in this job I am in now in an earlier period. I could never have tolerated the attitude that I encountered when I went in there—the don't-rock-the-boat attitude in the personnel field.... So I admit that I did a little stirring up of my own. I felt that that was what I was paid for—I wasn't paid to be a yes-woman—and that it was my job to find out what was wrong with the utilization of women, and I set at it."

Holm was promoted to brigadier general on 1 August 1971 and continued serving as WAF director until 30 May 1973. On 1 June 1973 she was promoted to major general, the first female two-star general in U.S. history, and was assigned as director of the Secretary of the Air Force Personnel Council. Major General Holm retired from the air force in 1975.

Holm has continued her involvement in military issues, serving on DACOWITS, contributing to professional journals, testifying to Congress regarding gender discrimination in the military, serving as special assistant for women to the president, and speaking publicly about the important role of military women. She is the author of *Women in the Military: An Unfinished Revolution,* widely regarded as an essential text on U.S. servicewomen. Holm's many awards include the Distinguished Service Medal with oakleaf cluster, Legion of Merit, and the National Defense Service Medal.

Bibliography: Brokaw, Tom, *The Greatest Generation* (New York: Random House, 1998); Friedl, *Women in the United States Military,* 1996; Holm, *Women in the Military,* 1992.

—Vicki L. Friedl

HOPPER, GRACE BREWSTER MURRAY (born 9 December 1906, New York City, United States; died 1 January 1992, Arlington, Virginia, United States), also known as "Amazing Grace," "the Mother of COBOL." Retired rear admiral, United States Navy, mathematician and computer science pioneer.

Grace Hopper's two loves, computers and the United States Navy, led her to a life of extraordinary achievement. Her work in computer science included programming languages, software-development concepts, compiler verification, and data processing—the foundation for modern computing. Hopper was a visionary who wanted to write computer programs that would allow other scientists (and someday nonscientists) to use computers directly, without relying on computer specialists. She resisted conventional thinking, believing that there was always more than one way to solve a problem. Hopper once said that the most dangerous phrase uttered by computer programmers was "But we've always done it that way." To illustrate flexible thinking, she hung a wall clock in her office that ran backwards. She had the distinction of serving forty-three years in the United States Navy and at her retirement asked those in attendance, "Do you realize I'm the last of the World War II **WAVES** to leave active duty?"

Hopper attributed her early interest in mathematics to her maternal grandfather, the senior civil engineer for New York City. This interest was encouraged by her parents, and she excelled in mathematics at school, especially geometry. In 1928 Hopper graduated from Vassar College with a degree in mathematics and physics. She continued her studies at Yale, completing a master's degree in 1930. She earned a doctorate in 1934 in mathematics and mathematical physics, writing a thesis titled "A New Criterion for Reducibility of Algebraic Equations." Hopper then taught mathematics at Vassar, where she became an associate professor. She married Vincent Foster Hopper in 1930; they divorced in 1945, but Hopper kept her married name. Hopper never remarried and had no children.

Grace Hopper, originator of electronic computer automatic programming for the Remington Rand Division of Sperry Rand Corporation. (AP Photo)

Hopper's great-grandfather and childhood hero, Alexander Wilson Russell, had been a rear admiral in the United States Navy, and when the United States became involved in the Second World War, she wanted to join the navy as well. Hopper was sworn into the United States Naval Reserve in December 1943 and attended the Naval Reserve Midshipman's School in Northhampton, Massachusetts. She graduated first in her class and received a commission as a lieutenant, junior grade, in June 1944. After commissioning, Hopper went to lay flowers on the grave of her great-grandfather, Rear Admiral Russell, telling him that it was "all right for females to be Navy officers." Hopper was assigned to the Bureau of Ships Computation Project at Harvard University, where, along with two navy ensigns, she worked on the Mark I. The Mark I was the first programmable computing machine made in America, to be used to compute firing tables for the new naval guns. Hopper became only the third person ever to program the Mark I and wrote the manual on operating the machine. It was during the summer of 1945, while working on the Mark II, that she is credited with coining the term "bug" to refer to strange computer malfunctions. One summer day it was hot in the computer building, so the windows were left open. Suddenly the Mark II stopped working. As Hopper told the story, "Finally someone located the trouble spot and, using ordinary tweezers, removed the problem, a two inch moth. From then on, when anything went wrong with a computer, we said it had bugs in it."

At the end of the Second World War Hopper requested a transfer from the

Reserve to the Regular Navy, but this request was denied because as her age of forty exceeded the navy's maximum age limit of thirty-eight for staying on active duty. She remained in the Navy Reserve and continued her work, now as a civilian, at the Harvard Computational Laboratory on the Mark II and Mark III computers. In 1949 she joined the Eckert-Mauchley Computer Corporation in Philadelphia as a senior mathematician. The company was merged into the Sperry Corporation, where Hopper made many remarkable contributions to the computer-programming field. She developed the first practical compiler, a program that translates instructions written by human programmers into more specific codes that can be directly read by a computer. A later compiler, the Flow-Matic, led to the development of COBOL, the first business-oriented, English-like higher-level programming language. Hopper was instrumental in persuading the navy to use the new programming language.

Hopper retired from the Naval Reserve with the rank of commander on 31 December 1966, a date she called "the saddest day of my life." Within a year the navy recalled her to active duty to help standardize the Navy's computer programs and languages. She remained on active duty for nineteen more years with the Naval Data Automation Command, working to refine computer language techniques. When Admiral Hyman Rickover retired from active duty in January 1982 at the age of eighty-two Hopper became the oldest person on active duty in the navy. She was promoted to the rank of rear admiral on 8 November 1985. At the age of seventy-nine Hopper was involuntarily retired from the navy; the ceremony was held on 14 August 1986 in Boston, Massachusetts, on the USS *Constitution*, ending her forty-three-year naval career. During the ceremony she was presented with the highest award given by the Department of Defense, the Defense Distinguished Service Medal. Hopper continued working as a senior consultant to Digital Equipment Corporation until her death in her sleep on 1 January 1992 at the age of eighty-five. She was buried at Arlington National Cemetery.

Grace Hopper's many awards include honorary degrees from more than forty colleges and universities. She received the first Computer Sciences "Man of the Year" Award from the Data Processing Management Association (1969); was the first American citizen and first woman of any nationality to be named as a distinguished fellow of the British Computer Society (1973); and was the first recipient of the National Medal of Technology in 1991, presented to her by President George Bush. The USS *Hopper*, a navy Arleigh Burke Class destroyer named for Admiral Hopper, was commissioned in San Francisco on 6 September 1997.

Bibliography: Billings, *Grace Hopper*, 1989; Lee, J.A.N., *Computer Pioneers* (Los Alamitos, CA: 1995); Slater, Robert, *Portraits in Silicon* (Cambridge: 1987).

—Vicki L. Friedl

HUA MULAN (ca. late sixth century, China), also known as Fua Mulan or Wei Mulan. General and "filial-staunchness." Soldier during the Sui dynasty (581–618).

Dressing as a man, Hua Mulan took her father's place in battle for a dozen years. The inspiration for the popular Disney film *Mulan,* Hua Mulan was a legendary female warrior whose historical existence has been a point of argument among scholars for centuries. However, since the anonymous poem that celebrates her comes from a ballad tradition, it might well be based on some similar incidents of its time. Further, the legendary Mulan has become familiar to Chinese households for centuries through vernacular stories, literary works, and dramatic performances; whether she existed or not, the image of Mulan has been more influential than that of any other woman warrior in Chinese history. Her filial piety and bravery have been an inspiration to generations of Chinese scholars.

> After I grew up, I heard the chant of Fa Mu Lan, the girl who took her father's place in battle. Instantly I remembered that as a child I had followed my mother about the house, the two of us singing about how Fa Mu Lan fought gloriously and returned alive from war to settle in the village. I had forgotten this chant that was once mine, given me by my mother, who may not have known its power to remind.... she taught me the song of the warrior woman, Fa Mu Lan. I would have to grow up a warrior woman.
>
> —Maxine Hong Kingston, *The Woman Warrior: Memoirs of a Girlhood among Ghosts* (New York: Vintage Books, 1977), 24.

During the Sui dynasty (581–618) the khan of the north often came down and skirmished with the Sui court. The court drafted soldiers, and Mulan's family was on the draft list. Because her father was old and her brother was still young, she decided to go in their place. She bought armor and saddles and asked her father's permission to go on his behalf. She served in the army for twelve years, her true identity never suspected by others. After she returned to the capital, the emperor rewarded her with the position of an imperial secretary, but she did not accept. Some friends from the army accompanied her home. After she changed into her old clothes and came out to receive them again, her companions were shocked to learn that she was a woman.

When the emperor learned the truth, he wanted to have Mulan as his consort. She claimed that it would not be appropriate for a minister to marry the emperor, and that she would rather die than comply. The emperor was surprised by her reaction but then understood and conferred upon her the title of general. She was posthumously given the honorific title Filial-Staunchness.

The earliest extant description of Mulan's life is found in a ballad, "Mulan shi" (The poem of Mulan), collected in *Yuefu shiji* (Collection of music-bureau poetry), compiled by Guo Maoqian (twelfth century). Most scholars date the poem to either the Northern dynasties (386–581) or the Sui dynasty (581–618). Her place of birth is usually considered to be Shangqiu of today's Henan Province. Thus Shangqiu's local gazetteer has her biography, and there was even a temple for her as early as the Qing dynasty (1644–1911). Her townspeople always commemorate her on her birthday, the eighth day of the fourth month.

Bibliography: Chen Menglei, *Gujin tushu jicheng (Qin ding)*, 1993; Liu Dechang, *Shangqui xian zhi*, reprint ed. (Taipei: 1968); "Mulan shi er shou," *Yuefu shiji*, comp. Guo Maoqian (twelfth century) (Beijing: 1979); Ren Shoushi, *Bo zhou zhi* (1825).

—Sherry J. Mou

HUNGARY, WOMEN IN 1956 REVOLUTION. During the night of 23 October 1956 women fighters took up arms against the Hungarian as well as the Soviet protectors of the Stalinist regime in Hungary. They continued their resistance up to 10–11 November 1956. The price for their short-lived rebellion was death or, if they managed to survive the military terrors of the Red Army, lifelong persecution during the ensuing phases of communism. Until 1989 their activities were labeled by the authorities as counterrevolutionary, disturbing the peace, and, therefore, a simple crime against common law. Accordingly, the legal charge brought against them was robbery, looting, or inciting to riot. As a result, the only legal defense open to them was to try to mitigate the severity of the charge, without, however, being able to reclassify it as rightful, legitimate protest. The list here is far from complete and is limited to women freedom fighters active in Budapest, as a listing of all women participants in all parts of Hungary has never yet been compiled.

An armed female on guard in a bomb damaged Hungarian town, November 12, 1956. (Hulton/Archive by Getty Images)

Representative Participants

Hrozova, Erzsebet (born 5 May 1938). Hrozova was born into a poor family of seven children. Although she was a member of the Hungarian Military Alliance, where she was taught to use weapons, she worked as a nurse in a hospital. Both she and her first husband, a pilot (lieutenant) posted to the Soviet-Hungarian Military Air Base in Taszar, joined the revolutionary group from Budapest's Eighth District, and between 23 October and 4 November 1956 she fought in street engagements in various parts of Budapest. She was captured and imprisoned in June 1957 and charged with manslaughter and robbery. No hint of the political intent of her activities was allowed to surface. The Court of First Instance sentenced her to death on triple charges of attempts to kill (*a*) a State Security Authority (A VH) man who would not allow food from a burned-out shop to be distributed to the citizens, (*b*) a sniper, and (*c*) a Soviet soldier who jumped out of a tank. The Court of Appeals commuted the death penalty, but she remained in prison for thirteen years, until 1970. She is alive and resides in Hungary.

Magori, Maria (born 1913; died 28 October 1958). A cleaning woman, Magori was executed on 28 October 1958.

Salabert, Erzsebet. A crane operator, Salabert was executed on 28 November 1958.

Sebestyen, Maria (born 28 February 1935). Sebestyen's father, an officer of the Hungarian army, and her brother, a Lutheran priest, were both deported to the Soviet Union in 1945. Her family suffered severe discrimination under the Rakosi regime, making it impossible for her to become a registered nurse. During the 1956 street fighting Sebestyen joined the Revolutionary Red Cross (Voluntary Ambulance) attending wounded regardless of affiliation. On 4 November 1956 Sebestyen took the lead in women's demonstrations in front of the American Embassy in Budapest, protesting Soviet use of heavy military attacks on civilians. After being arrested in 1957 and beaten during interrogations, she was charged with agitating against the People's Republic of Hungary, based on her role in the demonstrations. Although she was released in 1958, she remained under surveillance and could only find sporadic employment as an extra in movie crowd scenes. After 1991 she was living on a meager disability pension.

Sticker, Katalin (born 26 December 1932; died 26 February 1959). A divorced textile worker, Sticker fought in street fighting alongside Maria Wittner (see below) until Wittner was wounded. Sticker fled to Switzerland, but was called back by her new fiancé on the promise of an official pardon. Instead, in May 1958 she found herself in the dock again alongside Wittner; the Court of First Instance sentenced Sticker to death, and the sentence was carried out on 26 February 1959.

Toth, Ilona (born 23 October 1932; died 27 June 1957). A medical student, Toth took part in the organization of the Revolutionary Volunteer Ambulance (RVA) service, which supplied revolutionaries with first aid and food. She became head of RVA in November 1956. Toth allowed the illegal organ *Our Lives* to operate in the hospital under her management. She was arrested on 19 November; hers was a show trial designed to demonstrate how intellectuals became criminals during the 1956 Revolution. She was executed on 27 June 1957.

Wittner, Maria (born 9 June 1937, Budapest). Abandoned as a child, at the age of two Wittner was placed in an orphanage run by nuns. When these nunneries were nationalized, Wittner was placed into various state-run, prisonlike educational institutions. In 1955 she gave birth to a son, who was also made a ward of the state, and at age eighteen she was allowed to leave the institution. On the afternoon of 23 October 1956 she joined a group of street fighters assaulting the Hungarian Radio Station, helping the young men staging the attack to load machine guns on rooftops. Together with Kati Sticker (see above), and as a member of the Corvin Group, she fought against State Security Authority (AVH) forces then stationed in the Ulloi Street barracks. Later, in November 1956, when Soviet heavy military forces staged the recapture of the barracks, she was hit by shrapnel in her thigh, leg, and spine as she and Kati Sticker were carrying munitions to their group. Her streetfighting days were over, but her troubles were not. After partial recovery from her injuries, she planned to flee from the country, but when she returned for her son, she was arrested and imprisoned on 16 July 1957. After long and drastic interrogations, the trial of Wittner and nine others began in May 1958. The Court of First Instance sentenced both Wittner and Sticker to death; however, the death sentence was commuted in her case because of her youth. She was, however, not released from prison until 1970. Wittner is alive and resides in Hungary.

Bibliography: Kapitany, Eva, *301* (Aura Publisher, 1990); Krasso, Gyorgy, *Maradj velunk* [Stay with us] (Maganzarka kiado, 1991); *Pesti Utca—1956* (Budapest Streets—1956), a selection of street fighters' memoirs (Budapest: 1994).

—Antal Leisen

HUSSITE WARS, WOMEN IN. Bohemia, 1420–1428.

Early-fifteenth-century Bohemia (home of the Czech peoples) was part of the Holy Roman Empire when Jan Hus, chancellor of the university at Prague, called for reform within the Catholic Church. King Sigismund of Hungary granted Hus a safe-conduct to answer charges of heresy at the Council of Constance in 1415, but after Hus arrived, he was arrested and burned at the stake. The subsequent revolt in Bohemia turned into the Hussite Wars, which would plague the empire until 1436. General Jan Žižka, the commander of the Hussite armies, was a military genius who achieved victory through disciplined training, innovative use of new weapons, and troop organization. He did not hesitate to employ anyone willing to fight, so that some writers have incorrectly described his force as "Žižka's army of women."

The activities of female soldiers seem to be concentrated in the years 1420–1421. Open conflict began in 1420 when the Hussites found themselves under increasing pressure from Sigismund, who was preparing to launch the first of five anti-Hussite crusades. Women played an important role in preparations for and participation in battle as the Hussite Czechs defended Prague against Sigismund's first crusading army. Our knowledge of women's participation in the Hussite military campaigns comes from Laurence of Březová, the author of the *Hussite Chronicle*. Historians consider Březová's work reliable. He tells us that on 8 April 1420 the people of Prague dug a protective moat, a task in which all residents, including women and children, participated. Women were also participating in revolts in the rural parts of the kingdom. Laurence tells us that on 23 April "a multitude of squires and common people, even women, adherents of the chalice, and a multitude of rowdies gathered.... They committed many unheard and strange scandals in the Czech kingdom, especially in the regions of Bechyn'e and Plzen' ... burning and destroying churches, monasteries, vicarages and conquering and burning castles, towns and fortresses." Some of these eventually went to Prague to help the capital city. Soon women from many parts of the land joined the struggle to defend their faith and their nation.

> *A description of medieval warring women from the Czech foundation myth:*
>
> They went forward, with Vlasta, the most powerful in the lead.
> She stationed herself in the middle of the finest,
> and they stood on the battlefield, eager for battle with the men,
> Vlasta, sitting on her horse holding her spear, spoke to her troops:
> "O maidens, noble creation! nothing in the world is more noble than you are.
> Preserve your nobility, attain something good for yourselves.
> Take pleasure in exerting yourselves
> and as a result you will have eternal peace and honor.
> If we crush them now, our deed will always be remembered and praised.
> And we will select our own husbands,
> and when we want to, we can beat them.
> And we will be like the Amazonian women,
> who ordered the men to do the plowing,
> while the women governed the land.
>
> —M. Bláhová, *Staročeská kronika tak řečeného Dalimila* (vol. 3) (Prague: Academia, 1995), 179 (submitted by John Klassen).

On 7 May the lord of Vartemberk, a man thought of as a friend of the Hussites, handed over the main Prague castle to their enemy, King Sigismund. In serious danger now, Žižka called upon the radical Hussite community at Tábor to join him on the march to Prague, where they would meet the army of King Sigismund of Hungary, defender of Catholic unity and heir to the Czech throne. According to Březová, 9,000 soldiers joined Žižka's march, many of whom were women and children. Equipped with lances and maces, female soldiers from the Bohemian towns of Tábor, Louny, Slany, and Zatec participated in the march to Prague.

Women worked and fought on their own initiative as well as under the leadership of men. The king and his European crusaders held two important approaches to the city: Hradčany Castle to the west and Vyšehrad Castle to the south. Their goal was to take a third spot leading northeast below Vítkův hill, held by a small contingent of Hussites. The city stepped up work on its fortifications immediately after Vartemberk's action. Women took a leading role in rebuilding the city's guard. On 25 May 1420 the Táborite women, acting on their own initiative, removed enemy sympathizers from areas where they could do damage by destroying the Monastery of St. Catherine's in New Town (after evicting the nuns who lived there). The Hussite leaders had tried to prevent such actions but nonetheless sympathized with the anticlerical sentiment of the female soldiers. A few days later a large group of women dug a moat leading away from St. Catherine's, about a kilometer in length, toward the Church of St. Appollinary under instructions from the city elders. A little later two women and a priest entered the Church of St. Michael's, and while the priest talked, the women carried off its benches in order to fortify the city's defenses.

Women continued their pivotal contribution as warriors in the first major battle of the revolution. On 14 July 1420 troops led by the margrave of Meissen attempted to surmount a defensive wall protecting the small contingent stationed on Vitkův hill. It is impossible to judge how many women fought at Vitkův, but Březová's most famous anecdote from this battle involves three Hussite women who saved the day against the larger army. One of them ignored the advice of her male colleagues, who wanted to retreat, and went out from behind the wall. The words of the chronicler describe her action and fate: "And when the enemy wanted to ascend the wall made of stone and earth, two women and a virgin with sixteen men, who remained in the fortress, manfully defended it with stones and pikes, because they had neither arrows, catapults nor powder. Therefore one of those women, although unarmed, overcame the strategy of the males. Not wanting to withdraw her feet from where she stood, she said: 'It is not fitting for a faithful Christian to retreat before Antichrist'. And fighting manfully she was killed and gave up her spirit" (*Laurence of Březová*, 387–388).

The Hussites went on to win the ensuing battle, and Sigismund had to retreat. The unnamed woman had saved a small garrison and determined the tide of the battle, which was one of many Hussite victories against their foreign enemies throughout the fifteenth century. In the same year Hungarian troops reported that they had captured 156 Hussite women dressed as men with their hair cut and armed with swords and stones. In another battle in 1422 Hussite women fought openly alongside the men, and in general they were equal to males in ferocity and cruelty, refusing to spare the lives of their enemies in what they saw as a holy war. The last reference to women in battle was in 1428.

Women's roles in Žižka's military campaigns found their way into the nationalist legends of the Czech people. Alois Jirásek, who wrote the beloved *Old Czech Legends* in the 1890s, told the story of one battle in which the Hussites were rendered victorious by the participation of women along with Žižka's military genius. Žižka led his troops into a recently drained pond and ordered his female soldiers to lay their veils and cloaks on the swampy ground at the bottom of the pond. When the heavily armed enemy descended upon the small, weak Czech army in the pond, they began to sink into the mud and consequently got entangled in the cloth of the women's veils. The knights were unable to keep their balance or use their swords, and the enemy retreated. While this story is most likely apocryphal, it—as well as the more reliable information from Březová's chronicle—served to inspire Czech feminists in the nineteenth century. Women demanded their rightful place in the Czech national movement partly on the grounds that their foremothers had defended their nation during the Hussite Wars.

Bibliography: David, "The First in Austria," 1991; Haselholdt-Stockheim, *Herzog Albrecht IV. von Bayern und seine Zeit,* vol. 1, part 1, 1865; Heymann, *John Zizka and the Hussite Revolution,* 1955; Klassen, "Women and Religious Reform in Late Medieval Bohemia," 1981; Kolářová-Císařová, A., *Žena v hnutí husitském,* 1915; Laurence of Březová, "Laurence of Březová's Hussite Chronicle," 1873; Palacky, František, ed., *Archiv Český čili staré písemné památky české í morauské,* vol. 5 (Prague: 1862).

—John Klassen and Cynthia Paces

I

ISABEL DE CONCHES (fl. 1070–1100s), also known as Elizabeth of Conches; Isabel de Montfort, Tosny, Toni. Flanders.

Orderic Vitalis writes that in 1090 Isabel and her sister-in-law Helwise of Evreux instigated a civil war between their husbands Ralph, the lord of Conches, and William, count of Evreux. The antagonism between the two women was allegedly over insults to Helwise by Isabel. Helwise is accused of persuading her husband Count William to make war on Conches. However, as Marjorie Chibnall points out in *The World of Orderic Vitalis,* the real issue at hand was inheritance rights between the two brothers.

Orderic says that Isabel "rode in knightly armour when the vassals took the field." He also comments on her courage, but there is no other evidence that Isabel actually fought. Her primary military roles were encouraging her husband to keep the war going (which he eventually won) and, by her presence, helping to keep morale generally high. The story does, however, indicate that the presence of women on or near the battlefield (especially the wives of the nobility) was not unusual, although Isabel's mode of dress may have been. Orderic also provides a story showing Isabel as advisor and confidant to her household

> *Historical evidence of military roles for medieval women:*
>
> Scattered studies have indicated military actions by real women in various European countries. Such evidence was often preserved by local chroniclers: Alpert of Metz mentions Liutgarda, who, on her sister's death, took over her estates by force, then captured Elten on the Rhine. Thietmar of Merseburg writes of a certain Christianized Hungarian called Deviux around 1018. His beautiful wife "drank quite a bit, rode to battle like a knight, and even once killed a man in anger." . . . Orderic speaks of a number of women who openly played male roles, like Isabel of Conches, who fought "as fiercely as Camilla."
>
> —Gloria Allaire, " 'Babes' in Arms: Literary and Historical Antecedents to the *Guerriera* of Renaissance Epic," unpublished article.

knights when her son and two other knights discuss their dreams with her in the great hall at Conches.

Bibliography: Chibnall, *The World of Orderic Vitalis*, 1984; Echols and Williams, *An Annotated Index of Medieval Women*, 1992; McLaughlin, "The Woman Warrior," 1990; Orderic Vitalis, *The Ecclesiastical History of Orderic Vitalis*, 1854.

—Annette Parks

ISABEL, COUNTESS OF BUCHAN (born ca. 1287; died ca. 1314), also known as Isabel of Fife. Scottish-English wars, fourteenth century.

Isabel was the sister of Duncan III, the ninth earl of Fife, and the wife of John Comyn, count of Buchan. According to tradition, one of the three special privileges claimed by the earls of Fife was that of placing the king on the throne above the Stone of Destiny. Although the Stone had been removed from Scotland by Edward I in 1296, Robert Bruce and his supporters were concerned to make his coronation conform as closely as possible to tradition in order to give his reign as much legitimacy as possible. As the representative of her birth family, it was Isabel who crowned Robert Bruce Robert I of Scotland at his second coronation on 27 March 1306. Isabel was a staunch nationalist. Even though she was only about twenty years old, Isabel defied her husband in his support of the English and acted in a way that was certain to escalate the ongoing conflict between England and Scotland.

Robert Bruce's claim to the throne of Scotland was precipitated on 10 February 1306 when he murdered John "Red" Comyn, his chief political rival and the cousin of Isabel's husband. Based on Bruce's haste to assume power in the wake of the killing, many historians believe that a rebellion was already in the final stages of planning by February 1306. Isabel might have been aware of such a plot. In any event, Walter of Guisborough says that she slipped away from her home (it is unclear whether she was in England or Scotland at the time), stealing several of her husband's war horses, and raced to Scone, the traditional coronation place of the Scottish kings. Once there, she asserted her right as the representative of the senior Scottish earldom of Fife (her father was dead and her brother was being held in England) to assist in the coronation. Bruce's first coronation, held on 25 March 1306 at the traditional site at Scone, was attended by at least three Scottish bishops and four earls. However, as G.W.S. Barrow properly notes, without the traditional ceremony—in which the participation of the house of Fife was a crucial element—the coronation might have been challenged by the more conservative Celtic elements. Isabel's arrival and claim to her family's right were considered important enough by Bruce and his advisors to conduct a second ceremony on Palm Sunday, two days after the first coronation. By participating in the coronation, Isabel helped to add a much-needed aura of legitimacy to Bruce's coronation that bolstered his position in intangible, but important ways. Despite the murder and sacrilege that preceded his ascension, he was enthroned as the king of Scotland according to ancient Celtic tradition. It should be noted that not all historians accept the story of a second coronation as valid (cf. Archibald A.M. Duncan, *Acts of Robert I*).

Bruce's rebellion ran into trouble almost immediately with a rout of his army on 18 June 1306 at Methven. In the aftermath of this defeat Isabel and several of Bruce's female relatives went into hiding. They took refuge at Kildrummy Castle, were besieged there by Edward's forces, and managed to escape to the sanctuary of St. Duthac's Chapel at Tain, but were finally captured by Edward I's loyalists. They were removed from the sanctuary by the earl of Ross and sent to Edward in the summer of 1306.

Isabel, along with Bruce's sister Mary, was marked out by Edward for special punishment. They were both ordered confined in specially constructed cages "strongly latticed with wood, crossbarred, and secured with iron" (Palgrave). The cages were also to be equipped with privy chambers. Walter of Guisborough adds the colorful detail that because of her role in the coronation, Isabel's cage was to be built in the shape of a crown. During her confinement her needs were to be attended by Englishwomen who were not sympathetic to her, and she was not allowed to have contact with anyone of Scottish extraction. When the weather permitted, her cage was to be displayed outside so that all passing could see the occupant and note her punishment. Isabel was considered such an important prisoner that when Edward I died in June 1307, Edward II elected to continue her confinement, rejecting pleas for clemency on her behalf. She was confined at Berwick Castle for three years, until she was moved to a Carmelite nunnery in June 1310, probably due to ill health, and then to the custody of Englishman Henry de Beaumont. Her further fate is unknown. Although most of the women taken with her were released after the Scottish victory at Bannockburn in 1314, Isabel probably died in English custody.

Isabel of Buchan was not a soldier, but was willing to take actions of great political and military significance despite the risk to herself. After all, the previous rebellion led by William Wallace had led to his horrible execution by the English in August 1305. Isabel's courage as well as her subsequent harsh punishment made her an inspiration and martyr in Scottish history.

Bibliography: *Dictionary of National Biography*; Echols and Williams, *An Annotated Index of Medieval Women*, 1992; Fittis, *Heroines of Scotland*, 1889; Neville, "Widows of War," 1993; Powicke, *The Thirteenth Century*, 1991.

—Annette Parks

ISABELLA OF FRANCE (born 1292; died 1358, London), also known as Isabella of England, Isabelle de France, Isabelle d'Angleterre. Queen consort of England, 1308–1327, war leader, head of government, 1327–1330, Hundred Years' War.

Isabella of France

Queen Isabella (1292–1358) of France, wife of King Edward II of England, entering Paris. From *Chronicles of France*, by Froissart, 1400. (The Art Archive/British Library)

Isabella was the daughter of Philippe IV "le Bel," king of France. She married Edward II, king of England, in 1308 and later led a successful invasion to depose Edward in favor of their young son Edward (III). With her lover, Roger Mortimer, Isabella led the government during the minority of Edward III until they were overthrown by Edward in a coup at Nottingham in 1330.

One of the most colorful and controversial figures in fourteenth-century English history, Isabella was a formidable child of an equally formidable father, Philippe IV of France, who clashed with the papacy and engineered the suppression of the Knights Templar. Married to Edward II of England soon after his accession in 1307, Isabella was immediately alienated personally and politically when Edward allegedly gave her wedding jewels to his favorite, and probable lover, Piers Gaveston. She accompanied Edward on his ill-fated Scottish campaign of 1319 and had to escape by riverboat when York was attacked by Scots invaders. She played no apparent role in the political opposition to Edward in the middle years of his reign, which saw the overthrow and death of Gaveston.

Isabella may have first become an active participant in the political-military conflicts in 1321. While traveling, Isabella sought hospitality at the royal castle of Leeds, Kent. Edward doubted the loyalty of Lord Badlesmere, constable of the castle, who was away. It has been suggested that Isabella was sent to gain admittance to, and possibly control of, the castle. However, **Lady Badlesmere** refused to receive the queen. A fight ensued, and some of Isabella's attendants were killed. This incident became the pretext for Edward to besiege and occupy the castle.

After the defeat and death of Edward's leading adversary, Thomas of Lancaster, at Boroughbridge in 1322, Isabella and her young son Edward, Prince of Wales, increasingly became a focus of opposition to Edward's tyranny. Edward had allowed the political dominance of the Despensers, a father and son who had gained the ascendancy at Edward's court and had been enriched with confiscated lands seized from the rebels defeated at Boroughbridge. The seizure of Isabella's estates by Edward in September 1324, on the pretext of fear of a French invasion, may have helped push her into opposition.

Isabella's political maturity in these years was demonstrated by the decision of Edward II to send her to France to negotiate a treaty with her brother Charles IV over the disputed duchy of Gascony in 1325. In France she began to organize opposition to her husband and to the Despensers and was joined in exile by Roger Mortimer, a dissident lord who became her lover. When the king foolishly allowed their son Prince Edward to join her in France in order to perform homage for the duchy of Gascony on his father's behalf, Isabella had the trump card in her possession: the heir to the throne. She declared that neither she nor her son would return to England while the younger Despenser remained at court.

Isabella, Mortimer, and young Edward invaded England with the support of William II, count of Hainault, landing in Suffolk on 24 September 1326. With a force of only 700 Hainaulters, the queen, in the words of the chronicler Henry Knighton, "found favour with all." The people of London rose in her support, forcing the king and his supporters to flee to the west; Isabella pursued him through Oxfordshire and Gloucestershire. The people of Bristol forced the elder Despenser to surrender to Isabella, whereupon he was tried for treason and executed. The army then proceeded to Hereford, and Henry of Lancaster was dispatched to hunt down the king and the younger Despenser, who had fled into South Wales. They were captured at Neath, and Despenser was subsequently tried and executed as a traitor. Edward was deposed, at the instigation of the rebels, by an assembly of the "clergy and people" in January 1327 and was succeeded by his son as Edward III, although Isabella and Mortimer held de facto power. The deposed king died, presumed murdered, at Berkeley Castle in September 1327.

Isabella and Mortimer did not enjoy continued military success; the defeat of the English army at Stanhope Park led to the Treaty of Northampton in 1328, which finally recognized Robert Bruce as king of an independent Scotland. This treaty, dubbed a "shameful peace" by contemporary chroniclers, did much to undermine the Isabella-Mortimer government, especially in the eyes of the "disinherited," those English lords who had lost their lands in Scotland. The regime was also accused of corruption and favoritism. Isabella and Mortimer were overthrown in a coup d'état by the young Edward III and his supporters at Nottingham in 1330.

After her displacement Isabella was allowed to go into an honorable retirement on a pension of £3,000 a year at Castle Rising in Norfolk. There is no truth in the story that she went into exile in Scotland and led a war band against her son's armies. In her old age she took the habit of the religious order of the Poor Clares. She died in 1358 and was buried in the Franciscan church at Newgate, London.

Isabella has been much maligned by history. To contemporary chroniclers she was a "Jezebel"; to posterity she became the "She-Wolf of France." However, in 1327 she was, albeit briefly, seen as the "liberator" who had delivered her adopted people from the tyranny of Edward, and she was almost certainly guiltless of her husband's murder. Although her military leadership qualities were never tested in a pitched battle, she showed courage, resourcefulness, decisiveness, and the ability to inspire loyalty in her campaign to overthrow Edward.

Bibliography: Echols and Williams, *An Annotated Index of Medieval Women,* 1992; Fryde, *The Tyranny and Fall of Edward II,* 1979; Johnstone, "Isabella, the She-Wolf of France," 1936; McKisack, *The Fourteenth Century,* 1959.

—Michael Evans

ISABELLA DE LORRAINE (born 1410; died 28 February 1453). Queen of Sicily, duchess of Anjou, Lorraine, and Bar, countess of Provence.

While Isabella de Lorraine did not don armor or ride at the head of troops, she is credited with staving off a challenge to her inheritance of the duchy of Lorraine during her husband's long imprisonment and with valiantly defending his claim to Naples and Sicily. Isabella was the elder daughter and heir of Charles II, duke of Lorraine, and Margaret of Bavaria. In 1419 she was married to René of Anjou, second son of Louis II of Anjou and Yolanda of Aragon. The marriage produced at least four children, including **Margaret of Anjou** (1430–1482), who was married to King Henry VI of England in 1445 as part of a truce in the Hundred Years' War. When Isabella's father, Charles II, died in 1431, René became duke-consort of Lorraine; however, the succession was challenged by a nephew of Charles II's, Antony of Vaudémont. Later that year (2 July) Antony defeated René at Bulgnéville and took him prisoner. René was subsequently turned over to his great political enemy Philip the Good of Burgundy, who imprisoned him at Dijon.

Isabella immediately took up the defense of the duchy against Antony and helped to negotiate René's release on parole in May 1432. As part of the terms of his release, Isabella and René had to agree to give their sons John and Louis as hostages in his place. During this period of freedom René acquired the titles of duke of Anjou and count of Provence on the death of his brother, along with the family's titular claim to the kingdom of Sicily. In 1434, shortly before her death, Joan II, the reigning queen of Naples and Sicily, adopted René, drawing him and Isabella into a succession dispute with Aragon's king Alphonso. Alarmed by René's growing power and political influence, Philip the Good demanded his return to captivity in the same year, leaving Isabella not only to defend the duchy of Lorraine, but also to press his claims to Naples and Sicily. By all accounts, Isabella was successful in both endeavors.

In 1435 Isabella sailed from Marseilles to Naples. Once there, she took control of the troops loyal to René's cause and repelled Aragonese efforts to take the city. In addition to directing the city's defense, she also skillfully negotiated with both mercenary leaders and the Genoese for soldiers and naval support. Her political skills were further tested by recurring money shortages and the need for delicate diplomacy in the dispute that not only involved her husband and the king of Aragon, but also the king of France and Pope Eugenius IV. René was finally released in February 1437 and went to Naples in the following spring to join his wife. However, it was Isabella who successfully defended the city against a major offensive launched by Alphonso in 1438. Even after his arrival René continued to rely on Isabella in both political and military matters, and the two worked together in attempting to secure René's claim to Naples and Sicily. The city was once again under siege by Alphonso in 1441, and by 1442 René and Isabella were forced to abandon Naples permanently, along with René's ambitions to establish an Italian kingdom.

After their return to France Isabella continued to be actively involved in the administration and defense of Lorraine, where, with help from Charles VII of France, her right to the duchy was firmly established. Working with René, she was also involved in negotiating marriages for her children designed to secure peace with Burgundy and to bolster René's claims to territories in Maine. Isabella died on 28 February 1453 after a lengthy illness and was succeeded in Lorraine by her son John.

Bibliography: *Dictionary of National Biography;* Echols and Williams, *An Annotated Index of Medieval Women,* 1992; Hoefer, *Nouvelle biographie générale,* 1858; Kibler and Zinns, *Medieval France: An Encyclopedia,* 1995; Lalanne, *Dictionnaire historique de la France,* 1877; Ryder, *Alfonso the Magnanimous,* 1990.

—Annette Parks

ISRAELI WAR OF INDEPENDENCE, WOMEN'S MILITARY ROLES. Defensive and revolutionary activities from early twentieth century; open conflict, 1947–1949.

The history of women's military experience in Israel is important because it is studied by many other nations. For many in the West, it is the best-known example of women in combat, although women fought on a much larger scale in the Soviet Union and other places. Similarly, the fact that Israel removed women from combat roles immediately upon professionalizing its military implies a failure of women in the military role, even though this phenomenon was observable in the Soviet Union, Yugoslavia, and many other situations and had nothing to do with combat performance.

Israeli War of Independence, Women's Military Roles

An Israeli woman soldier, a member of the Haganah, 1948. (Hulton/Archive by Getty Images)

It was the political and cultural situation in the Middle East that led to numerous Jewish women taking up arms in the first half of the twentieth century. Five major waves of immigration (*Aliyah*) brought Jews to Palestine beginning in the late nineteenth century. The first émigrés were primarily Jewish families seeking a traditional lifestyle, but the second and third waves in the first decades of the twentieth century brought many revolutionary-minded Jews who hoped to create an entirely new kind of society. It was during this period that the frontier mentality became dominant, with communal and socialist philosophies inspiring the formation of collective settlements (*kibbutzim*) and collective farms (*moshavim*). These new settlements often faced hostile neighbors, and self-defense was as much a part of the communal life as farming. In addition, many of the new émigrés were strong believers in equality for women, and most of the settlements held an ideal of equal roles for women and men, though there was often a discrepancy between the ideal and actual practice.

Many among the *Yishuv* (the term used for the Jews who lived in Palestine before Israel became a recognized state) felt the need for structured defense. As early as 1907 a secret group called Bar Giora was formed; its model was the warrior Bedouin, and two women who later became famous were among its members (Ester Becker and **Mania Shochat**). In 1909 Bar Giora became Hashomer, whose policies were vague but whose purpose was to create "watchmen" who would guard the *Yishuv*. From the beginning, there was intense debate over whether women could actively participate in "guarding." The question was never formally resolved, and this would lead to a dichotomy in women's roles in Hashomer. Most women, wives or relatives of male members, took some part in Hashomer activities, but were not considered equal-status members and generally performed traditional "women's work" like caring for the men. Other women, like Becker and Shochat, were full and active participants who were intimately involved in organizing and carrying out Hashomer activities, including stockpiling and transporting weapons. This dichotomy would persist in other secret organizations and in fact is still found in the conflicting images of Israeli military women today.

Israeli War of Independence, Women's Military Roles

After the First World War control of Palestine was transferred from the Ottomans to the British, after the 1917 Balfour Declaration had supported the idea of a Jewish homeland in the region. These changes led to increased conflict between Jews and Arabs living in Palestine. In 1920 the Arabs attacked two outposts, and three Jewish women died defending the one at Tel Hai. The Arab success led the *Yishuv* to organize a new secret group, Haganah ("defense"), to organize the equipping, training, and defense of Jews in both settlements and towns. Haganah was a much bigger organization than its predecessors, but central control was weak, and women's roles varied depending on region. There were ultimately more than 10,000 women soldiers in the Haganah. Young women especially were encouraged to join, but local commanders often dictated the extent of their activities. Many women were involved in hiding, cleaning, and moving weapons. In an interview with Anne Bloom, Mina Ben-Zvi recalled that "we fought against separation and wanted mixed platoons of men and women. There were those [women] who disagreed, and they were sent to prepare sandwiches and do first aid. [Other women] wanted to take part in combat and fought for the same training.... I was then sent to Juara for officer training, where there were about four women to forty men" (Bloom, 143). Sara Kahn stated that "women took equal part with the men at that time" because the Haganah was mainly involved in defending its own positions. Later, she says, around 1937, when the Haganah began conducting raids into enemy territory, most local groups insisted that the women remain behind on defense while men went raiding.

By the outbreak of the Second World War the situation in Palestine had grown even more complex. The British had declared in 1939 that only a handful of additional Jews would be allowed to immigrate, but the British were also fighting Hitler, whose anti-Semitic policies were already clear. Most of the *Yishuv* decided to support the British in fighting the Axis powers (and call a truce with the Arabs in Palestine). The British extended their recruitment among the *Yishuv*, and some 4,000 women were taken into the **ATS** (Auxiliary Territorial Service) to serve in the region. A few Arabs and other women joined the ATS as well, but the recruits were predominantly Jewish. Women in the ATS were relegated to support roles, and these veterans would later serve as a counterweight to the Israeli women who fought on the front lines, as ATS veterans would dominate the postwar **CHEN** (Cheil Nashim [women's corps]).

> The Army is the supreme symbol of duty and as long as women are not equal to men in performing this duty, they have not yet obtained true equality. If the daughters of Israel are absent from the army, then the character of the Yishuv will be distorted.
>
> —David Ben-Gurion, quoted on the IDF Web site, http://www.idf.il/english/organization/chen/chen.stm.

Other Jewish women preferred to serve in Haganah. In May 1941 Haganah created a new secret organization, the Palmach (assault companies) to take on especially demanding missions. Palmach companies, like the Haganah, were based in both cities and rural areas. There was some financial support from the British for the Palmach in the early period, but it later was withdrawn. The question of women's roles immediately came up, and Haganah issued orders against them. Some local commanders disobeyed; the Jerusalem company created the first women's unit in the Palmach. There was some administrative flurry—a regional Palmach commander declared that the women were to be court-martialed for illegally joining the Palmach—but by May 1942 the national command decided to permit women to join, though it set a limit of 10 percent. Later, however, youth groups supporting the Palmach were typically half female, and ultimately women made up nearly a third of the Palmach.

Another secret group, the Irgun, targeted the British as the real enemy of the *Yishuv*. Of about 2,000 active members, 400 were women; they were mainly involved in anti-British agitation and courier duties, though some were involved with explosives. Again, it was mainly male reluctance rather than female ability or motivation that kept women away from the front lines; a ranking Irgun member later said that even though "there were some girls who were better in action than some of the men," most women were prevented from fighting by men in the local units (Bloom, 149).

After the Second World War was won, the United Nations refused to partition Palestine into separate Jewish and Arab states. The result was a sixteen-month war, ending on 10 March 1949, in which Israel fought for its right to exist. Women participated in combat during the Israeli War of Independence (1947–1949), and 114 were killed in action, 18 of whom were in the Palmach. Rachel Stahl commanded a Palmach company and led it in combat in the area of Tzfat (Safed)—one of five women to command a combat unit. Women soldiers fought in the Palmach's Negev and Harel Brigades. A female pilot, Yael Rom, flew combat missions with the tiny Israeli air force.

In the spring of 1948 an incident occurred that set the precedent for the future of Israeli military women. Two women officers who had served with the ATS during the war were asked to form a women's battalion in Jerusalem that would put women into the more restricted roles of the type they had held in the British women's auxiliary. Esther Herlitz, one of those officers, says, "One of the things we had to do was to take the Haganah girls out of the front line. It was a decision about which there is controversy till this day. ...The Haganah girls didn't like it, especially the radar girls.... They resented it."

With the creation of the IDF (Israel Defense Forces), women's roles became formalized, following the ATS model. In 1949 CHEN was formed, and women were thereafter excluded from combat roles. There was much conflict in the early years between ATS veterans (who stressed British-style military form, discipline, and order) and veterans of the secret organizations like Palmach, who reviled the lack of combat training for women, and who also rebelled against formal military trappings and courtesies. (See the entry on CHEN for a discussion of women in the postwar Israeli army.)

The official justification given for the combat exclusion in Israel is usually the fear that women in combat positions might be captured by Arab soldiers; they should be protected from direct contact with the enemy. Given the daily risk to civilian women in Israel from terrorist attacks and raids, this reasoning seems rather weak. The cultural basis for this fear is the belief that women are treated more harshly than male prisoners; reportedly, one female Palmach fighter in 1948 was captured and so badly mutilated by the enemy that women were pulled out of the fighting for a few days. This one incident, if true, appears to be the root of many horror stories about what could happen to female prisoners. However, Israeli men have also reportedly suffered gruesome fates as prisoners, including castration. Many question the oft-quoted myth of the tortured female prisoner; one veteran stated in print that "to the best of my knowledge those few Jewish girls, soldiers and civilians, who happened to fall alive into Arab armies' hands were treated in a respectable manner," noting that thirty women soldiers taken by Lebanese forces in the Six-Day War were treated well and returned unharmed (*Jerusalem Post* article, 22 June 1978, cited by Bloom).

Prominent historians and commentators often use the case of Israel to bolster arguments both for and against women in combat. The image of the brave Israeli woman fighting in the War of Independence is used to show that women can and have fought on an equal basis with men; the subsequent relegation of women to support roles in the IDF is used as proof that women should not be in combat. If Israeli women soldiers fought so well, the argument goes, why then are they no longer allowed to fight? Martin Van Creveld, a professor of history at Hebrew University in Jerusalem, testified in 1991 before the Presidential Commission on the Assignment of Women in the Armed Services in Washington, D.C., claiming that "women have never taken a major part in combat—in any culture, in any country, in any period of history." Creveld even tried to claim that no woman actually fought in the War of Independence, writing that he was "unaware of a single Israeli woman who has claimed, 'I was in combat in 1948.' There may have been a few such women, but for the most part they existed only in Arab imagination" (Gal makes a similar claim). Creveld went on to imply that the exclusion of Israeli women from combat roles is based on sounder historical experience, since "unlike most Americans, Israelis are familiar with combat." Like many aspects of the history of military women, this is a historical experience surrounded by mythology, assumption, and oratory, but few have attempted to systematically document what really happened, and why.

Few of the numerous books and articles produced in the West about the question of Israeli women in combat are based on any sort of historical evidence. Many, like Creveld and Gal, repeat stories and rumors of women who were mistreated, or claim that women did not actually fight. Israeli women veterans of this war have published memoirs in Israel (including those by **Netiva Ben-Yehuda** and **Rachel Yanait Ben-Zvi**), but few, if any, have been translated or are ever cited. Even in Israel itself, real inquiry into this issue has been neglected until recently. There is solid evidence that women did indeed participate in many combat and other violent actions in Israel, but no scholarly study, with the exception of Anne Bloom's well-researched article, has attempted to systematically document what really happened.

Bibliography: Bloom, "Israel: The Longest War," 1982; Gal, *A Portrait of the Israeli Soldier*, 1986; Katz, *The Lady Was a Terrorist, during Israel's War of Liberation*, 1953; Saywell, *Women in War*, 1985; Segal, "Women in the Armed Forces," 1993; Van Creveld, "Women of Valor: Why Israel Doesn't Send Women into Combat," 1991.

—Reina Pennington

JEANNE DE MONTFORT (born ca. 1315–1320 France; died after 1373, England), also known as Jeanne de Flandres, Joan of Flanders, Joan of Montfort. Military leader, France, Breton Civil War, 1341–1365.

Jeanne de Montfort was the daughter of Louis de Nevers, count of Flanders, and was married to Jean de Montfort in 1329. The death of Jean III of Brittany in 1341 led to a succession dispute for the duchy of Brittany between the rival claimants Jean de Montfort, half brother of the deceased duke, and **Jeanne de Penthièvre,** Jean III's niece by his younger brother Guy. The two rival factions were backed by the kings of England and France, respectively.

Jeanne de Montfort was a tough and ambitious woman and appears to have been the prime mover behind her husband's campaign. She and her husband moved swiftly to occupy Nantes and Rennes and secure the ducal treasury in May–June 1341. She was also a key military leader of the Montfort faction in her own right. After her husband's capture at Nantes on 18 November 1341, Jeanne organized the defense of the Montfort possessions in Brittany against the Penthièvre faction and their French allies through the winter of 1341–1342 and sent at least two embassies to England seeking support. These secured a formal alliance with Edward III of England in February 1342 and the dispatch of English armies to Brittany, including one led by Edward himself in October 1342. With English support the Montfort faction was able to seize and hold a series of fortresses around the Breton coast. Jeanne's most celebrated military feat was her alleged defense of Hennebont against the army of Charles de Blois in May 1342, when, according to Jean le Bel and Froissart, she donned armor, mounted a horse, and, showing "the heart of a lion and the courage of a man," rode through the streets, encouraging her forces and leading a rout of the opposing camp, burning its tents.

Following the Anglo-French truce of Malestroit in 1343, Jeanne and her two children, Jeanne and Jean (the future Jean IV), went to England with the returning Edward III. The struggle in Brittany was subsequently waged by Edward III's lieutenants; Jeanne took no further active role in the war. Froissart claimed that she led the pro-Montfort forces in a naval engagement off Guernsey, and that she continued to command her forces in battle after her husband's death in 1345; however, his account seems to be based on a confusion or conflation of different naval battles, in none of which Jeanne took part. She never returned from exile in England. Froissart, who was attached to the household of the queen of England, **Philippa of Hainault,** tended to glorify the role played by Jeanne de Montfort, so his descriptions of her actions need to be treated with some skepticism.

The experience of exile broke Jeanne's spirit, and she apparently lost her reason and was handed over to the care of keepers. She was probably briefly reunited with her husband at Easter 1345, when he escaped to England, but his death shortly afterwards in September probably proved the last straw. Several accounts relating to her keeping from 1343 until 1373 survive in the English exchequer records, while her children were brought up in the household of Queen Philippa. She died in England sometime after 1373, never returning to Brittany, where her son became Duke Jean IV in 1364.

Bibliography: Echols and Williams, *An Annotated Index of Medieval Women,* 1992; Froissart, *The Chronicle of Froissart,* 1967; Jones, "The Breton Civil War," 1981, and *The Creation of Brittany,* 1988; Sumption, *The Hundred Years War,* 1990.

—Michael Evans

JEANNE DE PENTHIÈVRE (born ca. 1320; died Guingamp, Brittany, 10 September 1384), also known as Joan of Penthièvre, Jeanne de Blois. Duchess of Brittany (1341–1365) and military leader, France, Breton Civil War.

Because a central feature of the Breton Civil War was the opposition between parties led by Jeanne de Penthièvre and **Jeanne de Montfort,** the war has sometimes been called the "War of the Two Jeannes." Jeanne de Penthièvre is said to have led her forces in person on a number of occasions, though reports of her presence during combat may be apocryphal. Like her opponent Jeanne de Montfort, Jeanne de Penthièvre proved an able and vigorous military leader.

Jeanne de Penthièvre inherited a strong claim in her own right to the duchy of Brittany in the disputed inheritance following the death of Duke Jean III in 1341. She was the daughter of Jean III's younger brother Guy (died 1331) and of Jeanne d'Avaugour (died 1327). She was opposed by her father's half brother Jean de Montfort until his death in 1345, and subsequently by his son Jean de Montfort the younger, later Duke Jean IV. The Parlement of Paris adjudicated in favor of the French king's nephew Charles de Blois, whom Jeanne had married in 1337, and who was the favored candidate of most of the Breton nobility and clergy and probably of the deceased Jean III himself. Consequently, Edward III of England, who was disputing the French throne with Philippe VI de Valois, backed the Montfort faction. The Blois-Penthièvre faction succeeded in maintaining control of the key towns of Rennes and Nantes, while the Montfort faction, aided by the English, secured the coasts and drew support from western Brittany.

Following the capture of her husband Charles de Blois by the English at La Roche Derrien in June 1347, Jeanne took over the military leadership and administration of the duchy. She was also engaged in diplomatic activity, attempting between 1348 and 1353 to negotiate a treaty with Edward III that was almost successfully concluded. During these years she also organized the defense of her lands against the English, who besieged Nantes in 1354–1355.

Charles de Blois was released in 1356 on the agreement to pay a huge ransom of 700,000 écus; his defeat and death at the Battle of Auray in September 1364 was a disastrous blow to the Blois-Penthièvre party. Jeanne conceded defeat and entered into settlement negotiations under the auspices of the archbishop of Reims. By the terms of the Treaty of Guérande of 12 April 1365, Jeanne was forced to surrender her claims to the duchy and recognize Jean de Montfort the younger. However, she was allowed to retain the title of duchess during her lifetime, the Penthièvre lands of her father, and rents to the value of 10,000 livres outside the duchy. She and her heirs were granted the County of Penthièvre, based on her estates in northern Brittany, and the recognition that the ducal title would revert to them in the event of a failure of the Montfort line to produce a male heir.

Ironically, Jeanne de Penthièvre helped the Montfort duke Jean IV to regain Brittany following an invasion by the French king Charles V in 1373. Having confiscated the duchy from Jean in 1378, the king also ignored Jeanne's claim to Brittany. Fearing that Charles planned to absorb Brittany into the royal demesne, Jeanne's former supporters, with her tacit approval, appealed to the exiled Jean to return in 1379. She spent her last years in comfortable semiretirement on her estates and died in September 1384.

Bibliography: Echols and Williams, *An Annotated Index of Medieval Women*, 1992; Jones, *Ducal Brittany*, 1970, "The Breton Civil War," 1981, and *Recueil des actes*, 1996.

—Michael Evans

JIMÉNEZ, ANGELA (born 1886, Jalapa del Marques, Oaxaca, Mexico; died ?), also known as "Angel." Soldier, Mexico, 1910 Revolution.

Angela Jiménez was the daughter of a Zapotec woman and the Spanish political chief of Tehuantepec. In 1911 Federal soldiers took over her family home looking for rebels. Angela's sister shot both a Federal officer and herself rather than submit to being raped. Angela decided to join rebel forces to seek revenge on Federal soldiers. At times she dressed as a man and called herself "Angel." She joined her father's command and with him served as a *soldadera*, banner carrier, explosives expert, spy, and, upon occasion, a cook. She also served in disguise under Villista and Zapatista officers. Jiménez left Mexico and settled in San Jose, California. She was one of the founders of the organization Veteranos de la Revolucion de 1910 and became a civil rights activist in the Chicano movement of the 1960s.

Bibliography: Jaiven and Escandon, *Mujeres y revolución,* 1993; Perez et al., *Those Years of the Revolution,* 1974; Salas, *Soldaderas,* 1990.

—Elizabeth Salas

Jeanne d'Arc. (Perry-Castañeda Library)

JOAN OF ARC (born ca. 1412, Lorraine, France; died 30 May 1431, Rouen, France), also known as Jeanne d'Arc, "the Maid of Orléans." Battle leader, strategist, and standard bearer, France, Hundred Years' War.

The military career of Joan of Arc was relatively brief: for some eighteen months Joan directed the French armies against the English in 1429–1430. In camp with the French soldiers, Joan advised the officers and planned battle strategy. Armed with a sword for self-defense, Joan led the French troops on horseback as a noncombatant.

As a child, Joan of Arc experienced mystical visitations from angels and saints, who, she claimed, commissioned her to serve as the Lord's soldier for French liberation. Joan confessed a vision of holy war. Throughout her public career Joan claimed that God intended to restore a united France to the French dauphin (Charles VII), and that she as "the Maid" was responsible for fulfilling this mission. In 1429, when she was seventeen, she set out to persuade various influential people to help her and gained an audience with the dauphin and the support of important military leaders.

Her first military activity was to relieve the siege of the city of Orléans, which had been dragging on for months, since October 1428. On 22 March 1429 Joan issued a letter to the English forces at Orléans with the following warning: "Believe for certain that the king of heaven will grant such power to the Maid, both to her and to her good men of arms, that you will not even know how to defend against all the attacks by her." After weeks of fighting at Orléans, on 8 May 1429 Joan led the French to a decisive victory over the English. Some of her contemporaries compared her to Alexander the Great and Caesar. The following month of June saw a series of French victories led by Joan over retreating English forces. On 17 July 1429 Joan accompanied the French dauphin Charles to his coronation at the cathedral in Reims, the holy site of French coronations. Joan carried her battle standard in the procession.

On 8 September 1429 Joan launched a military attack on Paris—probably her most ambitious effort—that proved unsuccessful. She led forces throughout the winter and spring and successfully besieged Saint-Pierre-le-Moutier and beat the English and Burgundians at the Battle of Lagny. However, in May 1430 Joan was wounded and captured at Compiègne, north of Paris, after attempting to escape pursuit by leaping from a tower. Joan was turned over to Burgundians, her political enemies, who put the nineteen-year-old girl on ecclesiastical trial for heresy beginning in January 1431. Joan's life ended with her condemnation as a relapsed heretic who had led soldiers to "exercise inhuman cruelties in shedding human blood." One of the main charges against her was wearing men's clothing, which was deemed "a thing abominable to God." She was burned at the stake in Rouen on 30 May 1431. Nearly five hundred years later, in 1920, Joan of Arc was canonized a virgin saint by the Roman Catholic Church.

Joan's efforts were not sufficient to bring the war to an immediate end. The Hundred Years' War between the throne of England and the throne of France for rule over French territories ultimately ended in French independence two decades after Joan's death. But Joan's methods continued to have influence on later commanders. Though her use of frontal assaults was not subtle, it proved successful for leaders such as La Hire and Arthur de Richmont in their attacks on the English in the 1440s. By 1449 the armies of the French king Charles VII had recovered from the English nearly all the disputed French territories, including Normandy. Joan's mission was fulfilled posthumously, and the Hundred Years' War drew to a close.

Joan's life inspired imitators, both during her own life and in following generations. Catherine de la Rochelle, a rival mystic who also prophesied on behalf of the king of France, was dismissed as a fake by Joan. In 1436, more than five years after Joan's death, a woman called Claude des Armoises toured France, claiming to be Joan the Maid herself. Oddly, this woman enjoyed the endorsement of Joan's surviving brothers and some of her senior military companions, including Gilles de Rais. In 1438–1439 she fought for France in the marches of Poitou and Guienne. She was arrested in 1440 when the king reoccupied Paris and her true identity was exposed. No longer pretending to be Joan, she seems to have made her way to Rome, still dressed as a man, where she fought as a mercenary for the Holy Father Eugenius.

Joan's achievements as a military leader have been seriously underestimated. Historian Kelly DeVries notes that in a single year Joan of Arc had "relieved the siege of Orléans, conquered Jargeau, Beaugency, Meung-sur-

Loire, and Saint-Pierre-le-Moutier, won the battle of Patay and numerous other skirmishes, attacked Paris and Le-Charité-sur-Loire, participated in the defense of Compiègne, saw the capitulation of too many towns to list, and stood next to King Charles VII of France as he was crowned.' " Any man who had done what she did—mapped out strategy, organized forces, led troops into battle, and been wounded—would have been called a general.

Bibliography: DeVries, *Joan of Arc: A Military Leader,* 1999; Doncoeur, P. Paul, *La minute française des interrogatoires de Jeanne la Pucelle, d'après le réquisitoire de Jean d'Estivet et les manuscrits de d'Urfé et d'Orléans* (Paris: 1952); Duparc, Pierre, ed., *Procès en nullité de la condamnation de Jeanne d'Arc* (Paris: 1977–1988); Pernoud, *Joan of Arc by Herself and Her Witnesses,* 1966; Pernoud and Clin, *Joan of Arc: Her Story,* 1998; Tisset, Pierre, ed., *Procès de condamnation de Jeanne d'Arc* (Paris: 1960–1971).

—Jane Marie Pinzino

JOANNA OF ENGLAND (born 1165; died 1199). Queen of Sicily and countess of Toulouse.

The daughter of King Henry II of England and **Eleanor of Aquitaine,** Joanna was married to King William II of Sicily in 1177, cementing the good relations between England and the Normans of Sicily. Around 1181 she was reputed to have given birth in Normandy to a son, Boamund, who was made duke of Apulia by his father; however, nothing else is known about Boamund or any other children of her first marriage.

Joanna was widowed in 1189 at age twenty-four. Two years later, like her mother before her, Joanna went crusading; she accompanied her brother as part of the Anglo-Norman contingent on the Third Crusade, a disastrous journey to Cyprus and the Holy Land. One reason for Joanna's presence was to act as chaperone to Berengaria of Navarre, who had married Joanna's brother Richard the Lionheart at Limassol, Cyprus, in May 1191.

In October 1191 Richard proposed that Joanna should marry Saladin's brother Safadin, with whom Richard was on good terms. Richard's proposal would have granted Joanna all of his lands in Palestine, which included Acre, Jaffa, and Ascalon; in turn, Saladin would grant his brother all of his lands in Palestine, and Safadin and Joanna would then be crowned king and queen of Jerusalem. Safadin thought Richard's plans feasible, and according to one Arab chronicler, they were approved by Saladin, "knowing quite well that the King of England would never agree to them and they were only a trick and a practical joke on his part." A furious Joanna refused to consider the idea on religious grounds; she declared that she would not sit on the throne of her great-grandfather Fulk of Anjou with an infidel and would not marry one even to secure peace in Jerusalem. Richard ignored her indignation, informing Safadin that the difficulty might be surmounted if Safadin became a Christian, but Richard's plan never came to fruition.

Back in France, as part of the settlement that ended the long feud between the dukes of Aquitaine and the counts of Toulouse, Joanna was instead married in 1196 to Count Raymond VII of Toulouse and gave him an heir, Raymond VIII. In 1199, while Raymond was fighting elsewhere, Joanna was called upon to put down a revolt in her husband's lands. She began a siege against the rebellious vassals, but was betrayed by some of her own knights and barely escaped with her life. Pregnant and suffering fatigue and despair after her narrow escape, she took refuge first with her mother Eleanor of Aquitaine at Niort and then, after securing special dispensation, at the Abbey of Fontevrault, where she died in 1199.

Bibliography: Barlow, *The Feudal Kingdom of England,* 1955; Chronicles of Pierre de Langtoft, Richard of San Germano, and Roger of Hoveden; Gabrieli, *Arab Historians of the Crusades,* 1969; Kelly, *Eleanor of Aquitaine and the Four Kings,* 1952; Runciman, *A History of the Crusades,* vol. 3, 1954.

—J.M.B. Porter

JOHANNA OF ROŽMITÁL (died 1474), also known as Johanna of Bohemia. Queen of Bohemia, fifteenth-century Hussite Wars.

Johanna of Rožmitál was one in a long line of militant **Hussite** women. She played a key role in the government of her husband King George, "the Hussite king." Together they faced imposing challenges as they sought to control a land split by two religious faiths and an ambitious nobility. Johanna participated energetically in public affairs, attempting to mediate between the warring factions and appearing in public as a symbol of the land's government.

In addition, Johanna of Rožmitál personally commanded her own army in battle. In 1461 she had offered to send her army to help the duke of Bavaria against his rivals. At a time before standing armies were common, she quickly fielded an army that helped fight King Matthew of Hungary, who invaded Bohemia in 1470. The Polish chronicler John Długosz described how she attacked Matthew from the front while her husband George pressed from the rear, and together the Czech armies drove him from the land. In September 1474 she indicated that one of the things she enjoyed about retirement was not leading troops in war. She ended her career by coming out of retirement to help her husband's successor rule.

Bibliography: Długosz, *Historiae Polonicae,* Book 13, 1878; Klassen, "Women and Religious Reform in Late Medieval Bohemia," 1981; Odlozžilik, *The Hussite King,* 1965.

—John Klassen

JUDITH (ca. sixth century BCE). Old Testament, possibly apocryphal figure. Military leader and strategist in conflict between Israelites and Assyrians.

Judith

The Return of Judith to Bethulia, Botticelli, 1472. (Dover Pictoral Archive)

The story of Judith is set in the era of King Nebuchadnezzar (ca. 605–562 BCE). In the Old Testament the first part of the Book of Judith (chaps. 1–7) describes the expanding power of Nebuchadnezzar II, king of Assyria. About to declare war on the king of Medea, Nebuchadnezzar summoned the armies of his vassal states in Syria, Lebanon, the anti-Lebanon, Judea, and Egypt, but they ignored the king's call for aid. After defeating the Medes, Nebuchadnezzar sent his chief general, Holofernes, to take revenge on those who had refused to obey his summons. On receiving word of Assyrian military successes to the north, the high priest and senate in Jerusalem directed the town of Bethulia to fortify the mountain passes into northern Judea. Nevertheless, Holofernes gained entrance into the plains around Bethulia, encamped near the hill town, and cut off its water supply. The Israelites became demoralized and prepared to surrender the town.

The second part of the Book of Judith (chaps. 8–16) describes Judith's role in turning back this Assyrian invasion. Judith, a righteous and beautiful widow, rebuked her fellow citizens and the elders of the town for failing to trust in God; she promised to save the town with God's help. She and her handmaiden, posing as informers, left Bethulia and under cover of darkness walked to the camp of Holofernes. They were stopped at the first Assyrian outpost and offered to give information on how Holofernes could capture the town without the loss of any Assyrian lives. The sentries took Judith to the tent of Holofernes, who promised to reward her for her information. Smitten by her beauty, Holofernes invited her to dine in his tent. After he passed out in a drunken stupor, Judith took his sword and cut off his head. Escaping from the Assyrian camp, Judith and her servant returned to Bethulia with the head of Holofernes in a sack. She summoned the people and the soldiers to see it. Holofernes' head was stuck on the town's main entry gate, and the Israelites, inspired by her act of courage, immediately attacked the Assyrians.

When the Assyrian soldiers learned of their general's death, they panicked and begin to flee northward. The Assyrian army was destroyed by the Israelites. With its defeat and the salvation of the Israelites assured, Judith became the most celebrated woman in Judea.

Whether an actual historical event lies behind the story of Judith is unknown, and historical inaccuracies appear; for example, Nebuchadnezzar was king of the Babylonians, not king of the Assyrians; he ruled from Babylon, not Nineveh. The Book of Judith is accepted as canonical by Catholics and Orthodox Christians. It forms part of the Apocrypha in the King James Bible and is not included in the Hebrew Bible. Generally thought to have been written down in the second century BCE by an unknown author at the time of the Maccabean wars, the story of Judith honors the tradition of pious and strong women. Like **Deborah,** Judith, through her faith in God, was able to inspire military action to save her people in a time of crisis, and like Jael (see the entry on Deborah), she was willing to kill to do so.

Bibliography: Bellis, *Helpmates, Harlots, and Heroes,* 1994; Lacocque, *The Feminine Unconventional,* 1990; Lamsa, *Old Testament Light,* 1964.

—Ralph F. Gallucci

JULIEN, SOPHIE (born ?; died ?). Volunteer in the French revolutionary armies, 1792–1793.

A native of Beauvais, Sophie Julien followed her father into the military. Trumpeter in the 2nd Regiment of Artillery, 2nd Battalion of the Pas-de-Calais, she was ultimately recognized as female after her father's death. Upon her forced discharge, which she challenged, she was granted 150 livres for provisory aid. Records noted that she had neither "women's clothes nor bread" to allow her to subsist.

According to petitions for aid and references from her peers in the military, Julien had been "inspired by patriotism" to wear the uniform of a male soldier and to enlist as a volunteer. Her courage in combat situations was well known, and when her commander was mortally wounded, she had sworn to avenge his death. Laws that excluded women from the military after 1793, however, precluded her remaining in the French armies, even though a clerk had noted that she "had been born with male traits" and that military service suited her.

Julien was granted various forms of aid from the French government in July and again in August 1793. Later the National Convention voted special recognition for her service, which had always been "with honor," and the Committee of Public Aid granted an additional award of 200 livres to assist her in her transition to civilian life.

Bibliography: Dossier XR48, "Julien," Service historique de l'Armée, Vincennes, France; France, *Archives parlementaires,* vol. 20, and *Procès-verbal de la Convention nationale,* 15 brumaire III.

—S.P. Conner

JULIENNE DU GUESCLIN (born ca. 1333; died 1405). Nun. Defender during siege, France, Hundred Years' War.

Julienne was the sister of Bertrand du Guesclin, constable of France. She entered religious orders as a Benedictine nun in a convent at Pontorson, Brittany. She later was to become abbess of the convent of Saint-Georges at Rennes. In 1370, in what may be an apocryphal episode, Pontorson is said to have come under attack from an English force led by Sir Thomas Felton, the future seneschal of Aquitaine. Julienne is said to have defended the convent, sword in hand, helping to push the attackers from the walls, holding them off until her brother arrived to break the siege. The incident is not easily verifiable. It is not mentioned by Froissart in his chronicles or Cuvelier in his verse biography of Bertrand. Felton was certainly militarily active at this time—the Hundred Years' War had resumed in 1369—but there appears to be little evidence to suggest that he was involved in a raid on Pontorson. It may be that the convent was attacked during Julienne's time there and that the incident became entwined with the legends that developed around the exploits of her brother. However, other militant nuns have existed in history, and women have often been active in defending besieged castles and towns, so Julienne may indeed have led a spirited defense of her convent against the English.

Bibliography: *Dictionnaire de Biographie Française,* vol. 11 (Paris: 1967), 1526; Gribble, *Women in War,* 1917.

—David S. Green

K

KĀHINAH, DAHYA BINT TATIT (died 693, North Africa), also known as al-Kāhina, Dihya-ibn Khaldūn. Military leader of Berber resistance during the expansion of Islam, seventh-century Africa.

As Islam spread across the North African coast, it met serious resistance in the area of modern Morocco from the Berbers, who were dominant in the area. There was also a population of Jews who had taken refuge among the Berbers during the reign of the emperor Justinian. Some Berber tribes had adopted Judaism and had later converted to Christianity. Most Berbers initially opposed Islam during its expansion in the late seventh century.

One prophetess who opposed Islam led a coalition of Berber tribes that included Romans, Greeks, Christians, and Jews. She was called "al-Kāhina" by the Arabs; "Kāhin" was a central Arabian term from pre-Islamic times that referred to the spiritual and intellectual guide of a tribe. The true name of al-Kāhina appears in some sources as "Dahya bint Tatit" and in others as "Dihya—ibn Khaldūn," with many variants. She has been described in legend as a Jewish queen, although the earliest local source (eleventh century) said that she was a Byzantine Christian. It appears that she was of mixed blood, as were her children, and although her tribe had at one point practiced Judaism, al-Kāhina and her people were practicing Christians at the time of their conflict with the Arabs. She has been compared to **Deborah** as a political and military leader. One source claims that she lived to be 127 years old and governed the Djarāwa tribe for 65 of those years, assisted by her sons. Kāhina was described as having a "mysterious influence" over her followers. She was an "ecstatic" who had visions regarded as prophecies by those around her. She appears to have been a tolerant leader in some ways: during the conflict she treated Arab prisoners well and adopted one young Arab prisoner, giving him equal status with her two natural sons. However, Jewish folklore of Algeria includes a poem that tells of her ruthlessness, even to her own people: "This cursed woman, more cruel than the others combined / She gave our virgins to her warriors / She washed her feet in the blood of our children."

In 689 a Moslem fleet under Hassān bin al-Nu'mān dislodged Byzantine forces from Carthage but then ran into difficulties with al-Kāhina. She was an important target, as Hassān was informed that she was infallible in her predictions, and if he could defeat her, all resistance would cease. Hassān's first attempt to defeat her was unsuccessful, and her forces drove his back into Tripolitania. Later offensives were equally fruitless. In desperation against his perseverance, al-Kāhina apparently instituted a scorched-earth policy to deprive the Arabs of supplies. Hassān finally defeated the forces of al-Kāhina on the fourth attempt, around 693, in what both regarded as a fight to the death. In fact, al-Kāhina was killed—some say that she was betrayed by one of her own people. She died near a well that still bears her name (Bir al-Kāhina).

Like many other tribal groups, al-Kāhina's coalition was weakened by disunity, and many of the people found Islam attractive; the great majority converted, and the Berbers became zealous Islamic warriors. Moslem legend has it that she ordered her sons to desert the night before her defeat and prophesied success for them as Moslems, for they joined Hassān and rose to high rank. After her death resistance to Islam in North Africa evaporated.

Because of the legendary quality of al-Kāhina's actions, there are many contradictions in the sources regarding her heritage, religion, and even her sex (some scholars see "kahinah" as a deformation of Kahya, a male proper name). All sources agree, however, on her influential role as a spiritual and political leader who led a determined armed resistance against outside invaders.

Bibliography: Baron, *A Social and Religious History of the Jews,* 1952; Cooley, *Baal, Christ, and Mohammed,* 1965; Donzel, *The Encyclopaedia of Islam,* 1993; Fage, *The Cambridge History of Africa,* 1978; Hitti, *History of the Arabs,* 1970.

—Paul K. Davis

KANG KEQING (born September 1911, Jianxi Province, China; died 22 April 1992, Beijing, China), also called The Girl Commander. Soldier, the Long March, China.

Kang Keqing was born to poor parents who gave her away as an infant. She joined the Communist Youth League when she was fifteen years old and soon afterward ran away to join the guerrillas at Jinggangshan, thus escaping the marriage that had been arranged for her. A year later she became the fourth wife of Zhu De (Chu The), then commander of the Communist forces. In 1934 the Chinese Nationalist forces succeeded in driving the Communists out of the area they controlled. Kang became known as "the girl commander" when she assumed command of a battalion of 800 men whose leader had been killed in one of these actions. The defeated Communists set out to begin again in the northern city of Yanan, 6,000 miles away. The retreat, known as the Long March, crossed several major mountain ranges, forded dozens of rivers, and fought a battle roughly every other day. Kang traveled with the 1st Front Army, a strictly disciplined unit of the top leadership. Only healthy women were allowed to travel; most had previous experience in Party cadres; and all were required to leave their children behind with adoptive families. They often marched at night to avoid air attacks. The survivors—8,000 of the 100,000, including 35 women, who had set out from Jiangxi—had truly been through the fire. In her later life Kang played down the rigors of the Long March, saying, "It was just like going out for a stroll every day," but it was a stroll on which she carried a rifle and two pistols and, at times, wounded comrades. She was one of the thousands of women and children who participated in the Long March; most served in support roles (collecting and preparing food, caring for the sick and wounded) and in functions such as propaganda and recruitment. Some women participated in fighting, and there were a few all-female units, such as the Women's Engineering Battalion of the 4th Front Army in Sichuan (some 2,000 women served in that army).

After the success of the Communist revolution in 1949, Kang held a number of government positions as leader of the Women's Federation and a member of the National People's Congress. In 1977, after the death of her husband, she became a member of the Communist Party Central Committee, retiring in 1985. Afterward, Kang avoided political power struggles, but continued to make ceremonial appearances to meet children or to congratulate the Women's Federation on its achievements.

Bibliography: *New York Times*, obituary, 23 April 1992; *The Times* (London), obituary, 5 May 1992; Young, "Women at Work: Chinese Soldiers on the Long March, 1934–1936," 2000.

—Valerie Eads

> *On Chinese women's role in resisting Japanese occupation:*
>
> Women made a remarkable contribution to the war effort, in the front line—in the CCP [Chinese Communist Party] fighting alongside men—as well as on the home front. During the period after the invasion in 1931, guerrillas led by women were particularly active in the Liaoxi area. More than three hundred women soldiers were in the CCP's Fifth Army, among them the heroine Zhao Yiman, who joined the party in 1926 and was captured and killed by the Japanese in 1936. The Guangxi women students' army, which fought in south China, numbered 130 soldiers; the women's battalion of Zhejiang, formed in May 1938, had more than sixty; and in CCP-controlled areas, women formed self-defense teams to undergo military training for local security.
>
> —Pan Yihong, "Feminism and Nationalism in China's War of Resistance against Japan," *International History Review* 9.1 (1997): 123.

KHAN, NOOR INAYAT (born 1 January 1914, Moscow, Russia; died 13 September 1944, Dachau, Germany), also known as Noor-un-nisa, "Madeleine." Ensign, WAAF/FANY/SOE, United Kingdom. Resistance radio operator, France, Second World War.

During the German occupation of France during the Second World War, Noor Inayat Khan was the first female radio operator sent to France by the British SOE (Special Operations Executive).

Khan had an interesting family history; she was related to Mary Baker Eddy on her mother's side, and her father, a Sufi leader, was descended from Islamic royalty in southern India. The family was living in Moscow as guests of Tsar Nicholas II when Khan was born in 1914. She grew up in France, graduated from the Sorbonne, and worked for Radio Paris. Her family evacuated to England during the German invasion of France, and in late 1940 Khan joined the WAAF (Women's Auxiliary Air Force). Her linguistic abilities were an asset, and she was in the first group of women to be trained as clandestine radio operators. She was recruited by the SOE and dropped into France in June 1943. Her cell was almost immediately betrayed, but Khan refused the opportunity to escape. She evaded capture for more than three months, continuing her work of transmitting messages between Britain and French resistance groups. Some believe that she took too many risks in her efforts to rebuild her network, and she was arrested.

Khan reportedly never broke during interrogation and made two unsuccessful escape attempts. In November 1943 she was placed in solitary confinement in chains; on 13 September 1944 she was executed at Dachau along with three other women trained by the SOE (Yolande Beekman, Madeleine Damerment, and Elaine Plewman). Khan was posthumously awarded the Croix de Guerre and was made the first female saint of the Sufis.

Bibliography: Foot, *SOE in France*, 1966; Fuller, *Noor-un-nisa Inayat Khan (Madeleine)*, 1971; Rossiter, *Women in the Resistance*, 1986; Weitz, *Sisters in the Resistance*, 1995, and "Soldiers in the Shadows," 2000.

—Reina Pennington

KNYVET, ALICE (born ca. 1420; died after 1491), also known as Knevet, Knevett. Defender of castle, England, Wars of the Roses.

Alice, the daughter of John(?) Lynne, married John Knyvet, Esq., a prominent Norfolk gentryman, around the year 1440. About twenty years later, in 1460, her husband and eldest son seized New Bokenham Castle in Norfolk following the death of its previous tenant. Their swift action flouted the finding of a royal inquisition, which had dismissed her husband's hereditary claim and ordered that the castle revert back to the English Crown.

In September 1461 John Twyer, J.P., was dispatched, along with a royal commission composed of local nobility, and charged by King Edward IV to take the stronghold into the custody of the Crown. They expected that Alice, holding the stronghold during her husband's absence, would immediately acquiesce to the royal order of eviction. When they passed through the outer ward, however, they were confounded to find the drawbridge of the inner keep raised and the battlements lined with fifty heavily armed men. Twyer demanded to be admitted and was bluntly refused by Knyvet, whose defiant response was reported in the king's Patent Rolls: "Maister Twyer, ye be a justice of the pees and I require you to kepe the peas for I woll nott leve the possession of this castell... and if ye begyn to breke the peas or make any warre to gete the place of me I shall defend me, for lever I had in suche wyse to dye than to be slayne when my husbond cometh home, for he charget me to kepe it." Faced by such determined opposition, the king's representatives were forced to withdraw. Knyvet and her husband were formally pardoned and granted possession of the castle in December 1461. Knyvet's deft handling of the royal commission demonstrates not only the confidence with which English gentrywomen could advance the interests of their families, but also how astute women could manipulate medieval notions of a wife's obligatory subservience to her husband in order to justify their defiance of a greater authority—in this case, that of the English king.

Bibliography: *Calendar of Patent Rolls, Edward IV, 1461–1467* (London: HMSO, 1897), 67, 83, 135; Echols and Williams, *An Annotated Index of Medieval Women*, 1992; Gillingham, *The Wars of the Roses*, 1981; Mackenzie, *Dame Christian Colet*, 1923; Postan, *Medieval Women*, 1975.

—Nicolas Lemire

KONG, first name unknown (fl. late fourth century to 460s). Military commander during the Southern dynasties (420–589) in China.

Kong was good at archery on horseback and other types of martial arts. In 397 an ex-aide to the minister of education, Wang Xin, rebelled against the Jin government. He established a women's troop and made Kong the commander and his daughter the general.

Later, after the peasant rebellion led by Sun En (died 402) had devastated the eastern part of the country, so much so that there were even cases of cannibalism, Kong dispensed her family's grain and saved many lives. In commemoration, many people named children after her. Kong lived more than a hundred years.

Bibliography: Shen Yue, *Song shu* (Beijing: 1974).

—Sherry J. Mou

KOVSHOVA, NATALIA VENEDIKTOVNA (born 26 November 1920, Ufa; died 14 August 1942, Novgorod Region, USSR); **POLIVANOVA, MARIIA SEMENOVNA** (born 24 October 1922, Naryshkino, Tula Region; died 14 August 1942, Novgorod Region, USSR). Both privates, Red Army Ground Forces. Snipers, Soviet Union, Second World War.

Natalia Kovshova and Mariia Polivanova, a sniper team, were the first women snipers to receive the **Hero of the Soviet Union**. From October 1941 they served with the Comintern District Home Guard Battalion of the 3rd

Voluntary Communist Division (organized to defend Moscow and subsequently renamed the 130th Rifle Division on 15 January 1942). In addition to serving as snipers and training new snipers, Kovshova and Polivanova operated as their battalion's command post security, as messengers, and as scouts. From the beginning of their service they were considered model soldiers.

Despite divergent family backgrounds, they had become close friends prior to enlisting. While Kovshova was born into a family with strong revolutionary traditions, Polivanova came from the nobility and was related to the anarchist Prince Peter Kropotkin and the tsarist defense minister A.A. Polivanov (1915–1916). In Soviet sources published before the breakup of the Soviet Union, Polivanova was portrayed as having a peasant or working-class background, as it was unacceptable for a member of the nobility to be a Soviet war hero. Her true origins, however, have recently been proven.

In 1937 Kovshova took up paramilitary training and earned the title "Voroshilov sharpshooter." In the evening both women attended the Moscow Aviation Institute, which trained aeronautical engineers. After the outbreak of war they enrolled in the School for Snipers of the **Osoaviakhim**. Initially their battalion operated in the Mozhaisk and Volokolamsk sectors, where Kovshova and Polivanova started the so-called 'sniper movement' in their regiment: in two months they had trained twenty-six new, mostly female snipers. Soon after arriving on the northwest front, on 21–22 February 1942, they shot their first enemy snipers, "cuckoo birds" hiding in trees.

One of Kovshova's assignments, in her capacity as sniper-observer attached to the regimental commander, was to destroy a group of enemy soldiers in the bell tower of a church. After this she was assigned to help adjust a field gun. She proved successful in carrying out these missions. Kovshova and Polivanova especially distinguished themselves in the fight for the village of Velikusha on 1–4 March 1942.

It was often the practice in the Red Army to track collective, rather than individual, kills of sniper pairs like Kovshova and Polivanova. By summer their joint score of enemy killed stood at more than 300. In addition, they had trained an equivalent number of Soviet snipers. In recognition, they were both awarded the Order of the Red Star.

On 14 August 1942, as their division attempted to prevent the meeting of two enemy forces near the town of Staraia Russa, Kovshova and Polivanova were sent with a platoon of snipers to an important sector where the fighting was fiercest. When the counterattacking enemy cut off the platoon from the main elements and when the platoon commander was killed, Kovshova replaced him. The enemy succeeded in steadily reducing the snipers' ranks and eventually outflanked them. The platoon's sole surviving sniper, Novikov, who had been wounded in both arms and put out of action at the beginning of the engagement, related their story. He said that after crawling to an alternate position, Kovshova and Polivanova continued to resist until they ran out of ammunition. Wounded several times, with their last grenades they destroyed themselves along with a group of enemy soldiers. Kovshova and Polivanova received the Hero of the Soviet Union on 14 February 1943 (posthumously).

Bibliography: Aralovets, *Natasha Kovshova,* 1952; Cottam, *Women in War and Resistance,* 1998; Erickson, "Soviet Women at War," 1993; Miniaeva, *Srazhalas' za rodinu,* 1964; Polivanov, "Voiteli iz roda Polivanovykh," 1994.

—Kazimiera J. Cottam

KRASIL'NIKOVA, ANNA ALEKSEEVNA (fl. early twentieth century, Kazan, Russia), also known as "Anatoly Krasil'nikov." Infantry, Russia, First World War.

Anna Krasil'nikova was the daughter of a miner from the Ural Mountains. When the war began, she tried to secure permission to enlist in the army from her regional governor; she was reportedly twenty-nine years old at the time. He refused, and so she cut her hair, donned military attire, and in the guise of "Anatoly Krasil'nikov" boarded the first available military transport. When she reached Ivangorod in central Poland, she was accepted as a volunteer in the 205th Infantry Regiment as Anatoly Krasil'nikov. She initially served as an orderly and medic, but was soon sent to the trenches. During the fall of 1914 she participated in nineteen battles. Krasil'nikova's sex was discovered after she was wounded in an attack on 7 November 1914, and she was awarded the St. George's Cross for her courageous actions (see **Order of St. George**). While recovering from her wounds, she longed to return to the front lines, but it is not known if she achieved her goal.

Bibliography: "A.A. Krasil'nikova (Zhenshchina Georgievskii Kavalier')," *Zhenshchina i voina* 1 (5 March 1915): 11; Stites, *The Women's Liberation Movement in Russia,* 1978.

—Laurie Stoff

L

LA GUETTE, MADAME DE (CATHERINE DE MEURDRAC) (born 20 February 1613, Mandres, France; died ca. 1681, the Netherlands). Counterrevolutionary, Wars of the Fronde, France.

Madame de La Guette is much less known than other "muses of the Fronde" such as the duchesses of **Montpensier,** of **Chevreuse,** or of **Longueville**. Her *Mémoires,* however, allow her to escape the anonymity in which a number of their emulators remain. Catherine de Meurdrac displayed a martial disposition very early. Her father allowed her to study with a master of arms, and she learned to use both sword and firearms. In 1635 she married an officer, Jean Marius, lord of La Guette, who died in 1665. She lived on his lands in Brie for nearly forty years, from her marriage until 1672; then she followed her two sons, who had gone to Holland in the service of the United Provinces.

It was civil war in France that brought the thirty-five-year-old Catherine de Meurdrac to the forefront of military activity. While her husband fought in foreign countries on behalf of the king, she defended their holdings against looters pushed into France from German lands by the **Thirty Years' War** (1618–1648). During the Fronde, a revolt by some of the nobility against the power of the royal house, she remained faithful to Louis XIV, whereas her husband took the part of the rebellious Prince Louis II de Condé. In the War of the First Fronde (1648–1649) her castle became a refuge for the villagers from the surrounding countryside fleeing from the soldiers, and she continued to fight vigorously against the ill-disciplined soldiery of whichever side, whether frondist or royalist. Some of them, on seeing her suddenly appear with her weapon raised, even swore to her that they would henceforth pillage only in the adjoining regions. In 1653, during the Second Fronde, she went to

Bordeaux at the request of the queen mother to negotiate an accord with the rebellious princes. In order to do this, she traveled with a small escort and twice passed through enemy lines. "I have always been," she said, "of a temper more given to war than to tranquil endeavors. My liking is only for the beat of the drums and the fanfare of trumpets."

Madame de La Guette's eldest son died at the siege of Maastricht in 1676. Her *Mémoires* break off at this date; they were published in 1681 at La Haye in the Netherlands and aroused enough historical interest to be issued in several modern editions. A century and a half later a novelist in 1841 took up her story and had her die in the course of a single combat against three swordsmen while defending her younger son, but it is only a novel.

Bibliography: Berthelot Marcellin, *La Grande encyclopédie*, n.d.; La Guette, *Mémoires*, 1856; Pillorget and Pillorget, *France baroque, France classique*, 1995.

—Marie-Thérèse Lalagüe-Guilhemsans (trans. by Valerie Eads)

LANE, ANNA MARIA (died 13 June 1810). Soldier, Continental army, American Revolutionary War.

The story of another woman who fought in disguise, Anna Maria Lane, has recently come to light. Lane was granted a pension in 1808 because "in the revolutionary war, in the garb, and with the courage of a soldier, [she] performed extraordinary military services, and received a severe wound at the battle of Germantown." However, unlike **Ann Bailey** or **Deborah Samson,** Lane's achievements garnered little attention in her own time or later.

Anna Maria Lane appears to have followed her middle-aged husband, John Lane, when he enlisted in the Continental army. She disguised herself as a man and enlisted in his unit. Although few details are available about the years she spent with the army, it is clear that she participated as a soldier in several battles, including Germantown, where she received an unspecified wound that plagued her throughout her life.

Sandra Treadway has published the most comprehensive account of Lane's story, which she believes was overlooked because "Lane left no personal papers, and no compatriot or descendant ever wrote an account of her exploits." Treadway pieced together Lane's story based on pension and genealogical records; she also quotes from a letter written by a governor of Virginia confirming that Lane was "disabled by a severe wound which she received while fighting as a common soldier." Treadway's article exemplifies the way in which military women's history can be uncovered through diligent research.

Bibliography: Treadway, "Anna Maria Lane," 1988.

—Reina Pennington

LANZ, KATHARINA (born 21 September 1771, Maräo, Italy; died 8 July 1854, Andraz, Italy), also known as Caterina. Defender, Tyrol, Napoleonic Wars.

All her life Katharina Lanz earned her upkeep as a domestic servant to parish priests. She was born in St. Vigil, a part of the Ladin village of Maräo (also called Marebbe) (South Tyrol), and was in service in the ethnically German village of Spinges when, in the course of the War of the First Coalition, French troops under General Joubert attempted to force their way through the Puster Valley in order to penetrate further into Austria. Lanz became known for her militant actions in helping defend her community against French invaders.

Tyrol had been Austrian territory since the time of the Holy Roman Empire and remained part of Habsburg domains until 1918. Historically Tyrol is a trilingual borderland, inhabited by German, Italian, and Ladin ethnic groups, the latter being concentrated in what is now South Tyrol. For census purposes the Austrian monarchy included the Ladins among its Italians. Nonetheless, the majority of politically conscious Ladins have long considered themselves Tyrolean rather than Italian, and Tyrol traditionally had a measure of autonomy.

The Tyroleans had a long history of self-defense against incursions by foreign troops, calling as the need arose for volunteer *Landesverteidiger* (irregular infantry). They fiercely resisted the French "liberators," whose troops had a very poor image on account of their oppressive demands of food and money, their sacrilegious behavior in churches, and their execution of combatants not in regular army uniform. Thus the Ladins countered the French slogan "liberté, égalité, et fraternité" with their own "God, emperor, and homeland!"

Spinges occupied a commanding position at the entrance to the Puster Valley, and three attempts by fresh contingents of French troops on 2 April 1797 to dislodge the outnumbered and minimally trained irregulars (most of them farmers or farmhands) were unsuccessful. Lanz was observed taking part in the defense of Spinges, standing on the wall of the churchyard, repelling the enemy with a pitchfork, and encouraging her fellow Tyroleans. French casualties were at least 600 dead, while the Tyroleans lost about 100. However, having reserves neither of men nor of ammunition, the defenders had to retire at nightfall, and the French pillaged the village the following day. Still, the French were impressed by the Tyroleans' resistance.

Lanz became one of the symbols of that resistance, which was seen as a precursor to the successful ousting of the French and their Bavarian allies in 1809. Lanz was praised in her own time and during the revolutionary upheavals of the subsequent decades of the nineteenth century. Lanz herself had no wish for fame, later saying that her conscience was burdened by her part in the bloodshed. She lived to be eighty-two and was buried with military honors.

With the rise in the later nineteenth century of German and Italian nationalistic fervor in southern Tyrol, the Spinges incident took on lasting legendary proportions. A statue of Lanz, erected in 1912 in Livinallongo, was accorded great importance as a symbol in this context. During the First World War it was moved behind Austrian lines to Corvara, later the occupying Italians moved it as a possible focus of Tyrolean insurgence to Rovereto in the Trentino, and finally in 1964 it was restored to its place of origin. In 1986 Luciana Palla still referred to the statue as the "symbol of Ladin independence."

Italian author Sebastiano Vassalli attempted in his historical novel *Marco e Mattio* (1992) to reduce the symbolic dimensions of the Spinges incident through levity and specious reasoning. He suggested that it was not so much Lanz's pitchfork as her ugliness that repelled the French, and since Maräo rather than Spinges was her home, this was not a case of local patriotism at all. However, Lanz's physical appearance is unknown, and Tyrolean patriotism embraced much more than the individual villages or towns invaded by the French, as is shown by the fact that the *Landesverteidiger* beside whom Lanz fought were from northern Tyrol. Katharina Lanz herself is historically interesting as an ordinary woman who was willing to fight and kill to defend her homeland—typical actions that took on atypical significance due to the extraordinary times in which she lived.

Bibliography: Derrécagaix, Victor-Bernard, *Nos campagnes au Tyrol* (Paris: 1910); Kolb, Franz, *Das Tiroler Volk in seinem Freiheitskampf, 1796–1797* (Innsbruck: 1957); Palla, Luciana, *I ladini fra tedeschi e italiani* (Venezia: 1986); Sparber, Anselm, "Wer war das Heldenmädchen von Spinges?" *Schlern* 22–23 (1948–1949).

—J. Keith Wikeley

LEBSTUCK, MARIA (born 15 August 1830, Zagreb; died 30 May 1892), also known as Hunter Charles, Lieutenant Charles, "First Lieutenant Maria." Revolution of 1848.

Maria Lebstuck fought in male disguise in the revolution of 1848 against the Austrian Empire. At the age of thirteen Lebstuck, the daughter of a Croatian merchant of substance, had been sent to live with her uncle in Vienna, Baron Balthasar Simunich, a brigadier in the Austrian imperial army. When she was eighteen, Lebstuck took refuge in the shop of a friendly shoemaker to avoid an excited street mob in Vienna (13 March 1848). While there, she learned about the "revolution against absolutism in Europe." She sold her earrings to the shoemaker's apprentice in return for his trousers and boots and, with the shoemaker's writ identifying her as his apprentice Charles, she immediately joined the Death's Head Legion of General Giron.

Despite her delicate appearance, she participated in many skirmishes; her adventures took her from Austria to Hungary, where General Dembinszky commissioned her as a lieutenant for shooting three Austrian officers in combat. As Lieutenant Charles, Lebstuck later enlisted in General Gorgey's Hussar Regiment, participating in the attack on Buda Castle, then in the hands of the Austrian general Heintzi. During the siege Lebstuck fired incendiary bullets with an accuracy that attracted the attention of Lieutenant Jozsef Jonak. Apparently Jonak and Lebstuck became friends, and she revealed her true identity to him. They became engaged, and Lebstuck attended a victory celebration as Jonak's fiancée, attired in female garments. General Gorgey realized the truth about "Lieutenant Charles," whereupon he put Lebstuck in jail for spying. Lajos Kossuth, the prime minister, personally intervened on behalf of Lebstuck. After her release she apparently married Jonak and continued to travel with the regiment, although it is not clear whether she still wore her male disguise or whether she continued to fight. What is known is that after 13 August 1849, when General Gorgey surrendered at Vilagos, Lebstuck was captured and imprisoned. In consideration of her pregnant condition, she was deported to Zagreb; she then moved to Hungary. Lebstuck lived for many years after the revolutionary upheavals of 1848. Two years after her death in 1892, she was reburied with honors in a grave in the Ujpest District of Budapest. Lebstuck's memory was best preserved in an operetta by Eugene Huszak, *First Lieutenant Maria*.

Bibliography: Hegyaljai Kiss, Geza, *Lebstuck Maria honvedhuszar fohadnagy emlekiratai 1848–49-bol* (Hadtorteneti Kozlemenyek: 1935).

—Antal Leisen

> Infantry veteran Saul Podvyshensky, who married a woman who served with a naval unit on the Baltic, on romantic relationships between soldiers:
>
> Those who got married at the front are the happiest people and the happiest couples.
>
> —Svetlana Alexiyevich, *War's Unwomanly Face*, trans. Keith Hammond and Lyudmila Lezhneva (Moscow: Progress Publishers, 1988), 79.

LESKA-DAAB, ANNA. Flight lieutenant, RAF; flight leader, Air Transport Auxiliary (ATA). Poland, Great Britain, Second World War.

Anna Leska-Daab qualified as a category A and B glider pilot in Milosna, Poland, when she was only eighteen, and as a balloon pilot at the Warsaw Flying Club. In the spring of 1939 she began flying aircraft at the Pomeranian Flying Club in Poland. In September 1939, after the Second World War had started, with only a few hours of independent flying to her credit, she was assigned to the Polish air force headquarters squadron to fly liaison and delivery missions. Like many military personnel in Poland, she was able to evade the Nazi and Soviet occupations and made her way to Great Britain via Romania and France. She initially worked at RAF headquarters and subsequently at the Air Ministry. She passed a flying test intended for pilots with 250 hours of flying experience, even though she had one tenth of the requirement, and was recruited by the Air Transport Auxiliary (**ATA**).

Leska-Daab started ferrying aircraft with the ATA in February 1941. One of three Polish women to fly with the ATA (along with Jadwiga Pilsudska and Barbara Wojtulanis), Leska-Daab stayed longest, until its disbandment in November 1945. At first she flew small airplanes, later qualifying to ferry larger, more maneuverable and diversified aircraft. Her favorite military aircraft were Spitfires, Mustangs, Tempests, and Mosquitos. She became flight leader in the spring of 1943, in charge of eight women ferry pilots, whom she instructed and assisted. Her subordinates included five British women and one each from the United States, Chile, and Argentina. Stationed at Hatfield and Hamble, she ferried a total of 1,295 aircraft, including 557 Supermarine Spitfires (I–XVII versions). She flew 93 types of aircraft and spent 1,241 hours in the air. Leska-Daab was awarded many Polish and British decorations, including the Polish Military Pilot Badge, and was the sole woman flying with the ATA to receive the Royal Medal.

Bibliography: Car, *Kobiety w szeregach Polskich Sil Zbrojnych na zachodzie, 1940–1948*, 1995; articles from *Skrzydlata Polska*: Leska-Daab, Anna, "Z lat wojny—wspomina Anna Leska-Daab" 20 (18 May 1980): 14; Malinowski, Tadeusz, "Z samolotu na samolot" 10 (8 March 1981): 3, 5; "Polki na wojnie" 10 (3 October 1985): 12.

—Kazimiera J. Cottam

LEVCHENKO, IRINA NIKOLAEVNA (born 15 March 1924, Kadievka, Voroshilovgrad Region, USSR; died 18 January 1973, Moscow). Lieutenant-colonel (ret.), Red Army Tank Troops. Tank liaison officer, Soviet Union, Second World War.

One of the few Soviet women to become a tank troop officer and to exercise command, Levchenko began her wartime career as assistant tank battalion chief of staff and ended it as tank corps liaison officer. Soviet tank liaison officers were responsible for delivering orders to tank units and were appointed from among those who demonstrated bravery and resourcefulness and had proved themselves in combat. In her capacity as liaison officer, all Levchenko's skills, including tank driving, firing the tank gun, handling small arms, and dispensing first aid, proved useful to her.

Levchenko was the granddaughter of Maria Zubkova-Saraeva, the first Soviet woman to receive the **Order of the Red Banner** for service in the First Mounted Army during the **Russian Civil War**. Levchenko's father, an electrician who became union deputy commissar of transportation, was a victim of Stalin's terror in the late 1930s. Like many Soviet teenagers, Levchenko learned to drive cars and motorcycles, fire small arms, and dispense first aid.

Levchenko experienced her baptism of fire in a first-aid platoon on 6 July 1941 near Smolensk. While serving as a company medic during the battle for Moscow, she was wounded for the first time. On recovery and due to her persistence, she was sent as a medical NCO to the 39th Tank Brigade in the Crimea, where she received her initial tank training in the small T-60 tank. Despite a serious wound in the right arm (which just escaped amputation), Levchenko secured permission to enroll at the Stalingrad Tank School in July 1942. Although her arm continued to trouble her, she became a competent tank driver. As tank platoon commander near Smolensk, she was wounded again and again returned to duty upon her recovery. Back at the front as a platoon commander and later liaison officer of the 41st Tank Brigade, 7th Mechanized Corps, on the 3rd Ukrainian Front, Levchenko advanced through

Moldavia, Romania, and Bulgaria into Hungary. Wounded in Budapest, she returned to action as a tank corps liaison officer on the 2nd Belorussian Front and ended up near Berlin.

Levchenko was present at the reunion of Soviet and U.S. veterans who met on the Elbe in 1945. She graduated from a course in engineering offered by the Academy of Armored and Mechanized Troops in Moscow in 1952. After obtaining a history degree from the Frunze Military Academy, she transferred to the reserves in 1958. A member of the Soviet Writers' Union, she wrote about the war and her comrades in arms. Her career could be summed up as follows: NCO at 17, senior lieutenant at 21, major at 28, and lieutenant colonel at 31. The first Soviet recipient of the international Florence Nightingale Medal, Levchenko was decorated twelve times and was awarded the **Hero of the Soviet Union** on 6 May 1965. She died at the age of 49 and was buried at Novodevich'e Cemetery.

Bibliography: Cottam, *Women in War and Resistance,* 1998; Erickson, "Soviet Women at War," 1993; Levchenko, *V gody Velikoi voiny,* 1955, and *Povest' o voennykh godakh,* 1965; Murmantseva, *Sovetskie zhenshchiny,* 1974; Toropov, *Geroini,* 1969.

—Kazimiera J. Cottam

LEWANDOWSKA, JANINA (born 22 April 1908, Khar'kov, Ukraine, Russia; died 1940, USSR), née Dowbór-Musnicka. Pilot officer, Polish air force, Poland, Second World War.

Janina Lewandowska was a pilot and reserve officer in the Polish air force when the Second World War broke out. She may have been the only female officer to have been executed in the infamous Katyn Massacre of 1940.

Lewandowska was the daughter of a Polish military officer, CGen Józef Dowbór-Musnicki. As a teenager, she joined the Poznan Flying Club and earned her glider and parachutist certificates. By 1937 she had learned to fly light aircraft; in the same year she attended a radiotelegraphy course in Lwów (now Lviv), probably attaining the rank of reserve second lieutenant. Shortly before the war began, she married instructor-pilot Mieczyslaw Lewandowski. Two months later Lewandowska was drafted in August 1939 for service with the 3rd Military Aviation Regiment stationed near Poznan, Poland. Her unit was evacuated by train, but on 22 September its members were taken prisoner by Soviet forces (reports that Lewandowska was shot down over Soviet territory were later proven false). Lewandowska was one of only two officers in the group; both were taken to the POW Camp for Polish Officers in Kozel'sk, about 250 kilometers southeast of Smolensk. There was some doubt about Lewandowska's identity; survivors reported that a woman officer there used an assumed name to disguise her family relationships from her Bolshevik captors for political reasons. However, other survivors later testified that the woman imprisoned at Kozel'sk was in fact Janina Lewandowska.

Lewandowska's subsequent fate is unknown, although it seems likely that she died in the Katyn Massacre. In the spring of 1940 the Soviets systematically executed almost all Polish officers, cadets, and NCOs who had been captured in 1939; most of them were buried in the Katyn forest near Smolensk. Lewandowska's name appears on the so-called Katyn List, compiled by Adam Muszynski in 1949, but is missing from the German "Katyn List" of exhumed bodies identified as the former inmates of the Polish officers' POW camp in Kozel'sk, Soviet Union. However, Lewandowska might have been buried in one of many other scattered graves.

Janina Lewandowska had a younger sister, Agnieszka, who also died in the war. She was arrested in Warsaw in April 1940 by the Gestapo and executed, along with a group of Polish prisoners, by a German firing squad on 21 June 1940. The Dowbór-Musnicki sisters, martyred daughters of a military hero, are thus a dramatic symbol to the Polish people.

Bibliography: Bauer, Piotr, "Wojenne Losy Janiny Lewandowskiej," *Skrzydlata Polska* 31 (30 July 1989): 7; Muszynski, Adam, comp., *Lista katynska*, 4th ed. (London: Gryf Publications, 1982; reprinted by the Omnipress Agency—Journalists' Employment Co-op and Polish Historical Society, Warsaw, 1989).

—Kazimiera J. Cottam

LI XIU (ca. 285, Tianshui, Gansu Province, China), also known as Li Shuxian. Governor and leader in early medieval China.

Li Xiu was the daughter of Li Yi, the regional inspector of Ningzhou, and the wife of Wang Zai. Li Xiu had been good at archery on horseback since she was young. At a time when Ningzhou was besieged by five tribal peoples, her father fell ill and died without a male heir. The subordinates supported Li Xiu as the new leader in charge of the prefecture. She led the troops and the people in fighting off their enemies. Afterwards Emperor Wu (r. 265–290) named Li Xiu the regional inspector of Ningzhou (now Qujing of Yunnan Province), placing her in charge of thirty-seven tribal divisions in the Yunnan area for more than thirty years. She proved adept at both political and military affairs; thus her people lived peacefully for more than three decades. After her death people erected temples in her memory.

Bibliography: Chen Menglei, *Gujin tushu jicheng (Qin ding)*, 1993.

—Sherry J. Mou

LIANG HONGYÜ (died 1135, China). Consort of State Peace and Consort of State Protection. Combat drummer and soldier in the Southern Song dynasty (1127–1279).

Liang Hongyü often fought side by side with her husband, Han Shizhong (1089–1151), one of the greatest generals of the Southern Song dynasty. Liang Hongyu was from Beichenfang. When she was young, she became a female entertainer in Jingkou (now in Jiangsu Province). Once, as she arrived at an engagement to entertain guests, she suddenly saw a tiger snoring in deep sleep next to a column at the main entrance. Terrified, she left in a great hurry and dared tell no one. Later, after more people arrived, she was emboldened to look again, but she found a guard instead. She woke him up with her foot and asked his name, which turned out to be Han Shizhong. She knew that Han must not be an ordinary man. Soon they fell in love, she gave him gold and silk, and they pledged to someday become husband and wife. Around 1121 Liang Hongyü became his concubine.

In 1129 Commander-General Miao Fu (died 1129) rebelled against Emperor Gaozong (1107–1162) in Hangzhou, where the emperor was on an inspection tour, and forced the emperor to abdicate to his son. As Han Shizhong and his troops approached the capital to rescue the emperor, his wife, Liang Hongyü, and one of his sons were taken hostage.

Liang Hongyü remained loyal to the imperial family during a process of political intrigue and mediation. She rejoined her husband Han Shizhong at Xiuzhou, and when he and other rescuing commanders defeated the rebel Miao Fu, the emperor rewarded the troops and bestowed on Liang Hongyü the title Consort of State Protection.

In 1130 Han Shizhong fought the Jurchen leader Wanyan Wuzhu, who came south to attack the Song dynasty. All the while, Liang Hongyü beat the drum to motivate the troops to stop the Jurchen from crossing the river. It is said that after the Jurchen left, Liang Hongyü wrote to the throne about Han Shizhong's guilt in letting Wanyan Wuzhu escape. The entire court was touched by her righteous action. Later that year Liang Hongyü also begged the court to pay the salary owed her and her husband from the previous year; the emperor granted her appeal and paid up.

In 1135 she went to Chu Prefecture (now Huanan of Jiangsu Province) with her husband and set up a military settlement. The place was not yet developed, so Han Shizhong established a military command and they lived in harsh conditions together. Liang Hongyü died later that year, and the emperor sent much silk and 500 pieces of silver in condolence.

Bibliography: Chen Menglei, *Gujin tushu jicheng (Qin ding)*, 1993; Deng Gongsan, *Han Shizhong nianpu* (China: 1944); Luo Dajing, *Helin yulu* (Beijing: 1983); Tuotuo et al., *Song shi*, 1977.

—Sherry J. Mou

LIBYA, MILITARY WOMEN IN. 1977–present.

The integration of women into the Libyan People's Army began in 1977; in a 1978 speech Libyan leader Muammar Ghadafi stated that "bearing arms is a right and duty of every Libyan woman" (Graeff-Wassink). In theory, no army jobs are closed to women, although the reality of this stated policy is unclear. Women primarily serve as regular army officers, enlisted and air support personnel, combat pilots, and instructors at the Women's Military Academy. Four women officers reportedly served in combat during Libya's ongoing conflict with Chad in the late 1970s and early 1980s, although the nature of their roles is unknown.

The main impetus to women's involvement in the Libyan military was the creation of the Women's Military Academy (WMA) outside Tripoli in 1979. Established by Ghadafi as a counterpart to the Men's Military Academy, the WMA has become a central symbol of Ghadafi's efforts to change the visible roles of women in Libyan society. In addition to careers in the regular military, the graduates of the WMA may also elect to take positions as paramilitary "Jamahiryan Guards." One of the functions of the Guards is to provide security for Ghadafi on his journeys within Libya and abroad. Other graduates serve as specialized secondary-school teachers charged with giving girls basic military instruction.

Libyan female soldiers salute as they chant the name of Libyan leader, Moammar Gadhafi, during a military parade Sunday, August, 31, 1997 in Sirte for the celebration of the 1969 coup which brought Gadhafi to power. (AP Photo/Mohamed El-Dakhakhny)

Bibliography: Addis, et al., *Women Soldiers,* 1994; Coughlin, "Women, War, and the Veil," 2000; Graeff-Wassink, *Women at Arms,* 1993; Kokan, Jane, "In Love with Gaddafi," *BBC Focus on Africa* 6.3 (1995): 4–8.

—William Henry Foster

LILLIARD OF ANCRUM (fl. 1545). Combatant, Scottish-English wars, sixteenth century.

Lilliard of Ancrum is a historical figure whose courage in battle is as much the stuff of legend as it is of fact; certainly it is the subject of nationalist mythmaking in Scotland. In the first half of the sixteenth century Scotland was frequently attacked in border wars by its more powerful neighbor, England. During an attack on 27 February 1545 Lilliard is said to have distinguished herself in battle in her village of Maxton, near Melrose and Jedburgh. The English had set Maxton on fire. Among the victims, according to legend, were the bedridden parents of Lilliard, an only child, who nearly died herself trying to rescue them; she was rescued in turn by her lover. As befits a tragedy, he soon thereafter died of wounds received from combat with the English soldiers.

With her parents and lover dead, an avenging Lilliard joined the fight. The Scots ambushed and savaged the English army. The victors were joined by Scottish soldiers who, before the present turn, had reluctantly collaborated with the enemy; this battle took on special significance as proving the rewards of Scottish unity. Lilliard is credited with inflicting a mortal wound upon the English field commander, Lord Evers. Thus were her efforts taken to be pivotal to the Scots' victory.

According to one legend, Lilliard had donned men's gear to do battle, wearing a helmet marked by a white plume. Although her flowing golden locks showed under the helmet, other soldiers were surprised to discover at the battle's end that she was a woman in men's costume. Her unorthodox dress and demeanor earned her the title "Amazoness" among some legendkeepers. The battlefield was named "Lilliard's Edge" in her honor, and in 1827 the novelist Sir Walter Scott, whose ancestor of the same name had fought on the Scottish side, paid a memorial visit to her grave in Ancrum Moor. He recorded the inscription on her tombstone: "Fair Maiden Lilliard lies under this stane / Little was her stature, but great was her fame / Upon the English loons she laid mony thumps / And when her legs were cutted off, she fought upon her stumps." Because of the legendary aura of her story, greatly enhanced by Scott's works, it is difficult to determine the true extent of Lilliard's military actions.

Bibliography: Child, *The English and Scottish Popular Ballads*, 1956; Fittis, *Heroines of Scotland*, 1889; Sir Walter Scott, *History of Scotland* (Philadelphia: 1827), and *The Journal of Sir Walter Scott*, ed. W.E.K. Anderson (Oxford: Clarendon Press, 1972).

—Alan Shepard

LITVIAK, LIDIIA VLADIMIROVNA (born 18 August 1921, Moscow; died 1 August 1943, Ukraine), also known as Lily, Lilya, Liliia, Litvak, Litvyak. Junior lieutenant, Red Army Air Force. Fighter pilot, ace, Soviet Union, Second World War.

Lidiia Litviak (who preferred to call herself "Liliia") was the first woman in the world to shoot down an enemy aircraft. With a score of eleven personal kills, three group kills, and one kill against an aerostat (artillery-spotting balloon), Litvak is the top-scoring female fighter ace to date. Short, beautiful, and blonde, Litviak attracted the attention of the Soviet press during her brief career. She flew in combat for less than a year before she went missing in action in 1943. Her mechanic, Inna Pasportnikova, helped sponsor a forty-year search for Litviak's body. After Litviak's remains were identified in the late 1980s, she was posthumously awarded the **Hero of the Soviet Union** medal by Soviet Communist Party leader Mikhail Gorbachev in May 1990.

From a very young age Litviak dreamed of becoming a pilot. At the age of fourteen she attempted to enroll in an **Osoaviakhim** flying club, but was told she could not begin training until she was sixteen. She attended the ground training classes anyway and hung around the airfield, washing and servicing aircraft. She earned a sport flying license in 1938 and an instructor's certificate in 1939; she then worked as an instructor pilot at the Kirov Flying Club in

Litviak, Lidiia Vladimirovna

Junior Lieutenant Liliia Litviak, pilot, 73rd Guards Fighter Aviation Regiment, in winter flying gear. (Courtesy of Inna Vladimirovna Pasportnikova and Reina Pennington)

Moscow until 1941. She was commended in the magazine *Samolet* [Airplane] on 5 May 1941 for setting a new record among Moscow instructor pilots by flying eight hours and forty minutes of training flights in one day.

When the Germans invaded the Soviet Union in June 1941, Litviak applied for military duty but was refused. In October 1941 she was permitted to join the Soviet Women's Combat Aviation Regiments formed by **Marina Raskova**. After completing training and graduating with the rank of sergeant, Litviak was assigned in January 1942 to the **586th Fighter Aviation Regiment**. The 586th, equipped with Yak-1 fighter aircraft, was based near Saratov on the Volga River and was assigned to air defense.

Litviak was always a maverick. Pasportnikova recalled one incident during the winter of 1941–1942, when the women were still in training. "When the cold season arrived, they gave out winter uniforms. At the morning formation, a command from Raskova rang out: 'Litviak, step forward!' Lilia took a step forward and the whole formation shouted. Instead of a brown beaver-lamb collar, she was showing off a white, fluffy one with ringlets. 'Litviak, what do you have on your shoulders?' asked Raskova. 'A goatskin collar. Why, doesn't it suit me?' Lilia had cut the goatskin from the linings of her winter boots. . . . I looked at her then: small, delicate, and beautiful; still completely a little girl. . . . I never imagined then that I would become the aircraft mechanic in the crew of Lidiia Vladimirovna Litviak, or that she would become the only woman in the world to have 15 enemy aircraft to her credit." Litviak was never one to live by the rules. Later, after each kill, she persisted in performing strictly forbidden victory rolls over the airfield despite her commanders' reprimands.

On 10 September 1942 Litviak was one of eight pilots sent from the 586th as a replacement to front-line fighter regiments at Stalingrad, where casualties had been heavy. Litviak and three other women pilots (Ekaterina Budanova, Maria Kuznetsova, and squadron commander Raisa Beliaeva) were initially assigned to the 437th Fighter Aviation Regiment. On 13 September 1942, on her second combat mission, Litviak shot down two enemy aircraft. Her first kill was a German Junkers-88 bomber; the second was made jointly with Beliaeva against a Messerschmitt-109 fighter. The pilot of the Me-109 was captured by the Soviets and brought to Litviak's airfield for questioning. The German pilot asked to be introduced to the pilot who had managed to shoot him down. When Litviak was brought to the dugout, he was certain that the Soviets were playing some sort of joke on him. He refused to believe that the tiny blonde girl before him could have defeated him in battle until Litviak began to relate the details of their air engagement, which only the two of them could know.

Most Soviet sources credit Valeriia Khomiakova as being the first woman to down an enemy aircraft. Khomiakova, who remained with the 586th, shot down a German bomber on the night of 24 September 1942. Archival documents show that Litviak's first kills were made two weeks before Khomiakova's night victory. Litviak's achievement was probably overlooked in the hectic battles of Stalingrad. She was no longer flying with a "women's" regiment (and so was less visible to the press); she made her first kills only three days after arriving at the 437th; and Litviak's squadron was transferred to another regiment within a month (giving the 437th chain of command little incentive to acknowledge or publicize her kills). These factors contributed to the lack of recognition for Litviak's victory.

On 1 October 1942 Litviak and the other women pilots were reassigned to the 9th Guards Fighter Aviation Regiment. There they flew with some of the most famous Soviet fighter aces. Colonel-General of Aviation Vladimir Lavrinenkov later wrote in his memoirs: "They disproved the erroneous opinion that the profession of air combat is unacceptable for women. Katia Budanova and Lilia Litviak were, for us, dependable comrades-in-arms."

On 9 January 1943 Litviak and Budanova were once more transferred, this time to the 296th Fighter Aviation Regiment (later redesignated the 73rd Guards Fighter Aviation Regiment). The other women pilots had been returned to their original regiment, the 586th, but Litviak and Budanova strenuously opposed this order and persuaded their commander to let them remain at the front. Boris Eremin was a regimental commander in the same air division. He described Litviak as "a remarkable girl, smart, with the true character of a fighter, and a daredevil."

Litviak made two more kills on 11 February 1943. A week later she was promoted to junior lieutenant. On 22 March 1943 she downed two more aircraft and was wounded during an engagement with enemy Me-109s. The shrapnel injury to her leg was serious enough to land her in a field hospital;

she was then sent to a military hospital in Moscow for recuperation. After a few days she was permitted to go home with orders to rest for one month. The inactivity palled on her, and she hopped a flight back to her regiment after only one week. By this time Litviak and Budanova had achieved some degree of fame; their picture was on the cover of the 20 April 1943 issue of *Ogonek*, a magazine for young people.

Much has been made of the relationship between Litviak and Alexei Solomatin, another pilot in the 73rd. The two flew together, and many of their friends knew that Solomatin was in love with Litviak. Despite many successful missions together, she transferred to another squadron when his feelings for her made their flying partnership too uncomfortable. Litviak's letters reveal that she did not know that she loved Solomatin until after he was killed in a crash in late May 1943.

Litviak made additional kills on 5, 7, and 31 May 1943 and on 6 and 16 July 1943. She was wounded again, but refused to take medical leave. She shot down two more Me-109s on 19 and 21 July 1943.

Litviak's accomplishments were achieved in the face of great personal hardship: her beloved father had been arrested during the purges of 1937. Part of Litviak's motivation as a pilot was to regain honor for her family name; she never stopped believing in her father and hoped that one day his innocence would be proven. But she carried the stigma of being the daughter of one of the "repressed." Litviak once told Pasportnikova, "More than anything, I am afraid of 'disappearing without a trace.'" Tragically, this was precisely her fate.

Litviak made her last two kills on the day she disappeared. On 1 August 1943 Guards Junior Lieutenant Litviak participated in a mass engagement against an enemy formation of Ju-88 bombers escorted by Me-109s. During the engagement her aircraft was badly damaged, and she disappeared over enemy-held territory. When Soviet forces recaptured the area, a search was conducted for her body. When she was not found, a cloud of suspicion began to form. Any pilot who went missing in combat was assumed to be a possible defector; such a pilot was guilty until proven innocent. There was a secondhand report that villagers had seen a blonde woman riding in a car with German officers; some investigators thought that it was Litviak. After all, her father had been an "enemy of the people." The only way to prove that Litviak had died honorably in combat would be to find her body. Until that time she would be listed as "missing without a trace," and no awards would be given.

As it turned out, Litviak had suffered a head wound and had tried desperately to return to friendly territory, flying some distance from the spot where she disappeared. She managed to land, but died soon afterwards. In 1979 her body was found in the village of Dmitrievka, Ukraine; in 1986 official confirmation was received that the body was indeed Litviak's. Her name was resubmitted for the Hero of the Soviet Union medal—the Soviet Union's highest military honor—which she received in 1990. A monument to Litviak now stands in the town of Krasny Luch, Ukraine.

According to archival records, Litviak made 168 combat sorties and achieved 12 personal victories and 3 shared kills. Her awards include the **Order of the Red Banner,** the Order of the Red Star, and the Hero of the Soviet Union.

Bibliography: Archival materials, Central Archive of the United Armed Forces of the Commonwealth of Independent States (TsAMO), Podolsk, Russia; Cottam, *Women in War and Resistance,* 1998; Eremin, *Vozdushnye boitsy,* 1987; Lavrinenkov, *Vozvrashchenie v nebo,* 1983; Murmantseva, *Zhenshchiny v soldatskikh shineliakh,* 1971; Pennington, *Wings, Women, and War,* 2001.

—Reina Pennington

LIU, first name unknown (fl. seventh century, northeast China). Mistress of Yanjun. Commander and guard of Fanucheng during the early Tang dynasty (618–907).

Liu defended a settlement against nomadic invaders. Liu was a skillful soldier, and her husband, Li Jinxing (died 682), was commander in chief of Yingzhou during the reign of Taizong (r. 627–649). One time, while Li Jinxing went in pursuit of some nomadic troops attacking Fanucheng, he left Liu as guard. When the nomads returned, Liu donned armor, mounted a horse, and led the soldiers against them, driving them off. Emperor Gaozong (r. 650–683) awarded her the title mistress of Yanjun.

Bibliography: Ou-yang Xiu, *Xin Tang shu,* 1975.

—Sherry J. Mou

LIU SHUYING (born 1621 ? Luling or Nanchang, Jiangxi Province, China; died ?), byname Jingwan, also known as Liu Shu. Leader of a private army and a loyalist of the Ming dynasty (1368–1644).

Liu Shuying was the daughter of a former prefect of Yangzhou who died when she was six years old. She was educated by her mother using her father's books, including Sunzi's *Art of War,* works on swordsmanship, and Buddhist scriptures. She was married and widowed by the age of twenty. Afflicted by the deaths of her father and husband and the downfall of the Ming dynasty in 1644, Liu sold her family holdings and recruited an army of 1,000 men under her command. Fighting against local bandits in Hunan Province in the year of 1646, Liu lost confidence in her cause after a quarrel with another local commander. She discharged her army and returned to her native place. Until her death she was said to have been a devotee of the Buddhist faith. According to legend, she died for her country, followed by the death of more than 8,000 soldiers.

Bibliography: *Ming yiminlu,* comp. Sun Jing'an, in *Qingdai zhuanji congkan* (QZC), ed. Zhou Junfu (Taibei: Mingwen, 1985), 68: 720–722; *Qingdai guigeshiren zhengluei,* comp. Shi Shuyi, in *QZC,* 25: 30–33; *Yishi zhiyi,* ed. Li Yao, in *Mingdai zhuanji congkan* (MZC), ed. Zhou Junfu (Taibei: Mingwen, 1991), 105: 233–234; Zou Yi, *Qi zhen yesheng,* in *MZC,* 127: 583–586.

—Doris Kehry-Kurz

LONGUEVILLE, ANNE GENEVIÈVE DE BOURBON-CONDÉ, DUCHESS OF

(born 29 August 1619, Vincennes, France; died 15 April 1679, Paris, France). One of the leaders of the Wars of the Fronde (1648–1653), France.

A figure symbolic of the Wars of the Fronde (1648–1653), Madame de Longueville was famous for proclaiming on several occasions, "I am not fond of innocent pleasures." Included among Madame de Longueville's more modest followers were women such as **Madame de La Guette** and Madame de Saint-Balmont.

A princess of the blood, she was the sister of the Grand Condé and the prince of Conti. She was born in the keep at Vincennes, where her mother had joined her father, who had been imprisoned there since 1616 by the order of **Marie de Médici**. She spent her childhood in an environment rife with conspiracies prior to her marriage in 1642 to Henry of Orléans, duke of Longueville. Nor was the duke a stranger to conspiracies, having taken part in several revolts against royal power, including the plot organized by **Madame de Chevreuse** in 1626.

Anne Geneviève de Bourbon, Duchesse de Longueville, ca. 1645. (Hulton/Archive by Getty Images)

Intelligent and ambitious, Madame de Longueville threw herself into the Fronde. She was the heart of the resistance during the siege of Paris by royal troops (1649) and set up her headquarters at the Hôtel de Ville, where she held a day to honor "the child of the Fronde," Charles-Paris d'Orléans. She succeeded in making her brother head of the troops raised by Parlement, and she helped negotiate the Peace of Rueil between the Crown and the Frondeurs.

She later used her influence to urge her husband, the governor of Normandy, to raise the province to fight against the royal forces. When her husband and her two brothers were arrested in 1650, she quickly outfitted herself as a man and went to Normandy at the head of a small band of mounted troops. She organized the defense of Dieppe against the royal army and tried to raise the province, but did not gain wide support. Again disguised as a man, she fled to take refuge in Holland, barely escaping closely following royal troops. Once there, she encouraged Turenne to negotiate with Spain.

Returning to Paris after the liberation of her husband and the princes, she quickly resumed her activities against the government. Condé, governor of Guyenne, was to organize resistance there while Conti and Madame de Longueville held Berry. However, confronted by royal troops, they were forced to evacuate Bourges and withdraw to Guyenne. Condé turned again toward Paris, leaving his sister and Conti at Bordeaux to continue the fight. However, the conspirators were ultimately forced to sue for terms, and the royalists entered Bordeaux in August 1653.

The Fronde over, Madame de Longueville essentially retired from political life, immersing herself instead in the Jansenist movement. After the death of her son in 1672 she spent the rest of her days in devotion to religious life, much the same as her old rival the duchess of Chevreuse.

Bibliography: Bluche, *Dictionnaire du Grand siècle*, 1990; Cousin, *Madame de Longueville pendant la Fronde*, 1891; Debû-Bridel, *Anne-Geneviève de Bourbon, duchesse de Longueville*, 1938; Erlanger, *Madame de Longueville, de la révolte au mysticisme*, 1977.

—Marie-Thérèse Lalaguë-Guilhemsans (trans. by Annette Parks)

LOZEN (born 1840?, southwest United States; died 17 June 1889 in prisoner-of-war camp, Mt. Vernon, Alabama). Chiricahua Apache warrior, holy woman, medicine woman.

As a girl, Lozen of the Warm Springs Chiricahua Apache tribe was taught the obligations expected of all Apache women and underwent warrior training as well (most Apache girls were taught both gender roles; boys learned only how to be warriors). However, Lozen balked at the women's mundane duties. Preferring to imitate her older brother, Victorio, she leaped from boulders, sliced the air with her knife, rode bareback on a stallion, and wore a bandanna to hold back her long black hair. In time, Victorio became a noted chief, and Lozen became one of his warriors and the only unmarried woman allowed to participate in warring and raiding.

Lozen was an excellent horsewoman and quite skilled at stealing horses, useful traits among the Chiricahua. Frequently acting simultaneously as a warrior and a woman, she was especially close to the other women and children in Victorio's band and often took responsibility for their well-being. In the 1880 breakout from the San Carlos Reservation, for example, Lozen led the women and children across a river to safety from the army's pursuit and then returned to the men to fight the soldiers.

During the years between 1860 and 1880 the U.S. government conducted extensive roundups of native peoples; those who resisted suffered death or incarceration. But they first had to be subdued, and the Warm Springs Chiricahua Apaches successfully resisted for many years due, in part, to Lozen's skills. Fearless, she could dash on horseback through hails of army bullets, knew the correct herbs to apply to warriors' wounds to effect quick healing, assisted young women giving birth in the middle of battles, and without hesitating helped slaughter more than 200 Mexicans in revenge for Victorio's death. Some say that had Lozen been near Victorio at the time of the massacre by Mexican soldiers at Tres Castillos, Mexico, he and more than 100 members of his band never would have perished.

For Lozen claimed a unique gift: through supernatural ability she could tell the direction from which the enemy was coming. Climbing to the top of a mesa or a similar high point, Lozen would lift her hands and arms high and pray: "Upon the earth / On which we live / Ussen has Power. This Power he grants me / For locating the enemy. I search for that enemy / Which only Ussen, Creator of Life / Can reveal to me."

Turning slowly in a clockwise direction, she halted when a tingling feeling or change in the color of her palms became noticeable. The direction she faced then was usually the place where the enemy was located. Occasionally the sensation became so intense that her palms turned purple, meaning that the foe was very near. Returning to the group, she quickly informed the leaders of her findings. Escape from harm was then facilitated.

Lozen eventually joined forces with Geronimo's band and rode with him until the end of their freedom. During one memorable battle she crawled through the line of fire to rescue an ammunition pouch containing more than 500 cartridges that had been dropped by an Apache warrior. Lozen retrieved the sack and returned to the battle unscathed. Honoring her, Geronimo chose Lozen and Dahteste—another Apache woman he held in high esteem—to initiate overtures to the American government that resulted in the negotiations leading to the Chiricahuas' final surrender on 4 September 1886.

Bibliography: Ball, *In the Days of Victorio*, 1970; Mathes, "Native American Women," 1982; Stockel, *Women of the Apache Nation*, 1991, and *Survival of the Spirit*, 1993.

—Henrietta Stockel

LU, first name unknown (fl. early fourth century, China). A military commander in the Jin dynasty (265–420).

Lu led troops to fight Shen Chong, a rebel who had killed her husband and three sons in 322. Lu was from the Wu Commandery, where her husband Zhang Mao had been governor. After he was killed by Shen Chong, Lu liquidated the family's assets to pay her husband's troops, whom she led against Shen Chong, preceding all other corps. After Shen Chong was defeated, she went to the capital and appealed to the throne to pardon her husband's failure. The emperor decreed honors for both the husband and the wife. A posthumous honorary title, Chamberlain for the Imperial Stud, was also conferred on her husband because of the loyalty the whole family displayed.

Bibliography: Chen Menglei, *Gujin tushu jicheng (Qin ding)*, 1993; Fang Xuanling and Baoyuan Du, eds., *Jin shu*, 1974.

—Sherry J. Mou

M

MABEL OF BELLÊME (born ca. 1030s; died 2 December 1079), also known as Mabille Talvas, de Montgomery, de Montgomerii. Countess of Shrewsbury, lady of Arundel. England and France, eleventh century.

Mabel was the daughter of William Talvas and heir to his estates in Normandy, Maine, and other parts of France, including the important castles of Domfort, Sees, Alençon, and Bellême. Mabel married Roger de Montgomery in 1050 and became an active partner in her husband's political and territorial ambitions as well as her own, including vigorously pursuing her birth family's feud with the Giorie family in France. While it remains unclear whether Mabel ever actually fought, she was involved in the planning and direction of military campaigns. She was typical of ambitious and powerful noblewomen of her time and did not hesitate to use military force or the threat of force to attain her goals.

Information about Mabel comes largely from the chronicler Orderic Vitalis, whose sympathies lay with the Giories, founders of his monastery at St. Evroult. Despite Orderic's hostility, he paints a fairly balanced portrait of Mabel as a vital political actor. Mabel was notably hostile to the monastery of St. Evroult, but could not openly harm the establishment because of her husband's affection for it. Instead, she practiced economic warfare and nearly brought the house to financial ruin by visiting it often, leading as many as a hundred men-at-arms and demanding that hospitality in the form of food and lodging be extended to her party for as long as she wished to remain.

Both Mabel and her husband were extremely effective in wielding political influence, especially with William, duke of Normandy, of whom Roger was a trusted supporter. Taking advantage of that influence, Mabel was instrumental in convincing the duke to banish her family's enemy Arnold d'Echaufflor (son of William fitz Giorie, who around 1048 had been treacherously captured and mutilated by her father, William) and to confiscate his lands. She later acquired large parts of those lands for herself and Roger. When Arnold, returning from exile, successfully petitioned for restoration, Orderic reports that Mabel decided to poison him, but accidentally killed her brother-in-law instead. She later managed to bribe Arnold's chamberlain into administering a new poison, successfully ridding herself of this rival on the second attempt.

Orderic says that Mabel was responsible for the disinheritance and ruin of many nobles, including one Hugh de la Roche, whom she deprived of his family castle. Along with his brothers, Hugh took his revenge on Mabel, who was staying at Bures on the Dive on 2 December 1079. Legend holds that Hugh, infiltrating the castle, found Mabel lying in bed after a bath and beheaded her there.

Bibliography: Chibnall, *The World of Orderic Vitalis*, 1984; *Dictionary of National Biography;* Echols and Williams, *An Annotated Index of Medieval Women*, 1992; Orderic Vitalis, *The Ecclesiastical History of Orderic Vitalis*, 1854; Thompson, "Family and Influence," 1985; White, Geoffrey H., "The First House of Belleme," *Transactions of the Royal Historical Society*, 4th ser., 22 (1940): 67–88.

—Annette Parks

MAHAUT D'ARTOIS (born ca. 1270s; died 27 November 1329, Paris). Countess of Burgundy and Artois. Defended counties of Artois and Burgundy, France, from internal and external enemies.

Mahaut d'Artois enforced the law, suppressed rebellions among her own vassals, and defended her claims to the county of Artois against the repeated assaults of her nephew Robert. Mahaut was left as sole ruler of the county of Artois after the deaths of her brother Philip in 1298 and her father Robert II of Artois at the Battle of Courtrai in 1302. She had already shared the rulership of Burgundy with her husband Otto since their marriage in 1285, and when he died in 1303, she became sole ruler of that county as well. In both territories her duties as ruler included enforcing justice and maintaining the peace, through force when necessary—although it appears that Mahaut did not normally lead armed expeditions herself but usually delegated these tasks to trusted subordinates. Thus when one of her vassals, the lord of Oisy, pillaged the property and mistreated the inhabitants of some lands under her protection (even going so far as to imprison one of her officials), Mahaut ordered the destruction of his castle, an order that was carried out by an armed band under the leadership of Jean de Ponthieu.

Mahaut's hold on Artois was contested several times by her nephew Robert. The difficulty of maintaining the peace was compounded by the fact that she had to defend her title and rights in Artois against Robert's repeated military and legal challenges. Although Mahaut received favorable rulings in the king's court, Robert was able to stir up rebellions against her by providing aid to some of her more disgruntled vassals; the problem became so acute in the first few years after the death of Philippe IV in 1314 that Mahaut was forced to spend five years under the king's protection, returning in 1319. Nevertheless, when her life is viewed as a whole, the countess emerges as an intelligent and competent ruler who successfully maintained her authority in difficult circumstances through her utilization of alliances and the delegated application of armed force.

Bibliography: Butler, *Women of Mediaeval France*, 1907; *Dictionnaire de biographie française*; Lehmann, *Le rôle de la femme*, 1952; Richard, "Les livres de Mahaut," 1886, and *Une petit-nièce de Saint Louis*, 1887.

—David Hay

MAKAROVA, TAT'IANA PETROVNA (born 25 November 1920, Moscow; died 25 August 1944, Poland); **BELIK, VERA LUK'IANOVNA** (born 5 August 1921, Crimea, Ukraine; died 25 August 1944, Poland). Both guards lieutenants, Red Army Air Force. Pilot and navigator, respectively, Soviet Union, Second World War.

Tat'iana Makarova and Vera Belik served in the **46th *Taman'sky* Guards Night Bomber Aviation Regiment,** one of the three Soviet women's combat aviation regiments formed by **Marina Raskova**. Makarova flew 628 combat sorties and Belik 813 aboard the Po-2 biplane. A very cohesive team, Makarova (pilot) and Belik (navigator) flew most of their missions as a crew. In September 1942 both were included in the first group of women to be decorated in their regiment: Makarova received the **Order of the Red Banner** and Belik the Order of the Red Star.

Makarova, the second daughter of an impoverished and disabled First World War veteran and an illiterate mother, was very much impressed with the famous 1938 nonstop flight to the Far East executed by Marina Raskova, Polina Osipenko, and **Valentina Grizodubova**. Admitted to a part-time flying course, she did well and was allowed to train as an instructor. At the age of nineteen she decided to become a professional pilot. In 1940 she was appointed an instructor at the Military School of Basic Training in Moscow and was awarded the rank of sergeant.

Belik was the oldest of six children of a master electrician. In 1939 she enrolled at the Karl Liebknecht Pedagogical Institute in Moscow, where she studied mathematics. Her sound judgment, good memory, and ability to calculate quickly were noted after she had volunteered for service in her combat aviation regiment.

Both joined the **122nd Aviation Group** in October 1941 and began flying combat missions with the 588th Night Bomber Aviation Regiment (later redesignated the 46th Guards) in mid-1942. In December 1942 the regiment was expanded, and Makarova was promoted to commander and Belik to navigator of the second squadron. After eight members of their squadron were shot down by German night fighters on 31 July 1943 over the Kuban' area, at their own request Makarova and Belik were demoted, reverting to duties of flight commander and navigator, respectively. They flew difficult sorties over Ukraine, the Kuban' area of North Caucasus, the Crimea, Belorussia, and Poland, often descending to very low altitudes to increase bombing accuracy. On 1 August 1944 they were singled out to fly the first bombing mission over East Prussia—the first aircrew of the regiment to fight over German soil.

Makarova and Belik died on 25 August 1944 when they were attacked by a German fighter near their home airfield in Poland. Their wood and canvas airplane caught fire, and like most Po-2 crews, they carried no parachutes; the aircraft went down in flames. Both were awarded the **Hero of the Soviet Union** posthumously on 23 February 1945.

Bibliography: Aronova, *Nochnye ved'my*, 1980; Chechneva, *Nebo ostaetsia nashim*, 1976; Cottam, *Women in War and Resistance*, 1998; Litvinova, *Letiat skvoz' gody*, 1983; Shamiakin, *Navechno v serdtse narodnom*, 1984.

—Kazimiera J. Cottam

MALASAÑA, MANUELA (died 1808). Spain, Napoleonic Wars/Peninsular War.

Manuela Malasaña's death during the *dos de Mayo* in Madrid in 1808 made her a martyr and symbol of Spanish unity against the French Empire, but historians debate whether she was a heroine or merely a victim. In March 1808 Napoleon's brother had been installed as ruler of Spain; on 2 May 1808 the people of Madrid rose up against the French occupation. Although imperial troops easily crushed the rebellion, the martyrs of the *dos de Mayo* served as a rallying cry for the Spanish resistance, which bled France for six years and prepared Napoleon's destruction.

Women played an important military and symbolic role in the events of 2 May. Contemporary accounts and illustrations of the uprising presented women in the thick of fighting, cutting down enemy troops, gutting the horses of the French cavalry as they galloped through the Puerta del Sol, and throwing tiles and boiling water down from their balconies and windows upon the French. The most famous of the civilian martyrs on 2 May was a young working-class woman named Manuela Malasaña, whose last name now describes the Madrid neighborhood where she died.

According to legend, French troops shot Malasaña as she carried cartridges to the men fighting in the artillery park of Monteleón, the focal point of the *dos de Mayo*. Early representations of the fighting at Monteleón show Manuela as an Amazon, actively participating in the combat. However, the reality on 2 May 1808 was rather different. The fifteen-year-old Malasaña took no part in the uprising; she simply had the misfortune to live near the Monteleón artillery park. Malasaña was surprised by French troops on 2 May as she returned from her work as an embroiderer, work that required her to carry shears. The French were shooting civilians holding anything resembling a weapon, and it was Manuela's fate to be shot down as a random victim of French excess.

The contradiction between the legend of Manuela and her senseless murder is emblematic of the distance between the popular image of women in combat and the real lives of most women during the long and bloody war with France. Women certainly took part in the resistance of 2 May. Indeed, a group of three women sparked the uprising, which started in the early morning in front of the royal palace. Women also took part in the fighting around the Puerta de Toledo on the western outskirts of Madrid. Of the 409 individuals killed by the French, 57 were women. The French garrison suffered just 14 killed and 150 total casualties, mostly in combat with the regular Spanish troops in the artillery park. The action of the mob in the streets had little impact. In particular, the horror of women pouring boiling water down upon French troops apparently resulted in only one superficial injury. Thus while it is clear that some women fought on 2 May 1808, "amazons" were neither typical nor effective; but then, neither were the men of the street mob.

The role of Spanish women in combat with France on 2 May and during the next six years was exaggerated by the small group of Spanish nationalists leading the resistance because female combatants served as potent symbols of a national unity and resolve that did not exist in reality. In fact, most Spanish women, like most men, did their best to remain aloof from the conflict, and very few women actively fought the occupation. In Spain the "nationalization of the masses," both male and female, is a twentieth-century phenomenon, except in the propaganda and panegyrics of nineteenth-century nationalists.

Bibliography: Alcaide Ibieca, Augustín, *Historia de los dos sitios que pusieron a Zaragoza en los años de 1808 y 1809 las tropas de Napoleon* (Madrid: 1830); García Nieto, José, *El dos de mayo en la poesía española del siglo XIX* (Madrid: 1983); *Madrid, el 2 de Mayo de 1808: Viaje a un día en la historia de España* (Madrid: 1992); Tone, "Women in the Resistance to Napoleon," 1998.

—John Lawrence Tone

MAO, EMPRESS (died 389, China). Military leader, early medieval China.

Empress Mao was the wife of Fu Deng (ca. 334–386), fifth emperor of the Former Qin dynasty (351–394), a short-lived nomadic reign in early medieval China. Toward the end of Fu Deng's reign the nomad chieftain Yao Chang (330–393) attacked and captured all of Fu Deng's military barracks. On horseback with bow and arrows, Empress Mao rode out with several hundred soldiers to attack Yao Chang and did considerable damage. She was said to be valiant and very good at archery on horseback. Nevertheless, since they were outnumbered, Mao was captured by Yao Chang. The bandit wanted to take her as his wife, but Mao cursed him, saying, "I was an empress, the wife of the son of heaven. How can I be violated by a bandit? Why don't you just kill me!" After that, she cried toward heaven, "Yao Chang has no sense of morality. He has harmed the Son of Heaven and is insulting the Empress now. Heaven above and earth below, witness it all!" Yao Chang was infuriated and killed her.

Bibliography: Chen Menglei, *Gujin tushu jicheng (Qin ding)*, 1993; Fang Xuanling and Baoyuan Du, eds., *Jin shu*, 1974.

—Sherry J. Mou

MARGARET OF ANJOU (born 24 March 1430, Pont-à-Mousson, Lorraine; died August 1482, Dampierre, Anjou), also known as Marguerite d'Anjou, Margaret of England. Queen of England, political and military leader, England, Wars of the Roses.

Margaret of Anjou, ca. 1640. (Hulton/Archive by Getty Images)

Warfare dictated the political career of Margaret of Anjou from its very beginning. She was the daughter of the militant **Isabella de Lorraine** and René, count of Anjou and titular king of Sicily; she was also the niece of Charles VII, king of France. Her marriage to King Henry VI of England was arranged at the Truce of Tours in May 1444, providing a brief respite in the ongoing conflict between England and France toward the end of the Hundred Years' War. One of her wedding gifts was a decorated manuscript of **Christine de Pisan**'s *The Book of Deeds of Arms and of Chivalry*. Margaret's influence was used to help persuade England to cede Maine to her father, a decision that was bitterly resented by the prowar faction at the English court.

Margaret soon became involved in the increasing factionalism at court as England lost further ground after the resumption of war in 1449. The year 1453 saw the final defeat of the English in France and the first attack of the mental illness that was to afflict King Henry, never a strong-willed ruler at the best of times, throughout his reign. The birth of a son, Edward, to Margaret and Henry strengthened her political position as mother of the heir to the throne. However, Richard, duke of York, who had emerged as the leader of the anticourt faction, was appointed "protector and defender of the realm" for the duration of the king's illness. The basis for the factions known to history as Lancastrian and Yorkist was being established. Margaret would emerge as the political and military leader of the Lancastrian faction, and the conflict would later be termed the Wars of the Roses.

Henry's recovery at Christmas 1454 removed power from York's hands, forcing him into open rebellion. York defeated the Lancastrians and captured Henry at St. Albans in May 1455 and was named Protector of the Realm. Margaret responded by rallying her supporters and was able to establish an effective power base on her estates in the northwest and west Midlands of England and in Wales. York fled to Ireland in 1459 after an armed face-off at Ludford Bridge. However, Margaret's regime was too narrowly based to bring stability to the country, and the Yorkists regrouped, capturing the king at the Battle of Northampton in June 1460. York returned from exile and for the first time stated his own claim to the throne as a descendant of Edward III; baronial opposition prevented him from seizing the throne, but he was recognized as heir apparent, supplanting Margaret's son Edward.

Margaret fled to Scotland, from where, having obtained the support of King James III of Scotland, she invaded England to regain her son's rights. York was defeated and killed at Wakefield in December 1460, where thirty-year-old Margaret led the army and probably had a hand in commanding the Lancastrian center on the battlefield. However, York's son Edward, earl of March, defeated Margaret's allies at Mortimer's Cross the following February. Marching at the head of her army, Margaret advanced on London and defeated the earl of Warwick's Yorkist army at St. Albans on 17 February, liberating her husband. Although Margaret did not formally take command on the battlefield, she was present at the battle and was probably responsible for masterminding a flanking maneuver that created the Lancastrian victory. However, the victory was nullified by the City of London's refusal to admit Margaret. Instead, March was able to enter the capital and have himself proclaimed King Edward IV. Retreating northwards, Margaret's army was defeated at Towton before she escaped to Scotland again.

Margaret did not accept defeat despite ten years in exile in Scotland and France and seized the opportunity to take up the Lancastrian cause once more when Edward alienated his most powerful supporter, the earl of Warwick. Believing that Henry would prove a more pliable king, Warwick allied himself with Margaret in July 1470 and restored her husband to the throne with the aid of Louis XI of France. However, Edward returned to England in April 1471 and at Barnet defeated and killed Warwick and recaptured Henry, just as Margaret and her son were landing at Weymouth. Despite this shattering blow, the queen was persuaded to continue her invasion, attempting to cross the Severn to join her allies in Wales. Edward advanced westwards from London; Margaret skillfully evaded his pursuing army, but was caught at Tewkesbury on 3 May, where her army was routed by Edward's attack, and her son the Prince of Wales was killed. Margaret, who this time did not lead her army on the field of battle, was captured attempting to escape. She was taken to the Tower of London on the very night that her husband was murdered there, extinguishing the direct Lancastrian line, and with it, Margaret's purpose and will to fight on. She remained in comfortable captivity for the next four years until she was ransomed out by her cousin Louis XI of France. Returning to France, she lived out the rest of her life in obscurity in Anjou and died at age fifty-two.

Margaret has often been decried as somebody whose arrogance and refusal to compromise condemned England to nearly two decades of civil war. A more sympathetic assessment, however, would suggest that almost alone of the main participants in the Wars of the Roses, she was steadfast in a cause—the defense of her husband's and son's rights to the throne—rather than being motivated by personal gain or the desire for power. The extent to which she commanded her armies on the field is a matter of debate, but she certainly led her armies on campaign in 1461 and 1471 and showed a strong sense of strategy and leadership in the victory at St. Albans and in nearly outmaneuvering the Yorkist army in the Tewkesbury campaign. However, she did not have the political skill to exploit her victories in war or to overcome factionalism in peacetime.

Bibliography: Bagley, *Margaret of Anjou*, 1948; Echols and Williams, *An Annotated Index of Medieval Women*, 1992; Green, *The Battlefields of Britain and Ireland*, 1983; Kendall, *The Yorkist Age*, 1962; Pollard, *The Wars of the Roses*, 1988.

—Michael Evans

MARIA OF POZZUOLI (fl. 1340s), also known as the knight of Naples. Fought in an unidentified territorial conflict near Naples, Italy.

A fourteenth-century example of a true virago, the otherwise unidentified Maria of Pozzuoli was seen by Petrarch on visits to Naples in 1341 and 1343. He describes her in detail, praising her strength, dexterity, mental discipline, courage, and unassailable virginity. In his letters he compares her, quite naturally, to Virgil's amazon heroine Camilla, whose legendary birthplace was in the same region of Italy. Maria was a native of Pozzuoli, a small town to the west of Naples. She was involved in a long-standing, bloody feud with a neighboring, but otherwise unidentified, enemy. She habitually wore armor, had the hardened body of a veteran soldier, and willingly endured the discomforts of military campaigning. Although she lived in the company of military men, none dared to violate her. Her fame was extensive, and men often came to see her and to challenge her to feats of strength. In one such weight-lifting contest she easily defeated a group of male challengers, and that despite an avowed pain in her arm. Sabadino degli Arienti, writing around 1483, mentions her death as the result of a wound (see Annecchino, 177). No other written records about her survive.

Bibliography: Annecchino, *Storia di Pozzuoli*, 1960; Boulting, *Woman in Italy*, 1910; Petrarca, *Rerum familiarium libri I–VIII*, 1975; Savi Lopez, *La donna italiana del trecento*, 1891.

—Gloria Allaire

MARIA THERESA (born 13 May 1717; died 29 November 1780, Vienna, Austria). Archduchess of Austria and queen of Hungary and Bohemia (r. 1740–1780), wife and empress of the Holy Roman Emperor Francis I (r. 1745–1765), and mother of the Holy Roman Emperor Joseph II (r. 1765–1790). Led the Austrian Empire during the War of the Austrian Succession (1740–1748), the Seven Years' War (1756–1763), and (with her son and co-ruler Joseph) the War of the Bavarian Succession (1778–1779), all against Frederick II (the Great) of Prussia (1712–1786) and his allies. Married Francis Stephen of Lorraine, with whom she had sixteen children, ten of whom lived to adulthood.

Just twenty-three at the time of her accession to the Austrian throne on the death of her father Charles VI, Maria Theresa began her reign in dynastic controversy and military conflict. The death of Charles's only son forced him to make his eldest daughter Maria Theresa his heir, but to do this, it was necessary that this "Pragmatic Sanction" be recognized by the crowned heads of Europe. This was duly granted, but placed her legitimacy under a cloud. King Frederick II of Prussia, exploiting what he perceived to be Maria Theresa's weak diplomatic and financial position (not to mention her youth and gender), invaded the Austrian province of Silesia in October 1740. Archduchess for only weeks, surrounded by her father's older advisors (most of whom counseled submission to Frederick's territorial demands), Maria Theresa made her most momentous decision: she and Austria would defy the invader and fight. She did so not for pride, but for principle: she believed that the legitimacy of her throne and, by extension, Austria's territorial and national integrity were at stake. But she was in desperate straits; by her own description, she was "without money, without credit, without an army, without knowledge or experience, and also without advice, because each of the counselors wanted first of all to wait and see how matters would turn out." Maria Theresa rejected advice that she surrender Silesia and accept Frederick's offer of recognition of the Pragmatic Sanction, believing that the Prussian could not be trusted and that he would eventually demand more territory. Almost 200

This portrait of the Empress Maria Theresa is in the reception salon of the Ghent Town Hall. The dress of Flemish lace that she is wearing was a gift to her from the states of Flanders. The dress was made by Ghent orphans. Out of gratitude for this expensive present, Maria Theresa allowed her portrait to be painted by her court painter, Martin van Meytens, after which the completed work was handed over to the members of the states. (Perry-Castañeda Library)

years before the word came into general use, she rejected "appeasement" of an aggressor.

Maria Theresa's decision soon proved costly, but she continued to insist on her prerogative to make war. She hired and fired generals, educated herself on the principles and strategies of warfare, and made pointed suggestions to her commanders on how to proceed against Frederick, much as the equally inexperienced Abraham Lincoln would do more than a century later. Undeterred by her inexperience, often using her gender as a weapon when necessary (her moral integrity and charm were much commented upon, even by Frederick), and bitterly reluctant to admit defeat ("My mind is made up! I will sacrifice and perhaps lose everything in order to save Bohemia.... All my soldiers, all of Hungary will I lose before I retreat even a step"), the empress waged war with a ferocity that astonished Frederick, and by 1748, although Austria was forced to cede Silesia, the Prussians had been fought to a draw.

The contest was renewed in 1756. The Seven Years' War was a "world war," but the duel between England and France for supremacy of the seas was of little moment to the protagonists in central Europe. Surrounded by the Grand Alliance of Austria and France, Russia, and Sweden, Frederick evaded disaster on the battlefield, and peace was concluded in 1763, leaving Prussia and Russia in the first rank as European great powers and Austria in the third rank at best. From this point a disillusioned Maria Theresa left affairs increasingly in the hands of her son and coruler, Joseph II.

Maria Theresa made a number of administrative and social reforms designed to further the war effort against the "Monster" Frederick, whom she personally hated. Upon being crowned queen of Hungary, against her ministers' advice she made a personal appeal to her restive Hungarian subjects, offering to arm them if they would provide troops. Defying expectations, the Hungarians fought for her enthusiastically. Maria Theresa attempted to improve morale in the officer corps by promoting commoners to the nobility in recognition of valor and initiative, culminating in the establishment of Austria's highest military honor, the Order of Maria Theresa. She thus anticipated Napoleon's "aristocracy of talent" and Legion of Honor. These innovations, together with the introduction of standardized national uniforms, militarized Austrian society in unwitting imitation of hated Prussia.

Maria Theresa lost all three of her wars against Frederick the Great of Prussia. Yet her far-seeing reforms of the army and administration laid the foundation for the "enlightened despotism" of her son Joseph II; her duel with Frederick launched more than a century of rivalry with Prussia that had lasting effects on the evolution of German nationhood; and her refusal to yield to Frederick's limited war aims and her dedication to the destruction of Prussia led to what she herself called "total war." In this respect, one could add to Maria Theresa's titles that of "protonationalist."

Bibliography: Crankshaw, *Maria Theresa,* 1970; Hochedlinger, "Mars Ennobled," 1999; Roider, *Maria Theresa,* 1973; Scott, Hamish M., "Verteidigung und Bewahrung: Österreich und die europäischen Mächte, 1740–1780," *Maria Theresia und ihre Zeit,* ed. Walter Koschatzky (Salzburg: 1980); Wandruszka, *The House of Habsburg,* 1965.

—Curtis F. Morgan

MARINER, ROSEMARY BRYANT (born 1953). First navy woman to fly a tactical jet, first to command a squadron of navy aircraft.

As a child, Rosemary Mariner dreamed of being a pilot, even after the death of her father, an air force pilot, in a plane crash when she was three years old. At age eight Mariner and her two sisters moved to San Diego, California, with their mother, a former navy nurse. There she spent many hours watching the planes take off from nearby Miramar Naval Air Station. Mariner earned money for flying lessons by cleaning houses and washing planes at Gillespie and Lindbergh Fields, earning her pilot's license at age seventeen.

Mariner attended college at Purdue University, where she was the first woman ever to graduate from the aviation program. In 1973, at age nineteen, she graduated with a degree in aviation technology. It was the same year that the navy admitted women to flight school, and Mariner joined the first group of women flight trainees at the Naval Aviation Training Command in Pensacola, Florida. Six of the women went on to win their wings, including Mariner, who graduated in June 1974. They were the first women pilots in any of the armed services since the **WASP** of the Second World War. Mariner was selected for jet pilot training in 1975, qualifying in the A-4L, A-4E, and the A-7E, all jet attack aircraft. She accumulated more than 3,500 flying hours in 15 types of aircraft during her navy career.

Mariner was Surface Warfare Officer qualified and in the early 1980s spent almost three years as ship's company aboard the USS *Lexington* (AVT-16), during which she stood thousands of hours of engineering and deck watches in order to qualify for her "water wings" (i.e., ship driver). At that time the *Lexington* had 120 enlisted women and 15 female officers in a crew of around 1,700. As a LCDR, Mariner was also a division officer who spent many hours overseeing young enlisted men and women aboard ship.

On 12 July 1990 Mariner accomplished another milestone when she became the first woman to assume command of a naval aviation squadron, Tactical Electronic Warfare Squadron 34, based at the Naval Air Station in Point Mugu, California. Mariner was selected for O-6 command (major aviation shore) and slated to go to Meridian Naval Air Station, but decided to retire instead. With an extensive background in joint warfare, she served as CJCS professor of military strategy at the National War College during her last two years of active duty.

Mariner explained her rise through the ranks of the navy when she said,

"When I hit a wall, I am going to get under it, over it, or around it. Put a wall in front of me and my reaction is to knock it down." Mariner also likes to acknowledge that she was mentored by several combat-experienced male officers and figures like Chief of Naval Operations Elmo Zumwalt, who supported the entry of women into military aviation. Mariner notes, "I think it is very important to give men credit, because the military can be a very hostile place for women and we must acknowledge the good guys who risk their careers by supporting women. 'Politically Correct' for most of my career meant 'No Women.'"

Mariner has spoken out on many issues involving women in the military. In her position as president of Women Military Aviators (WMA) in 1991, she helped fight to repeal the 1948 law banning women from flying aircraft engaged in combat missions. She has been vocal on other issues as well, including the role of women in the armed forces. In a *Washington Post* article (11 May 1997) she discussed integrated gender training in the armed forces, writing, "Internally, the military must resolve the outstanding issues of gender integration. Externally, we dare not squander our success by pandering to a few vocal critics pushing an anti-woman social agenda at the expense of national security. No amount of nostalgia over manly warriors protecting fair maidens erases the fact that this country cannot go to war without women on the front lines."

Mariner has published widely, especially in support of gender integration based on her own years of experience "in the trenches." Her article "A Soldier Is a Soldier" (*Joint Force Quarterly*, Fall/Winter 1994) was based on an award-winning paper for the National War College and discusses the basic principles for successful gender integration. She is working on a book tentatively entitled *No Excuses: Ending the Gender Wars in the Military.*

Bibliography: Ebbert and Hall, *Crossed Currents*, 1993; Holm, *Women in the Military*, 1992; Mariner, Rosemary, "The Military Needs Women," *Washington Post* 11 May 1997: 7C.

—Vicki L. Friedl

Inscription on tombstone at St. Nicholas Church, Hove, England:

In memory of Phoebe Hessel who was born in Stephney in the year 1713. She served for many years as a private soldier in the 5th Reg. of Foot in different parts of Europe and in the year 1745 fought under the command of the Duke of Cumberland at the Battle of Fontenoy (War of the Austrian Succession) where she received a bayonet wound in her arm. Her long life which commenced in the time of Queen Anne extended to the reign of George IV by whose munificence she received comfort and support in her latter years. She died at Brighton where she had long resided December 12th 1821 aged 108 years.

—*Minerva: Quarterly Report on Women and the Military* 1.2 (1983): 88.

MARKINA, TAT'IANA (born ?; died ?). Captain. Soldier, Russian infantry, eighteenth century.

Tat'iana Markina was a twenty-year-old Don Cossack from the village of Nagaevsk who, during the reign of **Catherine II,** disguised herself as a man with the surname of "Krutochkin" and enlisted as a soldier in an infantry regiment in Novocherkassk. She was promoted to the rank of captain and described as being "strong-willed, energetic, and militant." Someone apparently discovered her true identity and made a complaint, forcing "Captain Krutochkin" to appeal to the empress, who had physicians settle the question of "Krutochkin's" identity. Although Markina was forced to leave the service, she was given military retirement and a pension. She returned to her village near the Don River. Unfortunately, she left no writings or memoirs to describe her experiences and motivations. Her story is only beginning to be pieced together by Russian researchers from documents and archival materials.

A similar story from a slightly later period is that of Aleksandra Tikhomirova, who reportedly served for fifteen years in the Russian cavalry in the late eighteenth and early nineteenth centuries. She is said to have taken the place of her dead brother, a Guards officer whom she closely resembled. She gained the command of a company and was killed in 1807, whereupon her identity was discovered.

Bibliography: "Itak, ona zvalas' Tat'ianoi," *Nedelia* 1984: 11; Ivanova, "Zhenshchiny v istorii Rossiiskoi Armii," 1992.

—Reina Pennington

MARY I OF ENGLAND (born 18 February 1516, Greenwich, England; died 17 November 1558, Westminster, England), also known as Mary Tudor. Queen of England (1553–1558).

Mary Tudor was the daughter of Henry VIII and his first wife, **Catherine of Aragon**. She became queen on 19 July 1553 following the death of her half brother Edward VI and a brief effort to prevent her accession. Thus Mary, Henry's child of a repudiated marriage, became England's first woman ruler since the doubtful claim of Maud (**Matilda Augusta**) four centuries previous. A devout Catholic who rejected her father's creation of the Church of England, Mary set about attempting to reconcile England with the Roman Catholic Church. Her methods were extreme and included the burning of approximately 300 Protestants, earning her the epithet "Bloody Mary."

Mary had many opponents on both religious and political grounds. When she announced that she would marry her cousin Philip of Spain, the prospect of a foreign, Catholic king on England's throne provided a rallying point for Mary's enemies, and in 1554 a series of conspiracies and uprisings known as Wyatt's Rebellion erupted. Some lords and important gentry planned uprisings throughout the country, but when the government got wind of the plans, the uprisings failed or were aborted except for the one in Kent headed by Sir Thomas Wyatt. By late January soldiers from a royal force sent from London to subdue the rebels joined the rebellion, and Wyatt moved 3,000 men and even some of the queen's ships and artillery against the capital. The queen's council was said to have urged her to abandon her capital to the attackers, but the queen acted otherwise. On the first of February she entered the city and from Guildhall addressed the citizens. Her presence required defense, and

Mary Tudor (Mary I), 1544. (Hulton/Archive by Getty Images)

her courage inspired faltering resistance. Wyatt's forces entered the suburbs and even crossed the Thames River, but as London's gates remained closed to them, they withered to about 300 men at the final skirmish on 8 February. Mary's courage, resolve, and military decisiveness helped to secure the collapse and defeat of a dangerous rebellion. However, her forces managed to lose the port of Calais, the last English possession in France. Mary was willing to use violent military force to impose her policies, but was neither a political, nor a religious, nor a military success. When she died after only five years on the throne, England was more fiercely Protestant than when she took power.

Presbyterian author John Knox in 1558 spoke vehemently against the accession of Mary Tudor to power and the legitimacy of female rule in general in his famous "First Blast of the Trumpet against the Monstrous Regiment of Women." Knox wrote that "it is more than a monster in nature that a woman shall reign and have empire above man" and that "it is the subversion of good order, of all equity and justice." When a woman reigns, he said, "she hath obtained it by treason and conspiracy committed against God." Knox managed to gloss over the military role of **Deborah** in the Old Testament and in any event wrote an apology in 1559 when **Elizabeth I** succeeded her half sister Mary as queen of England, stating that his book should not be construed as applying to a Protestant queen.

Bibliography: Fletcher, *Tudor Rebellions,* 1983; King, *Women of the Renaissance,* 1991; Loades, *Mary Tudor,* 1989; Prescott, *Mary Tudor,* 1962; Wernham, *Before the Armada,* 1972.

—Marvin A. Breslow

MARY OF HUNGARY (born 17 September 1505, Brussels, Netherlands; died 18 October 1558, Cigales, Spain), also known as Marie de Hapsbourg, Maria von Habsburg. Ruling queen of Hungary, 1521–1527; ruling queen of Bohemia, 1522–1527; regent of the Netherlands, 1531–1556. Ruler of Hungary during Ottoman wars, ruler of the Netherlands during Habsburg-Valois wars. Military organizer, administrator.

In the immediate aftermath of the Battle of Mohacs (1526), where her husband, King Louis II of Bohemia and Hungary, fell to the Ottomans, Queen Mary organized political affairs and prepared for the election of her brother Ferdinand of Habsburg (the later Holy Roman Emperor Ferdinand I, ruled 1556–1564) to the thrones of Hungary and Bohemia. In 1531 she was named regent of the Netherlands and was personally involved in the military preparations and defensive improvements of the southern Netherlands front, including the construction of the defenses of Antwerp and the modern fortresses of Mariembourg, Charlemont, and Philippville. Mary was partially responsible for the importation of Italian fortification strategies and defenses into northern Europe and was involved in the reorganization of the Dutch navy. Queen Mary supervised the reorganization of the financial administration of the Netherlands during the intermittent wars with the French of the 1540s and 1550s. She also was responsible for the suppression of a number of domestic revolts during her reign.

Marble detail of the tomb of Mary of Hungary, Santa Maria Donnaregina, Naples, by Tino Da Camino, 1325. (Dover Pictoral Archive)

Mary was the daughter of King Phillip "the Handsome" of Habsburg (king of Castile, 1504–1506) and Queen Juana "the Mad" (queen of Castile, 1504–1555). She was raised by her aunt, Margaret of Habsburg (regent of the Netherlands, 1507–1515, 1518–1530), before moving to Austria in preparation for her marriage to the heir of the thrones of Bohemia and Hungary, Louis of Jagellon. From the age of nine until she left for Hungary and her coronation at the age of sixteen, Mary lived in an old armory in Vienna. She became a famed hunter, following the example of her grandfather, the Holy Roman Emperor Maximilian I (1493–1519). Her life in Hungary was marked by the ongoing wars with the Ottomans and political unrest in the realm. After her husband's death on the battlefield Mary organized resistance to the Ottoman troops from her headquarters in the castle at Bratislava, negotiating with the

Hungarian nobility while protecting her dowry incomes from Upper Hungary (Slovakia). The disastrous defeat at Mohacs, a battle in which much of the nobility of Hungary was killed, clearly showed the difficulties in relying upon personal ties and retinues to do battle in early-sixteenth-century Europe.

After the successful Habsburg defense of the city of Vienna during its siege by Ottoman troops in 1529, a siege that no doubt impressed on the young widow the importance of quality fortifications, Mary's brother, the Holy Roman Emperor Charles V (1519–1556), asked her in 1531 to replace her recently deceased aunt Margaret as regent of the Netherlands. There she built Renaissance castles, hunted extensively, and constructed a military power capable of defending itself against repeated assaults by the French. French propaganda portrayed Mary as an unnatural "Amazon" because of her inspection trips to the fronts and her reported choice to dress in male garb at times (all sanctioned portraits show her in formal widow's weeds).

French campaigns in the 1540s and 1550s failed to seriously breach the defenses Mary supervised, although her favorite château of Mariemont was sacked by French troops. These years were those of the so-called military revolution when the size of standing armies and navies and the costs of fortifications skyrocketed. Mary administered the transition to the new military system, innovating particularly in the financial realm and influencing naval strategy with concepts related to control of the seas.

Mary resigned her office as regent of the Netherlands in 1555 at the age of fifty, ending her administrative duties completely by August 1556. She retired to Spain and died shortly thereafter. She was buried next to her mother Queen Juana in the Church of Saint Claire in Tordesillas. Her body was later transferred to El Escorial.

Bibliography: Boom, *Marie de Hongrie*, 1956; Daniel, "Piety, Politics, and Perversion," 1989; Iongh, *Mary of Hungary*, 1958; Kerkhoff and Boogert, *Maria von Hongarije*, 1993; Tracy, "Herring Wars," 1993.

—Joseph F. Patrouch

MARZIA DEGLI UBALDINI (fl. 1330; died ca. 1374), also known as Marzia degli Ordelaffi, Marcia, Zia, and, most often, Madonna Cia. Defender of the castle of Cesena, Italy, against local and papal attackers.

Marzia degli Ubaldini grew up and lived in a period of continual warfare in Romagna, Italy. She was the wife of Francesco of the noble family Ordelaffi, tyrannical lord of Forlì and various towns and castles. This Ghibelline family strenuously opposed the pope, whose territorial ambitions threatened Ordelaffi holdings; a hundred years later the family would also be enemy to **Caterina Sforza**.

While her husband Francesco campaigned, "Madonna Cia" was often left to guard the town of Cesena. In August 1355 Count Carlo of Dovadola, with two Malatesta sons, 100 knights, and additional mercenaries, raided Cesena, capturing livestock and taking prisoners. Upon hearing of this, Cia armed herself, mounted a horse, and with great shouts mobilized her knights into pursuing the raiders. Matteo Villani, the only contemporary chronicler to record this incident, states that she acted "not like a woman but as a virtuous knight" (1:700). In the sortie the booty was retaken, the enemy was defeated, the two Malatesta were taken prisoner along with many of their soldiers, and Carlo himself was mortally wounded.

Pope Innocent VI placed Francesco and his lands under the interdict for various outrages committed against the Holy See. In September 1356 Francesco gave his wife command of 200 knights and many mercenary soldiers to hold Cesena while he held Forlì. In April 1357 papal sympathizers in the town revolted. The papal legate Cardinal Egidio Albornoz, supported by more than 800 Hungarian archers plus cavalry, laid siege to the town. When Cia learned that her former advisor Sgaraglino was negotiating with the enemy in order to end the siege, she had him and other traitors beheaded, and she remained sole defender of the fortress and captain of the soldiers. As the siege grew more desperate, she defiantly burned the cathedral's bell tower, the bishop's palace, and houses of rebel townspeople on 27 May. Facing overwhelming odds, Cia retreated into the citadel with her relatives, soldiers, and loyal citizens. The papal forces tightened their grip, repeatedly assailing and bombarding the town and mining its walls. Wearing armor day and night, Cia stoutly defended the walls and supervised the repair of barricades. At one point her own father unsuccessfully tried to convince her to yield. On 21 June she surrendered to the papal legate after personally negotiating a treaty that granted freedom to her soldiers, but spelled imprisonment for herself and her relatives. They were not released until 1359 when Francesco was pardoned. Her valiant actions won the respect of both allies and enemies and made her a legend in her own day.

Bibliography: Forlì, Biblioteca Comunale, Raccolta Piancastelli, sala O, ms. III/10, *Annales Caesenates* ad. an. 1357; Marchesi, Sigismondo, *Supplemento istorico dell'antica città di Forlì . . .* , 1678, reprint, Historiae Urbium et Regionum Italiae Rariores no. 34 (Bologna: Forni, 1968); Pecci, Giuseppe, *Gli Ordelaffi* (Faenza: Lega, 1974); Tuchman, *A Distant Mirror*, 1978; Villani, Matteo, *Cronica* (Parma: Fondazione Pietro Bembo, 1995).

—Gloria Allaire

MASARICO (fl. 1470–1545), also known as Macarico. Military leader, central Africa.

Various early Portuguese sources record that Masarico was a prominent courtier at the palace of Mandimansa, emperor of the Mande in the central sub-Saharan belt. She fell out of the emperor's favor and was asked to exile herself from the empire. As a leading matriarch, she left the empire with a massive retinue comprising friends, relatives, and dependants. She was also accompanied by a large detachment of soldiers under her command as she proceeded southwards toward the jungles of the sub-Saharan belt and the coast.

With what was ostensibly a marauding force of fortune seekers, Masarico and her followers, who became known as the Manes undertook a policy of subjugation and assimilation of all conquered people. An astute politician, she maintained a strict hierarchical order that meant the use of one of her followers as viceroy with strong kinship bonds to the main hub of the Mane force. The Manes extended their influence throughout what is now western Liberia, where they established the Manou Empire and their invasion of present-day Sierra Leone had a tremendous effect on the pattern of settlement and way of life of the traditional people. Their cultural, political, and social influence has been documented by a number of historians.

On the military front, Masarico's Manes introduced new weapons and tactics that accounted for their great success in subduing the people of the forest regions. Early Portuguese sources again describe a highly disciplined force equipped with bows and arrows such as those found among the Mande-speaking people in the area that straddled the Sahara and the forest regions to the south. In their documented attack on the Portuguese trading outpost at Elmina, the Manes are reported to have deployed and moved on their objective in a three-pronged formation, approximately forty-five miles equidistant from one another. Each prong comprised a vanguard, the main fighting force, and a rear guard. The vanguard was composed of scouts and received and gave ultimatums to prospective foes, requesting either acquiescence or defeat and death. Those who surrendered were conscripted into the conquering force, and those who resisted were vanquished and often cannibalized by the Sumbas. Portuguese accounts intimate that the Manes engaged the fortress at Mina in a bid to sack it. However, Mane fighting spirit withered as their forces were razed down by heavy artillery barrage from the fortress. They regrouped, formed a fourth army, and evaded the enemy by moving westwards along the Malaguetta coast toward present-day Sierra Leone. The Bulloms to the west, although politically fragmented, put up stiff resistance to the Mane onslaught. It is reported that in a fierce battle with the Bulloms Masarico's only son was killed. Grief stricken and aged, Masarico died shortly after in the Cape Mount region. The Manes continued to assert their hegemony over the native people and by the end of the century made large territorial acquisitions in Sierra Leone.

Bibliography: Alie, *A New History of Sierra Leone*, 1990; Alvares, Manuel, "Ethiopia Menor," manuscript at the Biblioteca da Sociedade de Geographia de Lisboa; Dornelas, Andre, "Relaçâo sobre a Serra Leoa," MS 51-VIII-25, Biblioteca de Ajuda, Lisboa; Rodney, "A Reconsideration of the Mane Invasions," 1967.

—Patrick Kagbeni Muana

MATILDA AUGUSTA (born ca. 1102, Winchester, England; died 10 September 1167, near Rouen, France), also known as Empress Matilda, Empress

Maud, Queen Maud. Military leader, besieger, and defender, England and France.

Matilda was the daughter and designated heir of King Henry I of England (r. 1100-1135). She was married to the Holy Roman Emperor, Henry V, from 1114 to his death in 1125, a marriage that conferred upon her the title of empress. Three years later she married Geoffrey, the count of Anjou. For nearly two decades beginning in 1135, Matilda fought bitterly for the crown of England against her cousin Stephen, count of Blois.

Although the magnates of England had promised her father that they would support Matilda's succession in 1127, 1131, and perhaps 1133, many apparently did not welcome the prospect of a female monarch. Consequently, when Henry died in 1135 and Stephen of Blois seized the crown, the majority recognized him as king. Shortly thereafter many in Normandy acknowledged him as duke, giving him control of substantial territory on the Continent, and the long struggle between the partisans of Matilda and Stephen commenced.

While adherents fought on her behalf in England, Matilda at first abetted her husband's efforts to secure

Matilda Augusta. Illumination from a medieval manuscript. (Perry-Castañeda Library)

several castles in southern Normandy. In October 1136 she led troops to Le Sap to bolster a siege that Geoffrey had mounted. Although Le Sap did not fall, the overall success that Geoffrey and Matilda enjoyed up to 1139 set the stage for Geoffrey's eventual conquest of Normandy. In 1139 Matilda turned her attention more directly to England, crossing the channel on 30 September. Her support, which had previously been confined largely to Gloucester and surrounding areas, increased considerably after Stephen was captured by her partisans at the Battle of Lincoln on 2 February 1141. On 2 March a council of magnates at Winchester acknowledged Matilda as "Lady of England and Normandy." Now clearly in the ascendant, Matilda moved to secure her position. By midsummer she was at Westminster attempting to win the support of the Londoners, but the burghers, angered at her demands and exactions, took up arms and forced her to retreat to Oxford. That flight marked the beginning of a decline in Matilda's fortunes.

In August 1141 she succeeded in taking the city and royal castle of Winchester and had laid siege to the castle of Stephen's brother, Henry, bishop of Winchester, located within the city's walls. But she and her forces were, in turn, beset by a siege directed by Stephen's wife, **Matilda of Boulogne**. The latter proved so effective that the empress, with a small escort, was forced to make a desperate escape. Riding, according to Florence of Worcester, "in the male fashion" (astride rather than sidesaddle), and then carried in a makeshift litter dragged behind two horses, Matilda was able to reach the temporary safety of Oxford. Meanwhile, her half brother Robert (earl of Gloucester and her most prominent supporter) was captured in the rearguard action fought to cover Matilda's escape. A prisoner exchange was arranged in November—Robert for Stephen of Blois.

The newly freed Stephen laid siege to Oxford Castle on 26 September 1142, leading to another daring escape by Matilda. One night near Christmas she and a few supporters lowered themselves to the frozen Thames outside the castle walls and, passing directly through Stephen's camp under the cover of darkness, made their way to Wallingford. However, Matilda was never able to regain the advantage against Stephen in England. In 1148 she departed for Normandy, which was by then in the hands of her husband, Geoffrey. The initiative in the struggle for the crown was taken up by her son, Henry, who married **Eleanor of Aquitaine** and ascended to the English throne as Henry II in 1154.

Some contemporary detractors described Matilda as "haughty." Although there may be some truth to this and similar charges, it seems that they were at least in part a reflection of a general prejudice against women operating in the traditionally male spheres of warfare and politics. For instance, the author of the *Gesta Stephani* asserts that she displayed "an extremely arrogant demeanor instead of the modest gait and bearing proper to the gentle sex." Several other chroniclers, both favorable and unfavorable, remarked upon Matilda's "masculine" qualities; for example, William of Malmesbury called her a "woman of masculine spirit." Such language was typically employed when writers of the period described women who wielded considerable authority or, more especially, played a role in military affairs.

Bibliography: Chibnall, *The Empress Matilda*, 1991; Florence of Worcester, *Chronicon ex chronicis*, 1848–1849; *Gesta Stephani*, 1976; Howlett, *Chronicles*, 1884–1889; William of Malmesbury, *Historia Novella*, 1955.

—David Balfour

MATILDA OF BOULOGNE (born ca. 1103, Boulogne, France; died 3 May 1152, Essex, England). Countess of Boulogne, queen of England. Military leader, besieger, and defender, England and France.

Matilda was the only child of Eustace III, count of Boulogne. In 1125 she succeeded him as countess and shortly after was married to Stephen, count of Blois. When King Henry I (r. 1100–1135) died in 1135, Stephen seized the English throne, and Matilda was crowned as queen of England on 22 March 1136. She played a leading military, political, and diplomatic role in the subsequent struggle for the crown between Stephen and his cousin, **Matilda Augusta** (Empress Matilda), the daughter and designated successor of Henry I.

The queen directed the successful siege of Dover Castle in 1138. The considerable support that Stephen enjoyed among the English magnates evaporated quickly following his defeat and capture at the Battle of Lincoln on 2 February 1141, but Matilda was instrumental in rallying Stephen's forces and soon regained the initiative for her husband. In midsummer 1141, while the empress was attempting to win the support of the citizens of London, Matilda led an army that ravaged the surrounding countryside. That helped to convince the Londoners to take up arms against the empress, forcing her to flee. When the empress took Winchester in August 1141, the queen mounted such an effective siege that the empress was compelled to take flight once again. Although she, along with a small retinue, managed to make her way to safety, her forces were defeated and her leading adherent, Robert, earl of Gloucester, was captured by the queen's army. It was the queen who then negotiated an exchange of prisoners, giving up Robert in exchange for her husband Stephen in November.

The efforts of Matilda of Boulogne were largely responsible for the fact that Stephen was able to retain his throne until his death in 1154. The author of the *Gesta Stephani* called her "a woman of subtlety and a man's resolution," who "bore herself with the valour of a man."

Bibliography: Davis, *King Stephen*, 1990; Florence of Worcester, *Chronicon ex chronicis*, 1848–1849; *Gesta Stephani*, 1976; Howlett, *Chronicles*, 1884–1889; William of Malmesbury, *Historia Novella*, 1955.

—David Balfour

MATILDA OF TUSCANY (born 1046, Mantua [?], Italy; died 24 July 1115, Bondeno, Italy), also called Matilda of Canossa, the great countess. Ruler, stateswoman, military leader, Investiture Controversy, Italy.

The Investiture Controversy (1075–1122) was, in modern terms, the first great war between church and state. The fundamental issues concerned the powers possessed by popes and kings in a Christian society. For more than twenty years Matilda of Tuscany provided the only reliable military support for Pope Gregory VII and his successors in their conflict with Henry IV, king of Germany and Holy Roman Emperor. The large landed possessions in Lombardy and Tuscany that Matilda inherited at the age of eleven were in a strategically vital position between the combatants and controlled access to the limited roads through the Apennines. Adroitly managing numerous fortified positions, she waged a classic campaign of active defense and outlasted a competent, popular, and more powerful opponent.

The sources for her early life are meager, but historians have not hesitated to fill the gaps. Matilda may have been present at various military actions from 1059 to 1064. Her mother, **Beatrice of Lorraine,** and her stepfather, Godfrey the Bearded, played an active role in the turbulent papal politics of Henry IV's minority. There are stories that Matilda took an active part in some of these actions. Writing in the mid-seventeenth century, Lodovico Vedriani described the fifteen-year-old Matilda leading a ferocious cavalry charge during a battle against the antipope Cadalus of Parma outside the walls of Rome. According to Vedriani, she shared the command of a small contingent of cavalry, 4,000 archers, and a similar number of pikemen with Arduino della Palude of Florence. This story, however, is not found in any eleventh-century sources, and Arduino della Palude first appears in Matilda's charters in 1101. It seems unlikely that a minor heir would have been exposed to the risk of death or capture on the field while her stepfather (and, in Vedriani's tale, her first husband) were present and battleworthy.

Matilda and Beatrice were at Rome with the army when Godfrey campaigned against the Normans of southern Italy on behalf of Pope Alexander II in 1067, but it is not clear whether the women accompanied the army into the Campagna. Bonizo of Sutri, writing about twenty years later, called this action the "first service" offered by Matilda to St. Peter. Although her stepfather still commanded in the field on her behalf, Matilda was now clearly past her majority, and Bonizo correctly credited the action to her. In 1074 Pope Gregory VII planned another campaign against the Normans, who were still encroaching on papal lands. Matilda and Beatrice brought troops to the pope's muster at Cimmino, but the army was torn by internal disputes and disbanded without action. Gregory also planned a campaign to the East against the Muslims on which Matilda and the dowager empress Agnes were to accompany him, but his letters of 1074–1075 do not specify Matilda's role, and the campaign never took place. No contemporary source describes Matilda commanding troops in any of these actions.

Also in the mid-1070s Beatrice and twenty-nine-year-old Matilda were active in the negotiations between Gregory VII and their kinsman, Henry IV, in a dispute that centered on the appointment of the archbishop of Milan. In January 1076 the king, possibly encouraged by his recent defeat of a rebellion in Saxony, reacted to a papal ultimatum by the not-unprecedented step of calling a council of German bishops and deposing the pope. The pope responded by excommunicating and deposing the king. King Henry had badly overestimated the strength of his position in Germany. His opponents pushed for the election of a new king. Matilda took an active role in negotiations to delay this action until the pope could go to Germany to confer with the princes.

Matilda's mother and her husband both died in 1076. Matilda, now ruling on her own, escorted the pope through her territory as he traveled to Germany for the meeting, and it is a measure of the turbulence of the times that she housed him in her great fortress of Canossa. King Henry, in a highly dramatic move, crossed the Alps in midwinter and stood before Canossa as a barefoot penitent. The pope had no choice but to give him absolution. Matilda's central role in this famous encounter was the culmination of her mediatory efforts. The conflict had been delayed, but was by no means forgotten, and neither Gregory nor Matilda could have failed to note how many Italian nobles came to meet with the king during his stay in Italy.

Henry's opponents in Germany quickly elected an antiking, Rudolf of Rheinfelden. For the next three years both men sent delegations to Rome asking for the pope's support. At the Lenten synod in 1080, Henry IV was excommunicated a second time on the grounds that he had not kept the promises he had made at Canossa. In June 1080 King Henry called a council, and many bishops, most from northern Italy, joined him in deposing Pope Gregory and electing Archbishop Wibert of Ravenna to replace him. Wibert returned to Italy, accompanied by Henry's heir, six-year-old Conrad, and possibly an older, illegitimate son, to gather support while Henry continued to campaign against Rudolf. In Italy Pope Gregory settled his differences with the Normans and began to gather troops for a campaign against Wibert. Among those to whom he turned for help was Matilda of Tuscany.

The first battle of the Investiture Controversy in Italy took place shortly after Rudolf of Rheinfelden was killed in battle on 15 October 1080. The troops of Henry's Italian supporters defeated Matilda's at Volta, above Mantua. Most Italian nobles supported Henry, partly in the hope of gaining some of the vast holdings of the widowed and childless Matilda, who clearly intended to oppose him. The citizens of the cities she controlled also saw an opportunity. The defeat at Volta was quickly followed by a revolt in Lucca, one of her most important cities south of the Apennines. Bishop Anselm, Matilda's spiritual advisor and Pope Gregory's legate, was driven out.

In March 1081 the pope wrote to his supporters in Germany, "But if you do not support our daughter Matilda, the loyalty of whose soldiers you have to consider, what else can she do, if her people refuse to resist him [Henry], but accept his terms of peace or lose all her possessions?" In April Henry IV arrived in Italy and gathered an army in Lombardy and Ravenna. The king rewarded the citizens of Lucca for their support with grants of privileges and immunities; he formally dispossessed Matilda. Unopposed, he traveled to Rome and laid siege to the city. The pope's nominal vassal and protector, the Norman duke Robert Guiscard, failed to answer the pope's summons, leaving Gregory VII defended largely by the formidable walls of Rome.

Matilda became the focal point of opposition to Henry. She gave refuge to the king's opponents and undertook operations against his supporters. In 1082 Henry's followers insisted that he lift the siege of Rome to campaign against Matilda. Matilda had a considerable strategic advantage; she did not need to defeat Henry in the field, but only to avoid being defeated herself. Given Henry's forces, battle posed an unacceptable degree of risk to Matilda, and she held to her Apennine strongholds. Henry, who had as much experience of battle as any eleventh-century general, conducted a campaign of systematic ravaging to deprive Matilda of resources and undermine the confidence of her followers. Benzo of Alba mocked Matilda as being stuck in Canossa, "wringing her hands and weeping for lost Tuscany," but apparently she could supply her troops while wiping her eyes. She probably did this at the expense of Henry's supporters, which is why they insisted on moving against her. Henry could neither bring her to battle nor contain her movements.

In the same year Matilda sent a substantial treasure of silver and gold, taken from local churches, to Rome to pay the pope's troops. The exact sequence of events is not clear, but one possibility is that Matilda undertook raids against Henry's supporters to force the king to leave Rome and campaign in the north, making it possible to get the precious metal into Rome. Henry's supporters also failed to dislodge the garrison loyal to the exiled Bishop Anselm of Lucca from the fortress of Moriana just above Lucca. The election in Germany of a second antiking, Herman of Salm, heartened Pope Gregory's supporters, but Herman never came to Italy. Employing both diplomacy and bribery, Henry IV entered Rome in 1083 and installed Wibert of Ravenna as Pope Clement III. Gregory VII took refuge in the Castel Sant'Angelo. In May 1084, when the pope's ally, Robert Guiscard, finally moved on Rome, Henry IV, newly crowned as Holy Roman emperor by Clement III, withdrew and left Italy by mid-June.

On 2 July at Sorbara Matilda surprised and soundly defeated an army of Henry's supporters. The sources, which include a possible eyewitness account, describe a dawn raid on a sleeping camp, the ignominious flight and capture of the leaders of the expedition, and the slaughter of the common soldiers who were left to the scavengers: "Now let them be consumed; let them learn by battle and blood that gentle faith when it wishes can be mighty in war." To Bishop Anselm's successor and biographer, Rangerius, Sorbara was a demonstration of the power of Anselm's blessing on Matilda's troops, but Rangerius also knew that the name Matilda means "mighty in war." Matilda's biographer, Donizone, understandably reverses the emphasis: "The renowned Matilda was a terror to them all." Matilda remained able to gather intelligence, mobilize rapidly, and act effectively.

After Pope Gregory's death in exile in 1085, several works emphasizing the righteousness of waging war in a just cause originated in Matilda's circle. Locally she occupied herself with recovering lost positions. In June 1087 Matilda, with her troops, supported Gregory's successor, Pope Victor III, in his second attempt to establish himself in Rome, but he remained only two weeks and died in September; Clement III's supporters held most of the city. Even so, her position was so strong at this time that she was able to support a joint campaign by the maritime cities of Pisa, which she ruled through a viscount, and Genoa against the Muslim trading city of Mahdia in North Africa. The emir was forced to pay tribute to St. Peter, that is, to the pope.

Pope Urban II was elected the following year and eventually entered Rome with his own troops. In 1089 Urban arranged a marriage between forty-three-year-old Matilda and Duke Welf, the seventeen-year-old heir of the duchy of Bavaria. The alliance put both sides of the Alpine passes under the control of Henry's enemies and threatened to cut him off from Italy. The emperor returned to Italy in 1090 and began a campaign of ravaging against the lands of Matilda and Welf. He took the castle of Gubernolo, but failed in an attempted siege of Mantua; in April 1091, however, the citizens surrendered Mantua in response to a generous bribe of imperial privileges.

Matilda fell back south of the Po; she had lost nearly everything north of the river. In the autumn she sent 1,000 soldiers to surprise Henry as he disbanded most of his army and moved to winter quarters near Verona, but the commander, Hugo del Mansi, warned Henry and delayed the march until the emperor could regroup and attack. Matilda's biographer, making it clear that she was not present at this defeat, says that she consoled the defeated troops, mindful of the morale of those who would face the same enemy again.

She clearly intended to keep fighting. The year 1092 was grim. Henry had made significant gains south of the Po; he won many positions through bribery rather than battle. A notable exception was the great fortress of Monteveglio, which cost Henry one of his sons in a failed siege. He now offered Matilda generous terms if she would only recognize Pope Clement III. At a council held at Carpineta in September, against the advice of many of her supporters, she decided to hold out.

In October Henry made a surprise move against Canossa itself. He apparently hoped to end the war by pinning her inside. When Matilda received word of his approach, she withdrew with some of her troops to Bianello, one of four smaller outlying castles. Henry's army, hampered by a sudden heavy fog, was caught between the two forces. Matilda's forces captured Henry's imperial standard and began a rout. Matilda harassed the emperor as he retreated back over the Po, preventing him from regrouping or resupplying his troops. After this defeat Henry's support from the Italian nobles melted away. His last recorded military action was a failed siege of Nogara in 1095. In 1095 Matilda escorted Pope Urban II when he went to France to preach the First Crusade at the Council of Clermont; upon his return she accompanied him southward and reentered her Tuscan possessions.

The political and religious issues of the Investiture Controversy did not end with Henry's final departure from Italy in 1097. Violent disturbances were put down at Parma (1102) and Ferrara (1104). Documents from 1107 were issued "during the siege of Prato." Matilda's last recorded military action took place in 1114, less than a year before her death. The city of Mantua, notoriously resistant to sieges because of its location in swampy terrain, revolted. As Matilda approached the city with her siege engines mounted on barges, the citizens sent a delegation to surrender.

Matilda of Tuscany appears regularly throughout meticulous studies of the Investiture Controversy produced by generations of scholars. However, just as she eluded the armies of Henry IV, the countess continues to elude biographers. The competent, twice-married aristocrat has been represented as a virgin saint, as an Amazon (complete with male sidekick), as a deluded dupe of the popes, and as a high-strung neurotic. Women's studies scholars have yet to apply their analysis to Matilda's military career. The most recent biography of Matilda in English, although competent and scholarly in its time, is now more than ninety years old. Studies of the strategic military actions of the Investiture Controversy, the sieges and scorched-earth actions, were until recently ignored by scholars occupied with decisive battles and thundering charges. Matilda as she was understood in her own time is shown in the frontispiece portrait from the *Vita Mathildis,* a biography that was prepared for presentation to her. She is seated in state; a monk holding an open book stands to her right and a soldier holding a sheathed sword to her left. A priest who was with Matilda in 1084 wrote, "Although she was most religious and devout in private, in the world she openly led the life of a soldier."

Bibliography: Cowdrey, *Pope Gregory VII: 1073–1085,* 1998; Duff, *Matilda of Tuscany, la Gran Donna d'Italia,* 1909; Eads, "Mighty in War: The Role of Matilda of Tuscany in the War between Pope Gregory VII and Emperor Henry IV," 2000, and "The Geography of Power: Matilda of Tuscany and the Strategy of Active Defense," 2002; Erdmann, *The Origin of the Idea of Crusade,* 1977; Goez and Goez, *Die Urkunde und Briefe der Markgräfin Mathilde von Tuszien,* 1998; Goez, "Markgräfin Mathilde von Canossa," 1983; Overmann, Alfred, *Gräfin Mathilde von Tuscien: Ihrer Besitzungen, Geschichte ihres Gutes von 1115–1230, und ihre Regesten* (Innsbruck: 1895).

—Valerie Eads

MAVIA (fl. late fourth century). Queen of the Saracens. Led troops into Phoenicia, Palestine, and Egypt in revolt against Rome.

Mavia was queen of the Saracens in the late fourth century, a time when the Roman Empire was disintegrating under the pressure of invasions and revolts. Around 373 CE she launched "one of the most serious attacks in the history of the Arabian frontier," focusing her assaults on Phoenicia and Palestine as far as Egypt. The local Roman commander, assuming that he could make quick work of this woman general, soon found himself sufficiently pressed that he had to call upon the general in charge of Rome's eastern armies for reinforcements. Mavia's forces continued to best the Romans, forcing the emperor, Valens (r. 364–378), to sue for peace and accede to her demands that a certain Moses be installed as bishop for her people. In 378 CE Mavia sent a contingent of troops to help defend the city of Constantinople, then besieged by the Goths. The initial encounter between the Saracens and the Goths proved so demoralizing for the latter that they were soon thereafter obliged to withdraw. This was the same year in which Valens died during the Visigoth revolt at Adrianople.

Interestingly, some later sources report that Mavia was a Roman by birth who had been taken prisoner by the Saracens, married their king, and upon his death ascended to the throne. However, the earlier tradition lacks these details, which has led some to question their authenticity.

Bibliography: Ammianus Marcellinus, 31.16.5–6; Bowersock, "Mavia," 1980; Jones et al., *Prosopography of the Later Roman Empire,* 1971; Parker, *Romans and Saracens,* 1986; Sozomenus, 6.38.1–5; Theophanes, 1.64–65 (de Boor).

—C. A. Hoffman

MCAFEE (HORTON), MILDRED HELEN (born 12 May 1900, Missouri; died 2 September 1994, New Hampshire). Captain, United States Women's Naval Reserve, Second World War.

Mildred McAfee was the first director of the **WAVES** (Women Accepted for Voluntary Emergency Service) during the Second World War. McAfee served in a series of university teaching and administrative positions from 1923 on, culminating in her appointment as president of Wellesley College. She helped to plan for the entry of women into a naval auxiliary program when the Second World War began and was appointed director when the WAVES was formed in 1942. Interestingly, she did not like the acronym WAVES and instead proposed the term "women in the navy," but the acronym was adopted.

McAfee was commissioned as a lieutenant commander on 3 August 1942 and was later promoted to captain. She established recruiting and training procedures for women in a wide variety of technical skills, primarily radio communications, during the war. She is remembered for her deep concern for the living and working conditions of women in naval service. She was particularly known for her "coffee klatsches," or daily informal meetings with women dispersed in various bureaus and departments of the navy, which she organized to keep informed of developments around the Department of the Navy.

McAfee was awarded the Distinguished Service Medal. Her postwar career included a series of public service appointments, including serving as a delegate to UNESCO and the National Women's Committee on Civil Rights. She married the Reverend Douglas Horton in 1945 and lived to the age of ninety-four.

Bibliography: Alsmeyer, *The Way of the WAVES,* 1981, and "A Preliminary Survey," 1983; Hancock, *Lady in the Navy,* 1972; Holm, *Women in the Military,* 1992.

—Patrick J. O'Connor

MCCAULEY, MARY LUDWIG HAYS (born ca. 1754, Trenton, New Jersey; died 22 January 1832, Carlisle, Pennsylvania), also known as "Molly Pitcher,"

"Captain Molly," "Molly McKolly." United States. Camp follower who operated an artillery piece in the Revolutionary War.

Mary Ludwig Hays McCauley followed her husband into service in the Continental army during the Revolutionary War, and when he was wounded at the Battle of Monmouth, she assumed control of his weapon and fired on the enemy. McCauley was granted an annuity by the state of Pennsylvania in recognition of her military service. Her wartime experiences inspired the popular legend of **"Molly Pitcher,"** a title sometimes applied generally to emphasize the vital military support of camp followers of the American Revolution who brought water to the troops.

Mary Ludwig was born to German parents who ran a dairy farm in New Jersey. In 1769 she entered domestic service in Carlisle, Pennsylvania. In that same year, while working as a maid for Dr. William Irvine, she met and married John Caspar Hays, a barber. In December 1775 John Hays enlisted as a gunner in the First Company of Pennsylvania Artillery under the command of Thomas Proctor. He enlisted again in January 1777, this time in the Seventh Pennsylvania Regiment. It is likely that Mary joined her husband in camp when her employer, Dr. Irvine, became commander of his regiment. She performed a variety of duties as a camp follower, including cooking, cleaning, nursing, and helping to maintain weapons.

When their regiment was engaged at the Battle of Monmouth on 28 June 1778, Mary crossed the line from camp follower to soldier. She carried water to her husband to cool his artillery piece (many accounts report, probably inaccurately, that she brought water to quench his thirst). Hays assumed her husband's place at the gun after he was wounded or collapsed from the heat, and she fired the gun at least once. Joseph Plumb Martin, a private in the 8th Connecticut Regiment, who was present at the battle, noted in his famous diary that a woman who is believed to be Hays had part of her petticoat shot off "while in the act of reaching for a cartridge."

Hays returned to Carlisle after the war and resumed her work as a servant. After John Hays died, she married another veteran named John McCauley in 1792. In February 1822 the Pennsylvania state legislature voted to grant Mary Ludwig Hays McCauley forty dollars and to give her the same amount each year for life. The original bill referred to "Molly McKolly" as the "widow of a soldier in the Revolutionary war," but subsequently the bill was amended to reward McCauley "for services rendered" in the war by herself. McCauley never received a congressional pension, but she collected the state's annuity until she died. Markers were placed on Mary Ludwig Hays McCauley's grave in 1876 and 1916, and the official monument commemorating the Battle of Monmouth depicts her.

Bibliography: De Pauw, "Women in Combat," 1980; Evans, *Weathering the Storm*, 1989; James et al., *Notable American Women*, 1971; Landis, "Investigation into American Tradition," 1911.

—Sarah J. Purcell

MÉDICI, CATHERINE DE (born 13 April 1519, Florence, Italy; died 5 January 1589, Blois, France). Queen of France during the Wars of Religion.

Widowed by the death of Henry II in 1559, Catherine de Médici controlled the destiny of France by right as the regent or by necessity when faced with a weak king. For thirty years during the reigns of her sons Francis II (1559–1560), Charles IX (1560–1574), and Henry III (1574–1589) and the crucial time of the Wars of Religion (1562–1598) she ruled France both directly and from behind the scenes.

Catherine was introduced to conflict at an early age. She had been held as a hostage in Florence in 1527 during a revolt that drove the Médici from power and was living in the city when it came under siege in 1530. Very intelligent, crafty, and without scruples, she used all means at her disposal to maintain her power and that of her sons after the death of Henry II. Rather than open warfare, she preferred more discreet means, like poison, occult practices, or discreet murder. Faithful to Catholicism, but not fanatical, she initially confronted the Protestants by playing on their divisions in an effort to keep the Crown in control of all the factions.

Undated portrait of Catherine de Mèdici. (Perry-Castañeda Library)

However, the Massacre of Vassy (1562) forced her hand since political circumstances assured those responsible that they had her support, as well as that of the young king. Catherine had to resort to war. With the royal army she besieged Rouen and ultimately was able to bring about a settlement with the Protestants, the Edict of Amboise, which put an end to the First War of Religion. It was said that during the siege at Rouen Catherine "was not at all concerned about the cannonades and arquebusades that rained all around her." From there she left to recapture Le Havre from the English; she united the Protestants and imposed peace in the province. Following her lengthy tour of France in the company of her son Charles IX (1564–1566), Condé and Coligny again rebelled and tried to abduct the king; the royal army stopped them. Catherine obtained a truce with the Treaty of Longjumeau (1568), but the fighting soon began again everywhere. She was forced to go to Metz to block the road to Paris to Dutch and German Huguenots hurrying to the aid of their coreligionists. The Peace of Saint-Germain (1570) restored a precarious peace. Faced with the growing power of the Protestants, Catherine decided to neutralize their leader, Henry of Navarre (the son of her rival **Jeanne d'Albret**) by marrying him to her daughter **Marguerite de Valois;** Catherine made Coligny the engineer of the alliance. But six days after the marriage at Paris of Henry and Marguerite came the St. Bartholomew's Day Massacre (24 August 1572). Spreading from Paris throughout all of France, it left thousands of Huguenots dead. Historians are still debating who was responsible for authorizing the massacre; some blame Catherine.

Although Catherine's power was greatly diminished under the reign of her son Henry III, she continued to wield a great deal of influence. With her favorite Henry securely on the throne, Catherine had to turn her attention to her last son, the duke of Anjou. Twice he fled the court, entering into open rebellion; twice she pursued and reconciled with him (1575 and 1578). Catherine then decided to negotiate with the king of Navarre, and after discussions during which she proved to be particularly obstinate, she obtained the Treaty of Nerac (1579). The duke of Anjou died in 1584, dimming her hopes of seeing her descendants retain the French throne, since Henry III had still produced no heir. But she fought up to the end for her family's interests. During the campaign of 1587 she remained at Paris with full power and as the chief officer of state. During the Days of the Barricades (May 1588) she was able to go out without fear of attack but was a beautiful and well-guarded hostage for some time in the capital. She was involved in negotiating the Treaty of Union (July 1588) that ended the crisis.

Catherine died seven months before the assassination of Henry III and the succession of Henry of Navarre to the French throne. Then began the dark legend of Catherine de Médici, amplified by Alexandre Dumas in the nineteenth century. At her death she was hated so much by Catholics and Protestants that no one wanted to bury her at Saint-Denis, in the necropolis of the kings of France; that was not done until twenty-one years later. Nevertheless, in quickly summing up her life, it can be said that like Jeanne d'Albret, Catherine was one of the most remarkable political personalities of her time who was intricately involved in the instigation, continuation, and cessation of warfare.

Bibliography: Cloulas, Ivan, *Catherine de Médicis* (Paris: 1979); Mariéjol, Jean-Hippolyte, *Catherine de Médicis, 1519–1589* (Paris: 1920); Médicis, Catherine de, *Lettres,* ed. Hector de La Ferrière and Gustave Baguenault de Puchesse, 10 vol. (Paris: 1880–1909); Roman d'Amat, Charles, *Dictionnaire de biographie française,* tome 7 (Paris: 1956).

—Marie-Thérèse Lalaguë-Guilhemsans (trans. by Annette Parks)

MÉDICI, MARIE DE (born 26 April 1573, Florence, Italy; died 3 July 1642, Cologne, Germany). Queen and queen mother of France. As regent during the minority of her son Louis XIII, Marie de Médici entered into a rebellion against him, seized power, and later died in exile.

Described as both proud and scheming, Marie de Médici became the second wife of Henry IV of France after he repudiated **Marguerite de Valois**. In 1610 she became the regent for her son after her husband's assassination, in which some believe she played a role. In embracing many policies directly opposite to those of the late king she was at least able to avert foreign wars. Domestically she had to deal with the unrest of the Protestants and with demonstrations of discontent by the "Great Ones." In 1614 she led the expedition of Nantes, where she obtained, at the head of 20,000 men, the submission of the duke of Vendôme, the bastard son of Henry IV. In 1615 she was again at the head of a small army, always fearful of a coup by the Protestants. When she went Bordeaux to celebrate "the Spanish marriages" of her son Louis XIII and of her daughter Elisabeth, Condé took advantage of her absence to attempt to raise part of the country against her. Although he formally submitted in May 1616, his continued plotting caused Marie to imprison him.

As queen mother, Marie was reproached for allowing herself to be exploited by her favorites, Concini and his wife. In 1617, when her son Louis XIII was sixteen, he seized power; he had Concini arrested and assassinated, then locked his mother up in the castle at Blois. In a famous incident Marie escaped in 1619 in the middle of the night by climbing out of a high window down to a terrace by a rope ladder; then, tied up in a coat, she was helped to slide along a wall to safety. It was about 130 feet of very risky descent considering her age (forty-six was regarded as old at that time) and her stoutness, evident in her full-length portraits and representations.

After her escape Marie alternated between open fighting and attempts at reconciliation with Louis XIII. Having taken refuge at Angoulême, she once approached her son as he headed threatening royal armies. Then, allying herself with the "Great Ones" in revolt, she took arms and entrenched herself in Anger; her army was beaten in battle at Ponts-de-Cé (1620). During a reconciliation she sought to overthrow Cardinal Richelieu, whom she accused of wanting to come between herself and her son. Following her failure to secure Richelieu's dismissal on the Day of Dupes (1630) and her refusal to renounce her schemes, she was locked up again, this time in the castle at Compiègne. She escaped in 1631 under less fantastic circumstances than twelve years earlier and left France for good. During eleven years in Brussels, The Hague, and London she found other sworn enemies of the cardinal, including **Madame de Chevreuse** and Madame de Cologne, and she continued to plot against the government of Louis XIII. She supported the attempted invasions of her second son, Gaston d'Orléans, also exiled, and financed his expedition at the time of the rising of Languedoc (1632). In May 1642 she believed that she could reenter France on the occasion of the plot of Saint-Mars, but the plan failed.

Marie de Médici died in Germany at the age of sixty-nine, deeply in debt, though not destitute, as is sometimes written. She was buried at Saint-Denis in the necropolis of the kings of France.

Bibliography: Batiffol, Louis, *La vie intime d'une reine de France au XVIIe siècle* (Paris: 1906); Bluche, *Dictionnaire du Grand Siècle*, 1990; Carmona, Michel, *Marie de Médicis* (Paris: 1981); Kermina, Françoise, *Marie de Médicis, reine, régente, et rebelle* (Paris: 1979).

—Marie-Thérèse Lalaguë-Guilhemsans (trans. by Annette Parks)

MEIR, GOLDA (born 3 May 1898, Kiev, Ukraine, Russia; died 8 December 1978, Jerusalem, Israel), also known as Golda Mabovitz Myerson. Zionist activist, founder of the State of Israel, Israeli diplomat, politician, and prime minister of Israel during the 1973 Yom Kippur War.

Photo of former Premier of Israel, Golda Mier, at an unknown location, 1968. (AP Photo)

Born Golda Mabovitz, Meir immigrated with her family from Russia to the United States in 1906, settling in Milwaukee, Wisconsin. She graduated from the Milwaukee State Teachers College and in 1917 married Morris Myerson. As a young woman she became a Zionist and dedicated herself to the establishment of a Jewish homeland in Palestine (present-day Israel). In 1921 the Myersons emigrated to Palestine, where they lived on a cooperative farm near Nazareth and raised a family. The Myersons separated in 1928 (although they never divorced), and in 1956 Golda Hebraized her surname to Meir.

Meir was active in the Zionist movement from the 1920s through the 1940s, and her work frequently took her to Europe and the United States. Just prior to Israel's declaration of independence in 1948 she crossed into Jordan dressed as an Arab woman in an unsuccessful attempt to persuade King Abdullah not to join in the impending Arab attack on the fledgling Jewish nation. Meir was among those who signed Israel's independence proclamation of that year, and after independence she remained active in Israeli politics and government. She served as Israel's first minister to the Soviet Union between 1948 and 1949. In 1949 Meir was elected to the Knesset (parliament) and served as minister of labor and social insurance. Between 1956 and 1966 she served as Israel's foreign minister. Meir also served as the secretary-general of the Mapai and Labor parties. Following the death of prime minister Levi Eshkol in 1969, Meir assumed Israel's most powerful office at the age of seventy. The appointment of an elderly woman to the office of prime minister concerned some in the Israeli government. However, Meir quickly assumed firm control over a fractious cabinet and gained a reputation for a strong yet personal leadership style in both domestic and foreign affairs.

As prime minister, Meir was commander in chief of Israel's armed forces. At the time she assumed power, Israel occupied large areas of Arab territory captured in the Six-Day War of 1967, and her nation remained surrounded by hostile Arab states. Meir's political leadership style extended to military affairs. She concentrated military decision making in a small group of advisors, the most important of whom was defense minister Moshe Dayan, a hero of previous Arab-Israeli wars. Once asked if she believed that Israeli democracy was threatened by its need for a strong defense, Meir stated that she did not want to have a "fine, liberal, anticolonial, antimilitaristic, *dead* Jewish people."

On 6 October 1973 Egyptian and Syrian forces launched a surprise attack against Israel in an attempt to regain territories lost in 1967. The attack was a shock for two reasons. First, it came on the holiest day in the Jewish calendar, Yom Kippur, the day of atonement. Second, many of Israel's military leaders, foremost among them Dayan, believed that the Arab states had yet to recover from the Six-Day War and were incapable of waging a coordinated war effort. During the war Meir appealed to the Israeli people's hearts and minds, and her leadership style proved to be a valuable asset. Dismissing the idea that the Arabs only sought to regain lost territory, she argued instead that the "war was launched . . . against the very existence of the Jewish state" and told Israelis that "our survival is in the balance." Meir's wartime posture was an inspiration to many; one diplomat described her as having "a singular strength of resistance." Within a week of the invasion Israeli forces counterattacked and gained the upper hand. By the third week of October the Israeli army had not only beaten back the Arab assault but threatened the capitals of Damascus, Syria, and Cairo, Egypt.

One of Meir's greatest challenges during the Yom Kippur War was the involvement of the two superpowers, the United States and the Soviet Union. The Arab-Israeli conflict, like other regional tension spots around the world, had become entangled in the global Cold War, with the Soviets backing the Arab states and the United States supporting Israel. The humiliating Arab defeat that was developing by late October threatened to destabilize the Middle East, and at one point the war almost sparked a major confrontation between the Americans and the Soviets. To restore stability to the region, the superpowers, working through the United Nations, pressured the warring parties into a cease-fire. The halt to the fighting was particularly galling to Israel, which was on the verge of another smashing victory. While Meir valued Israel's alliance with the United States, American pressure to end the war also disillusioned her about American-Israeli relations. "We have the right to demand that the United States policy should not be conducted at the expense of our vital interests," she once declared, although after the war she conceded regretfully that "when Washington and Moscow agree on a course of action, there isn't a lot of room for maneuver."

In her memoirs Golda Meir described the 1973 Yom Kippur War as "a nightmare . . . which will always be with me" and admitted that it was the most difficult period in her life. The war took a toll both on Meir's health and her political party. Despite Israel's military victory in the Yom Kippur War, controversy over Israel's apparent lack of preparedness for the Arab attack forced the resignation of Meir and the rest of the cabinet in 1974.

Bibliography: Allen, *The Yom Kippur War,* 1982; Howes and Stevenson, *Women and the Use of Military Force,* 1993; Meir, *My Life,* 1975.

—Mark D. Van Ells

MENG, first name unknown (fl. 504, Julu, China), also known as Consort Meng. Commanded defense of a prefecture during the Northern dynasties (386–534).

Consort Meng was the mother of Yuan Cheng (ca. 467–519), who was Prince Rencheng and the director of the Department of State Affairs. When Yuan Cheng led his troops south to quell uprisings, a general of the Liang dynasty (502–557), Jiang Qinzhen, gathered rebel groups to capture Rencheng. The administrator, Wei Zan, did not know what to do, so Consort Meng led the soldiers onto the ramparts. She encouraged both officers and enlisted men, comforted the new and the old soldiers alike, reminded people of rewards and punishments, and admonished them about the good times as well as the bad, inspiring them all. She also acted as a sentinel herself, unafraid of stray arrows and other weapons. As a result of all this, the rebels were unable to capture Rencheng.

Yuan Cheng submitted a report to the throne to honor his mother. But Emperor Shizong (r. 500–515) had died, so the matter was postponed. Later Empress Dowager Ling (died 528) sent down an order: "Such a great and splendid achievement should indeed be remembered forever" by a tablet to record and to honor Consort Meng's virtues.

Bibliography: Chen Menglei, *Gujin tushu jicheng (Qin ding),* 1993; Li Yanshou, *Bei shi,* 1974; Wei Shou, *Wei shu,* 1974.

—Sherry J. Mou

Undated portrait of Louise Michel. (Perry-Castañeda Library)

MICHEL, LOUISE (born 1830; died 1905, Marseilles, France), also known as "Red Virgin." Combatant for the insurrectionary Paris Commune in France, 1871.

Born out of wedlock to a maid in the household of the mayor of Vroncourt (Haute-Marne, France), Louise Michel spurned marriage, instead aspiring to be a writer and poet. In Paris during the 1860s she worked as a schoolteacher and moved in revolutionary political circles, becoming acquainted with many Blanquists and developing a lifelong commitment to anarchism.

During the Franco-Prussian War (1870–1871) Michel organized women's first-aid units during the siege of Paris (September 1870–January 1871). She was arrested for leading a group of women requesting weapons from military authorities and fought in several attempts to overthrow the moderate government in favor of a more vigorous and revolutionary republic. "The first time you take up arms in defense of your cause," she later wrote, "you enter into the struggle so completely that you yourself become a sort of projectile." Six weeks after the end of the war Michel helped spark the insurrection of 18 March 1871 that began the **Paris Commune** when she rallied Parisians against French government troops. During the Commune she donned a uniform and joined the 61st Battalion of the National Guard as a nurse and canteen worker, but also fought in many engagements. Michel once wrote, "Oh, I'm a savage all right, I love the smell of powder, grapeshot flying through the air, but above all, I'm devoted to revolution." She fought on the barricades after government forces entered the city on 21 May, but surrendered when she heard that her mother was to be shot in her place. At her trial she became a celebrity by proudly admitting her exploits and demanding the death sentence, but was instead deported to a fortress in New Caledonia with thousands of other Communards. After the amnesty of 1880 she returned to France and continued to work for revolution until her death.

Bibliography: Johnson, "Citizenship and Gender," 1994; Michel, *The Red Virgin*, 1981; Thomas, *Louise Michel*, 1980.

—Martin P. Johnson

"MOLLY PITCHER." American Revolutionary War.

"Molly Pitcher" is a composite mythical figure of the Revolutionary War, a woman who was said to have replaced her wounded husband at his cannon and to have fought against the British. Women who brought water in "pitchers" to cool artillery pieces during battle sometimes took over the weapons in emergency situations. The name "Molly Pitcher" was probably first used by early-nineteenth-century historian Samuel Thacher, who reported that an otherwise unidentified woman had fired an artillery piece during the Battle of Monmouth in 1778. The name has subsequently been applied to **Mary Ludwig Hays McCauley,** who replaced her husband at a cannon during the Battle of Monmouth. The name has also sometimes been applied to **Margaret Corbin,** better known as "Captain Molly," who fought at Fort Washington and received a pension from the state of Pennsylvania. Historians now believe that the term "Molly Pitcher" is best understood as representing a whole class of women who aided American troops during combat and who sometimes acted as combatants. Molly was a common nickname of the period, and the sobriquet "Molly Pitcher" was apparently used almost as freely as "Johnny Reb" or "GI Joe" would be in later wars.

Bibliography: De Pauw, "Women in Combat," 1980; De Pauw and Hunt, *Founding Mothers*, 1975; Landis, "Investigation into American Tradition," 1911; Mayer, *Belonging to the Army*, 1996; White, "The Truth about Molly Pitcher," 1977.

—Sarah J. Purcell

MONTPENSIER, ANNE MARIE LOUISE D'ORLÉANS, DUCHESS OF

(born 29 May 1627, Paris, France; died 5 April 1693, Paris, France), also known as La Grande Mademoiselle. Military leader, Wars of the Fronde, France.

Anne Marie Louise d'Orléans, duchess of Montpensier, could claim a family heritage of strong and militant women. Her maternal grandmother, Catherine of Lorraine, had without doubt inspired the murder of Henry III of France; her paternal grandmother, **Marie de Médici,** had raised troops against her son, Louis XIII; and La Grande Mademoiselle herself had the cannons of the Bastille fired against her cousin, Louis XIV. La Grande Mademoiselle left her mark on the history of the Fronde (1648–1653) by two notable deeds, at Orléans and at the gates of Paris.

The daughter of Gaston of Orléans, granddaughter of Henry IV, and heir to the greatest fortune in France, Anne Marie Louise was considered the most eligible young woman in Europe. She hoped to marry the young Louis XIV, but his minister, Cardinal Mazarin, opposed the match.

One of the first challenges to young Louis was the Wars of the Fronde, a series of revolts by some of the nobility against the power of the royal house.

Undated portrait of Anne Marie Louise d'Orleans Montpensier. (Perry-Castañeda Library)

Anne Marie was quickly favorable to the Fronde, but did not commit herself until 1651. Her father was an eternal plotter but also eternally indecisive, often leaving his daughter to act as head of the family. In March 1652 the twenty-five-year-old Anne Marie rushed to the defense of Orléans (part of her father's holdings), which was threatened by the royal army. Accompanied by her female "camp marshals," she roused the troops to action. But the people of Orléans refused to open the gates to her, fearing violent reprisals if they allowed her to enter the city with her army. She therefore used small boats and then a ladder to climb to the top of a barred gate that then had to be broken through. She was then welcomed in triumph. A month later she left for Paris, with Orléans firmly under her family's control.

A few months later, during the famous Battle of the Faubourg Saint-Germain (2 July 1652), she led a contingent of troops into Paris itself. The rumor spread that Anne Marie personally fired a cannon against Louis XIV. In reality, after having aimed the cannon of the Bastille in the right direction, she contented herself with giving the order to fire on the royal army. Her actions helped save the rebellious prince of Condé (hero of the Battle of Rocroi) by making it possible for him to enter Paris. The leader of the opposing royal army was the great general Turenne (who had joined the king after having followed **Madame de Longueville** for a while); as for Louis XIV, he watched the operations from the nearby heights. But the symbolic repercussions of the deed remained. Alluding to her former marital aspirations, Mazarin reputedly said, "That cannon killed her husband!" Historian Richard Vann assessed the significance of her actions when he wrote, "A woman commanded troops in the field in the Fronde," which was an example of "the responsibilities that a still half-feudal political order might throw on some women."

At the end of the Fronde, while waiting to obtain a pardon from the king, Anne Marie returned to her lands. There she began to compose her *Mémoires,* in which she narrated in detail her participation during the troubled years. After a belated unhappy love affair, she ended her days, like the duchesses of **Chevreuse** and Longueville before her, in religious devotion.

Bibliography: Barine, *La jeunesse de la Grande Mademoiselle,* 1905; Bluche, *Dictionnaire du Grand Siècle,* 1990; Bouyer, *La Grande Mademoiselle,* 1986; Le Moël, *La Grande Mademoiselle,* 1994; Montpensier, *Mémoires,* 1858–1859; Vann, "Women in Preindustrial Capitalism," 1977.

—Marie-Thérèse Lalaguë-Guilhemsans (trans. by Valerie Eads)

MOTTE, REBECCA BREWTON (born 1758; died 1815, South Carolina). United States, Revolutionary War.

Rebecca Motte provided aid to Continental army forces that were opposing Lord Rawdon's control of South Carolina after the Continental defeat at the Battle of Camden in 1780. Motte's position as the owner of a large and strategically located plantation involved her directly in the Revolutionary War. She helped the Continental army burn her own home, which the British had fortified as "Fort Motte," in 1781.

Rebecca Motte had married Jacob Motte, a planter, in 1758, and they had six children together. Her husband was killed early in the Revolutionary War, leaving Rebecca in charge of their plantation on the Santee River. She maintained control of the plantation, though it was encumbered by her husband's wartime debts, until she was ejected from her home by British forces following the capture of Charleston in 1780. A British captain named McPherson occupied Motte's mansion with 165 men and transformed the house into "Fort Motte" by building a high parapet to fortify the building. Just as Rebecca Motte removed herself to a farmhouse to the west of the mansion, American forces arrived under the command of Henry Lee and Francis Marion to contest McPherson. Henry Lee established headquarters in Motte's farmhouse as he planned a siege of the fort.

When the British commander in South Carolina, Lord Rawdon, arrived with reinforcements, Lee and Marion resolved that burning "Fort Motte" would be the only effective method to drive the British out. Though the commanders hesitated to destroy Motte's property, she insisted that they proceed to burn her mansion. To light the fire, Motte provided a suitable bow and arrows that her brother, Miles Brewton, had obtained from India.

After the British were driven from her burning plantation home, Motte entertained officers from both opposing camps at her table in the farmhouse. Her hospitality is an example of the codes of courtesy often extended between officers of opposing forces in the Revolutionary War.

After the war Rebecca Motte regained control of the property that she had helped to destroy. She grew rice on the Santee River and managed to pay off her husband's debts. Motte died in possession of a considerable fortune.

Bibliography: Ellet, *Women of the American Revolution*, 1848–1850; Lossing, *Harper's Encyclopedia of United States History*, 1901.

—Sarah J. Purcell

MOURON, MARIE MAGDELAINE (fl. ca. 1690–1696). Private soldier, France, War of the League of Augsburg (1688–1697).

Marie Magdelaine Mouron was one of the many lower-working-class women who, by dressing as a man and serving in the ranks of the army, sought to escape the limited opportunities that society usually offered women in the seventeenth and eighteenth centuries. At roughly age twenty Mouron fled her home in Picardy and enlisted in the French infantry about 1690. Her regiment then marched to Provence, where she left the army under somewhat mysterious circumstances. At Avignon she soon enlisted again as a dragoon in the Regiment de Morsan and served during the siege of Rosas. Upon the regiment's return to France she dueled with another dragoon, was wounded badly in the side, and had to seek treatment, which revealed her sex. She appealed to the army commander, who took pity and sent her to a home for girls. However, she could not fit in there as an obedient young lady; she physically assaulted the mother superior and fled. After further adventures she finally reached Picardy in 1696, where she joined the Regiment du Biez, a newly mustered unit. Fearing that her true identity would be discovered, she soon deserted that regiment and tried to enlist in another, but this time she was arrested as a deserter.

When an administrative officer at Saint-Omer realized that Mouron was a woman, he appealed to Paris for instructions, and thanks to his inquiry we know her story today. Mouron was never famous, and her story is of importance primarily because it stands as an example of many young women who masqueraded as men in hopes of a better life. Most we will never know; a few, like Mouron, Geneviève Grondar, and Madeleine Cauliern, appear on the records. Mouron's military career did not end in glory, as it appears that she was sent to prison for having deserted the king's forces.

The very real story of Mouron contrasts to the more famous tale of a woman who became an officer of regular troops. The Chevalier Balthazar, born Geneviève Prémoy in 1660, eventually told her tale in *Histoire de la dragonne*, although there is reason to doubt the validity of this account. She had wanted to dress as a man when only a youth and fled her home to enlist in male attire. She saw her first action at Condé in 1676, was promoted to cornet for bravery, and eventually reached the rank of lieutenant. Her service stretched into the War of the Spanish Succession. When she fell wounded, she had to reveal her sex, but instead of dismissing her, Louis presented her with a sword and elevated her to the order of St. Louis. After being discovered as a woman, she wore a skirt as part of her uniform. Similarly, Christine de Meyrac, a noblewoman dressed as a man, wrote a fictionalized biography, *L'heroïne mousquetaire*, which appeared in 1679. These sensational and popular, but probably fictional, biographies have overshadowed the unwritten stories of women like Mouron.

Bibliography: Dekker and van de Pol, *The Tradition of Female Transvestism in Early Modern Europe*, 1989; Lynn, "The Strange Case of the Maiden Soldier of Picardy," 1990, and *Giant of the Grand Siècle*, 1997.

—John A. Lynn

N

"NANCY HARTS" (1860s and 1952). Women's military groups, United States, American Civil War and 1950s.

Georgia's Revolutionary War heroine **Nancy Hart** provided a model for other women in Georgia who sought military accomplishment over the years, and two American military units composed of Georgia women have taken her name. During the **American Civil War** the Confederate women of La Grange, Georgia, organized themselves into an informal company to oppose Union troops who were closing in on their homes. Clifford L. Smith's *History of Troup County* notes that "during the Civil War, LaGrange had the unique distinction of having a company of women soldiers, under the captaincy of Mrs. J. Brown Morgan, who called themselves the Nancy Harts in honor of the revolutionary heroine of Georgia. This company was organized by Mrs. Morgan for the protection of the homes and the children in the absence of the men. When a detachment of Wilson's raiders under the command of Colonel LaGrange rode through the town that bore his name, the Nancy Harts lined up for action, but surrendered on the promise of the diplomatic colonel to spare the city from looting and destruction."

In 1952, according to local sources, the United States Marine Corps announced in Macon, Georgia, that Georgia women would be enlisted into a "Nancy Hart Squad." The marines had integrated women as a regular part of their military force in 1948. The Nancy Hart Squad was supposed to be sent for training at Parris Island, South Carolina. Further information is lacking, however, and it would appear that the twentieth-century unit never materialized. These groups, although they never performed significant military actions, show the influence of historical figures like Nancy Hart, their chosen role model.

Bibliography: Coulter, "Nancy Hart," 1955; Larson, "Bonny Yank and Ginny Reb," 1990; *Macon Telegraph*, 8 May 1952; Smith Clifford L., *History of Troup County* (Atlanta: 1933).

—Sarah J. Purcell

NATIVE AMERICAN WOMEN. North America, seventeenth century through Second World War.

> When the early Jesuit fathers preached to Hurons and Choctaws,
> They prayed to be delivered from the vengeance of the squaws.
> 'Twas the women, not the warriors, turned those stark enthusiasts pale.
> For the female of the species is more deadly than the male.
>
> —Rudyard Kipling, "The Female of the Species," *Rudyard Kipling's Verse: Inclusive Edition, 1885–1918* (London: Hodder and Stoughton, 1919).

The political power of Native American women in some tribes has been relatively well studied, but their military roles have been neglected or dismissed as anecdotal. The lack of traditional historical materials prevents historians from doing more than collecting anecdotes and oral histories and correlating them with available written accounts and anthropological studies. Even that, however, has not yet been attempted in a systematic fashion that would shed light on women's military experience in native cultures in North America as a whole.

Native women in North America were involved in warfare in various ways throughout millennia of tribal history and later as citizens of the United States and Canada. Most Native American tribes shared a culture that revolved around warfare. Women were intricately involved in preparations for war, as well as in violent actions such as defending their camps, torturing captives, and mutilating enemy corpses. The Iroquois, Huron, and Pawnee were particularly noted for their harsh treatment of prisoners. Native women often had the power of life and death over captives, deciding who might be adopted or enslaved, and who should be killed.

Although men were the primary combatants, many tribes had flexible gender roles that allowed for the occasional "feminine man" or "masculine woman." This allowed some women who were interested in warfare to live and act as warriors. More common, however, were Native American women who participated in fighting for limited periods, often for purposes of avenging a male relative. Stories of women who fought are found in the Lakota (Sioux), Cheyenne, Apache, Tapuya, Ojibwa, and Cherokee traditions, among others.

Carolyn Niethammer reports several Ojibwa legends of warrior women. For example, Chief Earth Woman was said to have followed warriors from her tribe on the warpath against the Sioux, claiming that she had foreseen enemy actions in a dream. Her predictions were reportedly true and led to victory for her party. Although Chief Earth Woman apparently did not fight, she did scalp a defeated enemy and was given war honors along with the male fighters. Another Ojibwa woman who went on raids reportedly collected enemy genitals along with scalps. Among the Lac Courte Oreilles band of the Ojibwa in Wisconsin, Hanging Cloud Woman was reported to have been a famed warrior in the years just before the **American Civil War**.

In many Plains tribes, Beatrice Medicine argues, a role for women warriors was institutionalized. The Piegan of Montana and Alberta recognized "manlyhearted women" who were political and military leaders. Similarly, the Blackfoot recognized the "manly woman" (*ninawaki*) who was bold and aggressive in hunting and sometimes warfare. These categories do not refer to sexual preference, but to social and economic behavior; most of these "masculine" women appear to have led a heterosexual life. What is important is that the very existence of these recognized categories indicates that it was accepted, if unusual, for women to pursue nontraditional roles, including combat. George Grinnell reports on Cheyenne women who fought, such as **Ehyophsta**. Among the Blackfoot, Elk-Hollering-in-the-Water, Running Eagle, and Throwing Down are women warriors whose names are prominent in oral traditions. Ruth Landes describes women among the Dakota who joined war parties, with some leading such expeditions. Landes believes that the female warrior role was relatively common among eastern Dakota but not reported in some western tribes (Teton), suggesting that there was regional variation, both within and among tribes. Women also fought in southwestern tribes, with **Lozen** of the Warm Springs Apache being one of the most famous.

In eastern cultures women served as tribal chiefs in historical times. Hernando de Soto took one such chief hostage in the mid-sixteenth century in Georgia, since her prestige helped guarantee his safe passage through hostile territory. During King Philip's War (1675–1676), several women held both political and military power. Wetamoo was a Wampanoag leader who played a military role in support of "King Philip" (her brother-in-law), who tried to force white colonialists out of New England. Wetamoo personally led her warriors on the field and was killed in 1676. In the same year a woman of the Narragansett called Magnus was killed by settlers while leading her warriors in combat. Among the Sakonnet in Rhode Island, Awashonks, whose brother had been killed by the Narragansett, allied her tribe with the colonialists and sent warriors to fight against the other tribes. See also **Molly Brant,** Ehyophsta, Lozen, **Woman Chief, Nancy Ward,** and **The Other Magpie** for individual biographies of Native American women who fought.

Few Native Americans served in the U.S. armed forces until after the end of the Indian Wars of the late nineteenth century, when American Indians were brought under governmental control and most survivors were forced into reservation life. Before the Second World War, many American Indians continued to struggle against a history of oppression as they fought for tribal and economic independence. They often earned subpoverty wages as manual laborers, and many lived in desperately poor conditions. Women worked mostly as maids, domestic employees, and waitresses, earning especially low wages.

Wartime brought opportunities not only to make better money, but also to prove their equality and integrity as Americans. In the United States during and after the Second World War, American Indians served disproportionately to whites, in part because military service provided an unparalleled opportunity to learn about life outside the reservation while proving their competence and patriotism. However, like other American women, Native American women had fewer military opportunities than men and were restricted from combat duties. American Indians were not segregated in the military; they served with and lived in racially integrated units. By the end of the Second World War, 70,000 men and women had left their reservations to enter the military or defense industries. Of these 70,000, about one-third served in the military; of these, about 800 Native American women joined the **WAC** and **WAVES**. As members of the WAC, American Indian women did the same tasks as all other members of the organization. They traveled around the world to all theaters of the war, supporting combat units and sometimes risking their lives in combat zones.

Not all American Indian tribes were represented in the WAC or the military as a whole. The Seminoles were still officially at war with the United States in the mid-twentieth century. The Iroquois, an independent nation, were at war with Germany, never having accepted the German surrender in 1918, and declared war on the Axis powers independently in 1942. Despite the lack of unified support for the U.S. war effort, as a whole, Native American participation in the Second World War was a turning point in white/Indian relations. American Indian servicewomen contributed to the changing climate of heightened respect and understanding.

Bibliography: Albers and Medicine, *The Hidden Half,* 1983; Bataille, *Native American Women,* 1993; Bellefaire, *The Women's Army Corps: A Commemoration of World War II Service,* 1993; Bernstein, *American Indians and World War II,* 1991; Davis, *Native America in the Twentieth Century,* 1994; Grinnell, *The Fighting Cheyennes,* 1915; Landes, *The Mystic Lake Sioux,* 1968; Mathes, "Native American Women," 1982; Niethammer, *Daughters of the Earth,* 1977; Williams, *The Spirit and the Flesh,* 1986.

—Reina Pennington

NEHANDA, also known as *Ambuya* (Grandmother) Nehanda, *Mbuya* Nehanda, Nehanda Nyakasikana. Religious leader of MaShona nation, name used by female military leaders against British. Rhodesia/Zimbabwe, wars for liberation, nineteenth and twentieth centuries.

The long-dead and possibly apocryphal Nehanda is the best-known *Mhondoro* or female ancestor in Zimbabwe. Her spirit is famous as having power over war and rain and usually possesses female mediums in the Mazoe and

Dande regions. Two of these mediums, supposedly possessed by Nehanda and often called by her name, played central political roles as military advisors and charismatic leaders in the Southern Rhodesian anticolonial struggle in 1893–1896 (a struggle locally known as Chimurenga I) and the Zimbabwean National Liberation War (Chimurenga II) against the white-minority colonial regime in 1964–1980.

Charwe (1862?–1898), a medium of Nehanda in the Mazoe (Zezuru) region, inspired local chiefs to become involved in rebellion against colonial rule in 1893–1896. "Nehanda" advised her people to use spears and sticks against European guns. This first Chimurenga was lost; Nehanda was captured in December 1897 and was sentenced to death for the killing of Pollard, the Native Commissioner of Mazoe. On 27 April 1898, the day of her hanging, she refused conversion to Christianity; on the scaffolding where she was to die, she danced, laughed, and screamed; and she prophesied that her bones would rise one day to regain freedom for her people. Charwe's final acts became legend, and "Nehanda" became a heroine of nationalist resistance.

In Chimurenga II Kunzaruwa, a medium of Nehanda of the Korekore, was conscripted by ZANU guerrillas in 1971. To protect her from Rhodesian soldiers, she was carried across to Mozambique, where she operated as advisor and spiritual leader to young freedom fighters. Nehanda advised them what to eat, when and where to fight, and where the enemy was, and she predicted victories. An operational zone was named Nehanda by freedom fighters. Kunzaruwa died in mid-1973 and was buried on the side of a road that led guerrillas back into Zimbabwe. (See **Zimbabwe, Women in the National Liberation War**.)

After independence was won, *Ambuya* (Grandmother) Nehanda was much praised in Zimbabwe. Her image was often seen above portraits of the new president Robert Mugabe. A maternity hospital and some schools were named after her, and she continues to be a potent symbol of militant anticolonialism.

Bibliography: British South Africa Company, *The '96 Rebellions,* 1898; Lan, *Guns and Rain,* 1985; Martin and Johnson, *The Struggle for Zimbabwe,* 1981; Ranger, *Revolt in Southern Rhodesia,* 1967, and *Peasant Consciousness and Guerrilla War,* 1985.

—Tanya Lyons

NHONGO, JOYCE MUGARI (born 1955), also known as Teurai Ropa (Spill Blood). Zimbabwe, National War of Liberation.

Joyce Mugari Nhongo became a heroine for thousands of black women who fought in the liberation struggles in Africa. She fought in Zimbabwe (formerly Rhodesia) against colonialism, racism, and oppression so that Africans would not have to be slaves on their own land. (See also **Zimbabwe, Women in the National Liberation War**.)

Joyce Mugari was born at Chawanda Village, Mount Darwin, Rhodesia. In 1973, after passing her Form 2 exams at the Howard Institute in the Mazoe area, she left Rhodesia to join the Zimbabwe African National Union (ZANU) in Zambia, which was waging a guerrilla war against the Ian Smith regime of Rhodesia. At the age of eighteen she was the only woman who, along with fifteen men, trained in Lusaka. She trained in light infantry for three months, using AK rifles and submachine guns. She also trained for six months as a medical assistant. Still only eighteen years old, she was promoted to the General Staff and given political instruction. By 1975 she had become the political instructor of two successive military bases (75 percent of all political instructors in ZANU were women). At age twenty-one she became camp commander at Chimoio military and refugee camp in Mozambique, where she met and married Rex Nhongo, deputy head of Zimbabwe African National Liberation Army (ZANLA) forces. She then became military commander of the ZANLA Women's Detachment, and in 1976 she was appointed director of politics in Chibawawa refugee camp.

At some point in her young career Joyce Nhongo adopted the revolutionary name of Teurai Ropa (Spill Blood). In 1977 in Mozambique, at the age of twenty-two, she became the youngest member of the Central Committee and was on the National Executive. She was then appointed secretary of women's affairs. As a politically active woman, Teurai Ropa became a prime target for the Rhodesian security forces, which unsuccessfully tried to capture her.

In 1978, when the camp she was in was attacked by Rhodesian soldiers, Teurai Ropa was an active combatant despite being in a late stage of pregnancy; two days later she gave birth. Her daughter, Priscilla Rungano, was sent away one month later and Teurai Ropa continued to fight until liberation was won in 1980. In newly independent Zimbabwe, while representing the Mashonaland Central Constituency, Teurai Ropa served briefly as minister of youth, sport, and recreation. In 1981 she became minister of community development and women's affairs.

Bibliography: Qunta, *Women in Southern Africa*, 1987; Stott, *Women and the Armed Struggle*, 1989; Uglow, *The Continuum Dictionary of Women's Biography*, 1989; Weiss, *The Women of Zimbabwe*, 1986.

—Tanya Lyons

NICOLAA DE LA HAYE (born 1160?; died 1230, Lincolnshire, England). Castellan and sheriff of Lincoln; defender, England, thirteenth century.

Nicolaa de la Haye was the oldest daughter of Richard de la Haye (died 1169), the hereditary castellan and sheriff of Lincoln. In the absence of direct male heirs she inherited both offices when her father died. Nicolaa won considerable praise from contemporaries for her able defense of Lincoln Castle during the struggle in 1191 between William Longchamp, regent for Richard I, and the king's brother, John, and during the Baronial Rebellion of 1215–1217, when John was forced to issue Magna Carta.

Nicolaa's first husband, William fitz Erneis, died in 1178; in the late 1180s she married Gerard de Camville. As was customary, Gerard administered his wife's inheritance, including the custodianship of Lincoln Castle and the shrievalty of Lincoln. In the spring of 1191 Gerard did homage to John in defiance of the fact that Lincoln Castle was held by John's brother, King Richard. This prompted William Longchamp to remove Gerard as sheriff and place the castle under siege in June 1191. By that time Gerard was with John, and Nicolaa defended Lincoln—"manfully" according to Richard of Devizes—until Longchamp lifted the siege in July. Later in the same month Gerard was reinstated as castellan and sheriff. Following his death around 1215, Nicolaa administered both offices in her own right. At the time it was unusual, though not unheard of, for a woman to hold an office with military responsibilities.

In her fifties Nicolaa defended Lincoln Castle again in 1216–1217, contributing significantly to the success of the royalist struggle against the rebellious English barons and their French allies. A siege mounted in the summer of 1216 ended when Nicolaa bought off the besiegers, but by March 1217 the insurgents controlled all of eastern England, with the exception of the fortifications at Lincoln and Dover, and the siege was resumed. Nicolaa's stout resistance on this occasion led her enemies to describe her as a "most ingenious and evil-intentioned and vigorous old woman" (Anon. de Bethune, quoted in Norgate, 37). Finally a royalist army came to the aid of the "matron conducting herself so manfully" (Walter of Coventry, 2:230) and defeated the rebels in the decisive Battle of Lincoln on 20 May.

Following these events, Nicolaa was rewarded with additional properties in Lincolnshire. Unfortunately, John I died in 1216, and his successor Henry III granted the shrievalty, city, and, by implication, the castle of Lincoln to his uncle William Longespee, the earl of Salisbury. Nicolaa protested to the king and, after a political struggle that lasted till late 1217, managed to regain custody of the castle and city, but not the shrievalty. She apparently spent much of the rest of her career as castellan resisting attempts by the earl to take control of the fortress. She resigned in 1226, probably due to her age, and died a few years later.

Bibliography: *Histoire des ducs de Normandie et des rois d'Angleterre* (Paris: 1840); McLaughlin, "The Woman Warrior," 1990; Norgate, *The Minority of Henry III,* 1912; Pollock and Maitland, *The History of English Law,* 1968; Powicke, *The Thirteenth Century,* 1991; Richard of Devizes, *Cronicon* (London: 1963); Walter of Coventry, *Memoriale* (London: 1873–1874).

—David Balfour

O

ODENA, LINA (born ca. 1914, Barcelona; died September 1936, Spain). Revolutionary, Spanish Civil War.

Lina Odena is considered a heroine of the Spanish Republic; she was one of the first women to die in combat in the **Spanish Civil War**. Odena was a Catalan seamstress and political activist in the Communist youth movement. She visited the Soviet Union and was instrumental in organizing the unification in 1934 of the Spanish Socialist and Communist youth wings into the Juventudes Socialistas Unificadas. Arrested, and later released, as part of the political repression associated with the Asturian uprising in October 1934, she campaigned with Dolores Ibárruri, "La Pasionaria," for the Communist Party in Asturias in the Popular Front elections in February 1936. Odena was also a member of the national committee of the movement of Women against War and Fascism.

When the civil war began in July 1936, Odena was in Almería and joined a militia unit at the front near Granada. While she was on a reconnaissance mission, she was cut off from her unit in ambush. Rather than face capture by the Nationalist forces, she committed suicide. Odena was transformed into a Republican heroine, particularly by the Spanish Communist Party. A military training school was named after her, as was a women's self-defense unit. Her courage and self-sacrifice were memorialized in popular ballad and song, and the story of her death is included in many of the patriotic writings from the Republican side during the war.

Bibliography: Ibárruri, Dolores, et al., *Lina Odena, heroína del pueblo* (Madrid: 1936); Scanlon, Geraldine M., *La polémica feminista en la España contemporánea (1868–1974)* (Madrid: 1976).

—Judith Keene

OKTIABRSKAIA, MARIIA VASIL'IEVNA (born 1902 or 1905, Blizhnee, Crimea, Russia; died 15 March 1944, Smolensk, USSR). Guards sergeant, Red Army Tank Troops. Tank driver-mechanic, Soviet Union, Second World War.

Mariia Oktiabrskaia served for only a few months as a tank driver-mechanic with the 26th Tank Brigade, 2nd Tatsinsky Guards Tank Corps, on the 3rd Belorussian Front before she was fatally wounded in combat, but she proved herself a fearless, confident, and expert tank driver who took good care of her vehicle and a full-fledged member of her tank crew, mostly young men approximately half her age. Oktiabrskaia was an officer's wife in the prewar years (1925–1941) and was accustomed to the harsh conditions of Soviet military life. When her husband was appointed political commissar of an infantry regiment stationed near the western border, Oktiabrskaia graduated from driving and marksmanship courses and earned the title "Voroshilov sharpshooter." Evacuated to Tomsk, Siberia, in 1941, she was notified in 1943 that her husband, parents, and two sons had all been killed. Reacting to her personal tragedy and encouraged by the success of a woman train engineer she had met, Oktiabrskaia decided that she would pay for a tank out of her own savings and become its driver. It took much effort on her part to secure the cooperation of the authorities and their permission to name the tank "Soldier's Helpmate," but she succeeded despite being a short, small woman already in her forties.

After five months of training, in September 1943 Oktiabrskaia was sent to a reserve T-34 tank company, part of the 26th Guards Tank Brigade. The unit was quickly sent to the front near Smolensk to join the 2nd Guards Tank Corps. The brigade commander, Colonel S.K. Nesterov, kept Oktiabrskaia out of combat for a month to allow her time to become accustomed to the battlefield, much against her wishes. On 21 October 1943, during an assault on Hill 208 near Novoe Selo of Vitebsk Region in Belorussia, Oktiabrskaia's tank was the first to break into enemy positions. A testimonial that recommended her for the Order of Patriotic War I Class reported that in her very first battle Oktiabrskaia had distinguished herself while maneuvering her tank onto enemy firing positions. Her tank took part in the capture of Novoe Selo on 17–18 November 1943, during which she received a slight wound while assisting in track repairs. In the fight for the village of Shvedy on 17 January 1944 Oktiabrskaia knocked out a self-propelled Ferdinand gun, destroyed a machine-gun

---

An RAF pilot in Russia during the Second World War comments on women's strength:

One British pilot who flew with an RAF unit near Murmansk wrote about a group of a dozen Russian girls whom he watched sawing up and loading timber. Stiff and bored after days on a transport ship, he decided to help the girls out:

The party was under the command of a girl about twenty, a qualified engineer. They worked away for three hours like absolute fury, the engineer-girl working the hardest of the lot.... Once a lorry was loaded up with sawn timber, the girls would sit down for a breather and start smoking and singing. And then, when another lorry would come up, they'd start in all over again. About halfway through the loading and sawing I felt that I had strained every muscle of my body ... but every time I sat down for a rest, there would appear another girl at the end of a log weighing half a ton ... and back one would have to go the treadmill of labour out of pure shame—explaining that one had been three weeks aboard ship, and one was not quite in one's natural athletic condition.

—Hubert Griffith, *R.A.F. in Russia* (London: Hammond, 1942), 29–30.

nest, and went on to flatten an earth-and-timber emplacement. On this occasion she was seriously wounded in the head by shrapnel while assisting in track repairs and died in a military hospital in Smolensk two months later. She was buried in Kutuzov Gardens in the center of Smolensk, along with other famous war heroes. Oktiabrskaia was awarded posthumously the **Hero of the Soviet Union** on 2 August 1944.

Bibliography: Cherniaeva, *Docheri Rossii,* 1975; Cottam, *Women in War and Resistance,* 1998; Erickson, "Soviet Women at War," 1993; Miniaeva, *Srazhalas' za rodinu,* 1964; Murmantseva, *Sovetskie zhenshchiny,* 1974.

—Kazimiera J. Cottam

OLK (*OCHOTNICZA LEGIA KOBIET*, OR WOMEN'S VOLUNTEER LEAGUE). Organized December 1918 in Lwów (Lemberg, Lviv, Lvov); dissolved by the Polish Ministry of Military Affairs on 1 February 1922. Functioned as an auxiliary service for the Polish army during the Polish-Soviet War, 1919–1920. Saw action at the front against Ukrainian and Bolshevik troops.

On 1 November 1918, in the vacuum left by the disintegrating Austro-Hungarian Empire, hostilities broke out between Polish and Ukrainian forces over the control of Lwów, a university town in Austrian Galicia. Among the groups rallying to the Polish side was the Citizens Committee of Polish Women, which set up a medical aid station when the fighting broke out. One member of the committee, Aleksandra Zagórska, concluded that this type of activity was too limited. Instead, she formed a courier service of seventeen women who conducted intelligence work, recruited Polish men to fight against the Ukrainians, and guided soldiers through hostile territory. By early December Zagórska had succeeded in creating the Women's Citizens Militia, which helped keep public order by guarding and patrolling secured areas. The militia also maintained a combat unit that was obliged to help defend Lwów.

Zagórska argued for more, however, insisting that the militia's fighting section should be recast as an authentic military formation, quartered and supplied according to military norms. In response, the Polish army command in Lwów established the Women's Volunteer League (*Ochotnicza Legia Kobiet*, or OLK) at the end of December. In April 1919 the commander in chief of the Polish army, Józef Piłsudski, confirmed a project for the organization of the OLK, and it was directly subordinated to the military high command in June. During this same period the OLK expanded its activities, ultimately forming six battalions in Polish-held territory. These included units in Kraków, Poznan, Warsaw, and Wilno (Vilnius).

The OLK's principal activity was auxiliary service, such as sentry duty, courier assignments, medical aid, and clerical work, yet the women also participated in combat. The fighting section of the Women's Citizens Militia briefly occupied a front-line position against Ukrainian forces in December

1918 until the arrival of a regular army unit. In the struggle for Lwów a number of OLK members served in the infantry and in field artillery units as observers and gunners. The group's most trying stint in combat came in July 1920 when the Wilno formation, a battalion of 250 women, was assigned to defend a twenty-five-kilometer-long section of the front against the advancing Third Cavalry Corps of Ghaia Dmitriyevich Ghai. Receiving no orders to pull back, OLK units held their position even after neighboring forces had already withdrawn. The battalion sustained heavy losses in clashes with Bolshevik forces, and approximately thirty women were taken prisoner during the subsequent retreat. In mid-August, at the height of the Red Army's offensive into Poland, OLK units also took up positions in the suburb of Grochów as the last line of Warsaw's defense.

Members of the OLK frequently earned praise from their superior officers. In addition, at the conclusion of the Polish-Soviet War the minister of military affairs cited the women for their service in the Battle of Lwów, during which they "performed their soldierly duty unswervingly and with the greatest sacrifice." He also commended OLK units for undertaking guard duty behind the lines, which "in large measure contributed to the protection of the state's resources."

The exact number of women who served in the ranks of the Women's Volunteer League is not known. One source puts the figure as high as 10,000, noting that at any given time the number did not exceed 2,500. Another claims that not many more than 3,000 women took part in the OLK, with roughly 1,500 serving throughout the period. Records of the Ministry of Military Affairs show that there were 2,693 women serving in the ranks of the Volunteer League in September 1920, a figure that included 17 officers and 103 NCOs.

By and large, the women who came forward to volunteer for service in the OLK were members of the intelligentsia (the educated urban class). According to the personnel records of the OLK Lwów Battalion for 1919, 18 percent of the 493 women were high-school and university students. The unit also had a large number of teachers, seamstresses, clerks, and store workers. Less than 3 percent had origins in the urban industrial proletariat. A relatively large number of women, 26 percent, declared no occupation, since they were still too young to have left their parents' homes. Figures from 1920–1921 confirmed the youth of the Volunteer League's participants. Of the personnel records from the Lwów Battalion that listed birthdates (312 out of 458 women), 39 percent cited ages of twenty-one or younger. The youngest member mentioned was Stefanja Pawecka, who would turn thirteen in February 1921.

Once the threat to Warsaw had subsided in the summer of 1920, the Ministry of Military Affairs issued orders to reorganize the OLK as a school to train women for auxiliary service. This directive reflected the assumption of a number of army planners who argued that modern warfare required a contribution from each and every member of society, including women. During the period 1919–1921 several officers advanced proposals that called for obligatory military training for women. Those who were physically capable of service would be assigned to sentry battalions, military police units, or other auxiliary formations.

The acute financial difficulties plaguing the newly created Polish army decided the fate of the OLK. At the close of the Polish-Soviet War not only the OLK, but many military institutions and programs were judged to be, if not unessential, then certainly unaffordable. In the early 1920s the army was forced to spend the largest share of its budget on food, clothing, and weaponry for its soldiers. Under such circumstances there was no room for the OLK. It was disbanded in early 1922.

Representative Participant

Zagórska, Aleksandra (1884–1965). Lieutenant colonel and commander of the Women's Volunteer League. Active from her earliest years in the struggle against Russian rule in Poland, Zagórska joined the Polish Socialist Party in 1906, making explosive materials for the party's Combat Organization in a home-based lab. Arrested in 1907 and sentenced to exile in Siberia, she escaped to Austrian Galicia, where in 1912 she joined the women's section of Józef Piłsudski's paramilitary group, the Riflemen's Association. Zagórska was a social activist in interwar Poland and participated in the Warsaw Uprising of 1944.

Bibliography: *Encyklopedja Wojskowa*, s.v. "Ochotnicza Legja Kobiet"; Kiedrzynska, Wanda, *Zarys Historji Wojennej O.L.K. (Ochotnicza Legja Kobiet)* (Warsaw: 1931); *Ochotnicza Legja Kobiet: Szkic Historyczny* (Lwów: 1921); Pilsudska, Aleksandra, ed., et al., *Służba Ojczyznie, 1915–1918,* vol. 2 of *Wspomnienia Uczestniczek Walk o Niepodleglosc* (Warsaw: 1929).

—Robert Ponichtera

OLYMPIAS (ca. 375–316 BCE), also called Myrtale, Polyxene, and Stratonice ("victor of the army"). Queen of Macedon and military-political strategist.

Olympias was one woman in antiquity who could not have shirked a military role had she wanted to. She was born in an age of military innovation in Greece and Macedon. In 357 BCE she married a man almost obsessed with the art of making war, Philip II of Macedon. Her son, Alexander the Great, was one of the greatest warriors in history. Olympias herself embarked upon a career that led at times to absolute rule and was always concerned with military strategy.

Olympias was the niece of the king of Epirus, with whom the Macedonians sought an alliance. We have some evidence that among certain groups to the north and west of Greece aristocratic girls were raised to play an active role in war. Olympias's stepdaughter **Cynane,** child of the Illyrian princess Eurydice I, is a well-known example. Although we have no evidence that Olympias

Olympias

Olympias, Queen of Macedon, mother of Alexander the Great, and her court. From the fourteenth-century Greek manuscript, *History of Alexander the Great*. (The Art Archive/Hellenic Institute Venice/Dagli Orti)

handled arms, the Greco-Roman biographer Plutarch does note the use of military equipment in the Bacchic rites Olympias brought from her native land.

We know little about Olympias during the lifetime of her husband, Philip (r. 359–336 BCE). Even if she merely observed her husband's frequent campaigns in the north and in Greece, she might well have gained insight into generalship. In a royal court full of soldiers this would have been hard to avoid. But it was after Philip's death, when her son Alexander set out to conquer the known world, that Olympias came into her own as a ruler.

From 336 to 331 Olympias was a decisive influence in Macedon while Alexander (r. 336–323 BCE) was in Asia. Alexander left Macedon in 335 BCE, never to return. As regent he left Antipater, who frequently came into conflict with Olympias; struggles for power in Macedon would continue for half a century. Olympias's importance is evident in the primary sources; she figured in the Greek political oratory of the period as a feared and redoubtable power. In a stream of letters to her son she discussed the grand strategy of the Macedonian empire. It is interesting to note the value she placed on the army in her political considerations. Olympias, who was hated by many Macedonians as a foreign queen, was forced out of the country in 331; she established herself in Epirus, from where she continued her efforts to regain ascendancy in Macedon. However, the support that the Macedonian army gave to an enemy of hers was decisive in causing her exile.

In 317, six years after Alexander's death, Olympias led an army into Macedon against Philip III Arrhidaeus and his martial wife **Eurydice II** (a granddaughter of Philip II). In the "first war between women," as the Greek historian Duris calls it, the two armies met at the border town of Euia. Eurydice wore a suit of Macedonian armor; Olympias was dressed as a devotee of Bacchus and marched her troops to the accompaniment of tambourines.

Eurydice's troops mutinied and defected to Olympias, who imprisoned, tortured, and killed Eurydice and her husband; she then began to rule Macedon with terror.

In the winter of 317 Olympias was threatened with an invasion by Antipater's son Cassander. Olympias assigned her generals to stations around Greece and Macedon but in a fatal error withdrew to Pydna. Diodorus of Siculus states that the city was poorly prepared for a siege; Olympias took people "mostly useless in war" in her personal entourage. She had with her the palace troops, a unit of cavalry, and some elephants. In a few weeks hopes for outside aid were crushed, and Cassander instituted a blockade. Starvation took the city as spring approached. Even in this situation Olympias's soldiers remained loyal to her; rather than desert, they approached her and asked for permission to withdraw. She granted the request and then surrendered to Cassander, ordering her generals to do likewise. She demanded a public trial in front of the Macedonian people. Probably in fear of her popularity with the army, Cassander sent 200 hundred soldiers to execute her. They entered the royal palace, but when Olympias confronted them, they refused to carry out their orders. Cassander then rounded up the relatives of Olympias's many victims, and they performed the deed.

Bibliography: Diodorus of Siculus, *Historical Library* 19.34–36, 50–51; Macurdy, *Hellenistic Queens*, 1975; Plutarch, *Life of Alexander*.

—Alexander Ingle

ONILOVA, NEONILA ANDREEVNA (born April 1921, Odessa Region, Ukraine; died 8 March 1942, Sevastopol, USSR), also known as Nina, Anka. Senior sergeant, Red Army Infantry. Commander of machine-gun crew, Soviet Union, Second World War.

"Nina" Onilova was a famed machine gunner who also trained dozens of fellow soldiers. Onilova joined the famous 25th "V.I. Chapaev" Division of the Independent Maritime Army on the Crimean front and served in the 54th Razin Regiment, which defended Odessa and Sevastopol. Her stamina, boldness, and bravery amazed even seasoned veterans.

Onilova was an orphan who worked before the war in a textile factory in Odessa. During those years she saw the famous film *Chapaev*, a movie about the **Russian Civil War** that spotlighted the achievements of a woman machine gunner nicknamed "Anka." Onilova determined to follow in Anka's footsteps and mastered the heavy machine gun in a course provided by the paramilitary **Osoaviakhim**.

Onilova served at the front from August 1941 as a medical NCO. She was transferred to a machine-gun platoon after proving her competence with the weapon. She had perfected her heroine Anka's machine gun technique, allowing the enemy to approach very close, then using aimed fire until the front rows were destroyed and those behind were forced to turn back.

She was seriously wounded in September 1941 and spent two months in a hospital. After her recovery, Onilova insisted on rejoining her regiment near Sevastopol. On 21 November, alone behind her heavy machine gun, she launched a Molotov cocktail that set fire to an enemy tank. She was recommended by her commander for the **Order of the Red Banner**. Army Commander General I.E. Petrov admitted that he had suspected that Onilova had been recommended on the basis of sexual favoritism rather than real achievement, since he found it hard to believe that a woman could be so brave and decisive in battle, so he summoned her to determine whether she deserved the award. He was convinced of her merit after discussing tactics with her and granted her the award and promotion to the rank of sergeant.

Onilova received her second Red Banner on 23 February 1942 for outstanding performance during the December enemy offensive. Now famous in the entire Maritime Army and Black Sea Fleet, she became a semilegendary figure like her prototype Anka and was given that nickname. When Petrov learned that she had been seriously wounded again on 28 February 1942, he immediately ordered his medical staff to do everything in their power to save her. When her condition deteriorated, Petrov himself came to say good-bye. The severe, smart-looking, grey-haired general, reportedly with tears in his eyes, said in a shaking voice: "You fought gloriously, daughter. Thank you, on behalf of the Army, on behalf of the nation. The entire Sevastopol knows about you and the entire country will learn about you too." It was only on 14 May 1965 that Onilova was posthumously granted the **Hero of the Soviet Union**.

Bibliography: Cottam, *Women in War and Resistance,* 1998; Erickson, "Soviet Women at War," 1993; Kolomiets, "Chapaevtsy stoiali na smert," 1967; Smirnova-Medvedeva, *On the Road to Stalingrad,* 1997; Toropov, *Geroini,* 1969.

—Kazimiera J. Cottam

ORDER OF GLORY, I CLASS, WOMEN RECIPIENTS. 1943–1948, USSR. Awarded to "the bravest among the brave" for skill in combat, Second World War.

The Order of Glory III, II, and I Class, created on 8 November 1943, was specifically intended for Red Army privates, NCOs, and junior lieutenants of aviation. Numbering about 2,500, recipients of the Order of Glory I Class received the three levels of the order on three separate occasions. Four of the recipients were women.

Representative Recipients

Necheporchukova-Nozdracheva, Matriona (Motia) Semenovna (born 1924). Sergeant (ret.), Red Army Ground Forces. Medical NCO. Served in the 100th Guards Regiment of the 35th Guards Division, which reached Berlin on 3 May 1945. She distinguished herself by dispensing first aid under heavy enemy fire, repelling enemy raids, and defending the wounded entrusted to her. She was awarded her third Order of Glory on 24 May 1945 for demonstrating bravery during the breakthrough of enemy defenses on the west bank of the Oder River and the fierce fighting in Berlin.

Petrova, Nina Pavlovna (1893–1945). Senior NCO, Red Army Ground Forces. Sniper-instructor. Graduate of a snipers' school and participant in the Soviet war with Finland, Petrova came first in a shooting competition for the entire Leningrad front. The oldest woman sniper in the army, she achieved a personal score of about 100 kills and trained 513 new snipers. With the 1st Battalion, 284th Regiment, 86th Tartu Division, Petrova advanced from her native Leningrad through the Baltic lands, East Prussia, and Poland to Schwerin, Germany, where she was killed on 1 May 1945. She was awarded the Order of Glory I Class posthumously on 29 June 1945.

Staniliéne-Markauskene, Danute Iurgievna (born 1922). Sergeant (ret.), Red Army Ground Forces. Machine gunner. Serving as the commander of a machine-gun crew, Company No. 3, 167th Regiment, 16th Lithuanian Division, she imitated the technique of Anka, the fictional machine gunner of Civil War fame (see **Neonila Onilova**). Waiting to open fire until the enemy made close approach demanded exceptional self-control. She was awarded the Order of Glory I Class on 24 March 1945 for her role in the fight for the Klaipeda-Tilsit Highway, the first Soviet woman soldier to be so honored.

Zhurkina-Kiek, Nadezhda (Nadia) Aleksandrovna (born 1920) Senior NCO (ret.), Red Army Air Force. Radio operator/gunner. She served in the 99th Independent Reconnaissance Aviation Regiment, the only woman air gunner in the entire 15th Air Army. She flew eighty-seven missions at maximum altitudes of 5,000–6,000 meters, during which she displayed an exceptional ability to spot enemy aircraft in time and to chase them away or shoot them down. She was awarded the Order of Glory I Class on 23 February 1948 for outstanding performance in late 1944 over Latvia, at which time she flew ten missions behind enemy lines and shot down a decorated German ace.

Bibliography: Barmin, "Kavaler ordena Slavy," 1975; Cottam, *Women in War and Resistance*, 1998; Eremin and Isakov, *Molodezh v gody Velikoi Otechestvennoi voiny*, 1984; Mikora, "Zhenshchiny kavalery ordena Slavy," 1976; *Sovetskaia Voennaia Entsiklopediia*.

—Kazimiera J. Cottam

ORDER OF ST. GEORGE, WOMEN RECIPIENTS. 1808–1917, Russia. Awarded for bravery in combat.

The four-level Order of St. George (alternatively referred to as the St. George Cross) was created exclusively for officers in 1769 by Empress **Catherine II**. Until 1856 it was also awarded to officers with at least twenty-five years of service or participation in eighteen to twenty naval campaigns. In 1807 a single variant and after 1856 a four-level variant also awarded to privates and NCOs were introduced. In 1808 **Nadezhda Durova** became the first Russian woman to receive the Order of St. George. **Mariia Bochkareva** was another famous recipient.

Representative Recipients

Ivanova, Rimma Mikhailovna (born 1895; died 1915, Stavropol, Russia). Nurse, acting soldier, Russia, First World War. Ivanova, a teacher in a small village, joined the active army as a nurse when the war broke out in 1914. Assigned to the 105th Regiment, she served with the men on the front line, attending the wounded in the trenches. When all of the officers in her company were put out of action, Ivanova led the Russian troops in a successful attack on enemy trenches on 9 September 1915. She was killed during this encounter. Her deed became widely known when German Military Red Cross authorities protested her participation in combat. The story of Ivanova's feat was released to the Russian press and caused something of a sensation in Russia. Archpriest Simeon Nikol'skii suggested that the young woman be canonized as a saint of the Russian Orthodox Church. Although this did not occur, the tsar, impressed by her courage and sacrifice, conferred the Order of St. George IV Class on 7 January 1916, posthumously.

Pal'shina, Antonina (born 1897; died ?), also known as Anton Pal'shin. Private, Russian army cavalry and infantry, First World War. Disguised as a man, Pal'shina joined the Cossack Cavalry at the beginning of the First World War. Her gender was discovered when she was wounded. On her recovery, Pal'shina was compelled to train as a nurse in a hospital in Lvov, from which she promptly ran away to join the Sixth Company of the 75th Sevastopol Infantry Regiment. Pal'shina was the only Russian woman to be awarded two Crosses of St. George.

Bibliography: Botchkareva, *Yashka*, 1919; Durova, *The Cavalry Maiden*, 1988; Ivanov, "I na Rusi byli Amazonki," 1992; Ivanova, "Prekrasneishie iz khrabrykh," 1994; Khristinin, "Ne radi nagrad," 1994; *Sovetskaia Voennaia Entsiklopediia*.

—Kazimiera J. Cottam

ORDER OF THE RED BANNER, WOMEN RECIPIENTS, CIVIL WAR. 1918–1928, Russian Soviet Federated Socialist Republic (RSFSR) and Soviet Union. Initially awarded for bravery in combat during the Civil War.

The first Soviet order and highest military award, the Red Banner was instituted on 16 September 1918 for the citizens of the RSFSR, after which several other Soviet republics adopted their own Red Banners. Following the establishment of the Soviet Union, a single Order of the Red Banner was proclaimed by the Executive Committee of the Soviet Union on 1 August 1924. There is no agreement among scholars as to how many Soviet women were awarded the Red Banner for their participation in the **Russian Civil War,** with cited numbers ranging from fifty-three to sixty-three (recent archival research indicates numbers in the higher range). Medics and military commissars predominated among those awarded the order. For example, Baturina lists one telegrapher, one commander of an unspecified company, two noncombatant executive secretaries, two army scouts, four partisans, and five machine gunners out of the total of fifty-three recipients of the order, with the rest being mainly medical and political personnel. (Many additional women received the order during the Second World War.)

Representative Recipients

Azarkh, Raisa Moiseevna (1897–1971). Senior political/medical officer, Red Army, Russia, Civil War. Her appointments from mid-1918 to mid-1920 included military commissar of the 1st Special Viatka Division; chief, Main Military Medical Administration of the Ukraine; chief, Medical Service of the 5th Army; and chief, Main Medical Administration of the Far Eastern Republic. She was awarded the Red Banner on 2 February 1928. During the **Spanish Civil War** she worked with world-famous Canadian medical doctor Normal Bethune.

Ianysheva, Aleksandra Aleksandrovna (1894–1983), née Permiakova. Served in the Red Guard and the Red Army, Russia, Civil War. One of the outstanding women Bolsheviks, she held the following key posts during the Civil War: chief of the Political Section of the Red Guards headquarters; chief of the Propaganda Section, Political Administration of the Red Army; and Chief, Political Section, the 15th Inzenskaia Infantry Division, subsequently renamed Sivashskaia following the division's unprecedented crossing of the Sivash Gulf, Crimea, with very heavy losses. On 2 February 1928 Ianysheva became one of the first recipients of the Order of the Red Banner in her division, awarded for her part in the legendary attack on enemy positions launched after the crossing.

Zemliachka, Rozaliia Samoilovna (1876–1947), née Zalkind, also called Demon and Osipov. Senior military commissar, Red Army, Russia, Civil War (and Soviet deputy premier, Second World War). A dedicated revolutionary, Old Bolshevik, and Lenin's close associate, Zemliachka held crucial military appointments during the Civil War, including those of chief of the Political Section of the 8th and 13th Armies, respectively. She was awarded the Red Banner on 23 January 1921 for enhancing combat effectiveness of the Red Army and facilitating its final victory. She served in important party posts until her death in 1947.

Bibliography: Baturina, *Pravda stavshaia legendoi*, 1964; Cottam, *Women in War and Resistance*, 1998; Johnson, "The Role of Women in the Russian Civil War," 1980; Levkovich, *Bez nikh my ne pobedili by*, 1975; *Sovetskaia Voennaia Entsiklopediia*.

—Kazimiera J. Cottam

OSOAVIAKHIM (*OBSHCHESTVO DRUZEI OBORONY I AVIATSIONNO-KHIMECHESKOGO STROITEL'STVA*, SOCIETY OF FRIENDS OF DEFENSE AND AVIATION-CHEMICAL CONSTRUCTION). 1927–1948, Soviet Union.

Osoaviakhim was the Soviet Union's mass voluntary society dedicated to educating and training its members in civil defense techniques. It was formed in January 1927 when two state-sponsored voluntary societies, OSO (Society for the Assistance of Defense) and Aviakhim (Society for Aviation-Chemical Construction) were merged. OSO organized military science circles and shooting clubs, while Aviakhim promoted the development and popularization of the aviation and chemical industries. Osoaviakhim combined the functions of the two agencies into one all-purpose mass civil defense organization, with a membership of more than three million in 1927. Thus Osoaviakhim was both a civil defense organization and a mass modernizing institution as well, seeking to inculcate its members and the Soviet population at large with militaristic, patriotic, and "modern" values, such as adopting a scientific worldview, promoting industrialization, and the like.

One of Osoaviakhim's stated tasks was to combat the alleged "backwardness" of the Soviet people. Since the Soviet leadership viewed women, particularly peasant women, as the most "backward" element of the population, Osoaviakhim made a concerted effort to enlist women in the organization. Women, in fact, composed a significant percentage of Osoaviakhim's membership. In 1928 there were 700,000 women in Osoaviakhim, making up 17.5 percent of the membership. By 1933 that figure had risen to 25 percent.

Besides promoting lotteries, press campaigns, and publicity stunts (such as balloon expeditions and other aviation exploits) to raise public consciousness about the organization and the need to construct modern aviation and chem-

---

Russia plans to reinstitute paramilitary training for teenagers:

For decades, Soviet children in secondary schools received training in military-related skills through the auspices of organizations such as Osoaviakhim and its descendants, including DOSAAF and NVP (Primary Military Training, 1968–1991). Basic skills including physical training, marksmanship, and first aid were combined with ideologically-oriented lectures and visits to military facilities. The program was dropped when the Soviet Union dissolved.

However, Russia's president Vladimir Putin recently announced that the program would be revived. Boys 15 and older will receive training in basic military skills and civil defense, to include a five-day orientation at local military facilities to learn drill and basic shooting skills. Girls will also receive training in marksmanship and other unspecified combat-related skills. Leaders in Russia believe the program might help to reinvigorate the youth's "moral compass." A senior leader in the Moscow Orthodox Church was quoted as saying that "children should be taught to love the smell of barracks and soldiers' boots."

—Reina Pennington (Source: 13 February 2000, "Putin to restart military training for schoolboys," by Guy Chazan in Moscow for the *Electric Telegraph* (United Kingdom).

ical industries, Osoaviakhim conducted numerous civil defense–related activities. The society established "military corners" in libraries and reading huts to educate the population about civil defense matters, including defense against air and chemical attacks. The society also sponsored lectures on these topics. In addition, Osoaviakhim organized "military circles" that taught such subjects as weapons familiarization, riflery, and small-unit tactics. In the summer members had the opportunity to train with active troops in the Red Army at Osoaviakhim summer camps. Finally, Osoaviakhim taught many Soviet citizens how to fly through the society's aerial clubs. Many of the Soviet Union's aviation heroes of the 1930s received their initial flight training in these clubs.

Women's participation in the society, its leaders hoped, would transform women into active participants in the Soviet modernization project. In the 1920s the Soviet Union's revolutionary approach to gender equality encouraged women to be as actively engaged in the defense of their country as men. The increasing militarization of Soviet society in the late 1920s and 1930s also helped to promote women's involvement in civil defense activities. Women, Soviet and Osoaviakhim leaders agreed, had an important role to play in the defense of their country.

Women participated in Osoaviakhim in a variety of ways. Women were primarily directed toward training in auxiliary and military support positions, such as intelligence gathering and analysis, map reading, first aid and medical care, communications, and air and chemical defense techniques. These functions reflected the continuation of customary gender roles in Soviet society that frowned on women engaging in combat, even though they were officially permitted (or even encouraged) to have military training. At the same time such specialists were urgently needed in the undertrained and undereducated support units of the Red Army. In the Second World War Osoaviakhim-trained women were an important part of the Red Army's rear-area support, auxiliary, and staff units. Moreover, many of these skills had peacetime uses applicable to the crash industrialization programs of the 1930s, enabling these women to play a more prominent role in the industrialization of the country.

Women also participated in most military activities sponsored by Osoaviakhim, such as shooting, parachuting, and flying. Like men, eligible women members received the GTO badge (Prepared for Labor and Defense), indicating that they had met the norms of basic military and physical training. Many women earned the marksmanship rank of "Voroshilov sharp shooters," and women trained in Osoaviakhim circles represented the Soviet Union in international shooting competitions. Participation in aerial clubs also exposed women to aviation and helped lay the groundwork for the aviation exploits performed by Soviet women in the 1930s and later. For example, **Valentina Grizodubova,** the commander of the 1938 record-setting nonstop distance flight (for which she was awarded the **Hero of the Soviet Union**) and later wartime commander of the 101st Long Range Bomber Regiment, first began her aviation career in an Osoaviakhim-sponsored aerial club, as did her colleague, **Marina Raskova**.

During the Second World War Osoaviakhim was a conduit for women seeking to join the armed forces. Osoaviakhim helped in the mass military training of civilians, including women. Many women had crash courses in parachuting and sabotage work and were sent behind German lines to blow up bridges, warehouses, and other such targets. Women trained in marksmanship became snipers on the front lines, while others provided essential support services for front-line and staff units as radio operators, nurses, and staff workers. Even women in the support ranks were exposed to the perils of combat. Radio operators and nurses, in particular, frequently came under enemy fire.

Osoaviakhim-trained women were active in other branches of the armed forces. In 1941 Marina Raskova convinced the Soviet air force leadership to form three women's air regiments, the **586th Fighter Aviation Regiment,** the 587th (**125th Guards**) Bomber Aviation Regiment (initially commanded by Raskova herself), and the 588th (**46th Guards**) Night Bomber Regiment. These units distinguished themselves in combat, the 587th and 588th earning the coveted "Guards" appellation. They were initially composed of women Osoaviakhim instructors, as well as personnel from the civilian and military air fleets.

In 1948 Osoaviakhim was disbanded into separate organizations. However, in 1952 a united civil defense organization reemerged with the establishment of DOSAAF.

Bibliography: Erickson, "Soviet Women at War," 1993; Odom, *The Soviet Volunteers,* 1973; Slepyan, "The Limits of Mobilisation," 1993.

—Kenneth Slepyan

THE OTHER MAGPIE (fl. nineteenth century). Warrior, Crow Nation.

All that is known about The Other Magpie is limited to oral testimony provided to Frank B. Linderman by Pretty Shield in the late 1920s. Pretty Shield's testimony is given in a book originally known as *Red Mother* first published in 1932. It has been reprinted more recently as *Pretty Shield: Medicine Woman of the Crows.*

The Other Magpie joined the Crow scouts working for Brigadier General George Crook during what is commonly called the Centennial Campaign of 1876. Crook's column was one of three designed to enforce a government order for all tribes to report to reservations. In May he moved north toward union with the other two columns, one commanded by Colonel John Gibbon and the other commanded by Brigadier General Alfred Terry, whose column contained the 7th Cavalry led by Lieutenant Colonel George A. Custer. Crook's movements were screened and scouted by a combination of Crow and Shoshoni scouts. On the morning of 17 June 1876 General Crook's 1,300 soldiers and scouts met 1,500 Lakota and Cheyenne warriors on the Rosebud

River. The battle involved more troops and Native Americans (on both sides), had fewer casualties, and was of eventual greater historical significance than any other battle in all of the Indian campaigns.

Pretty Shield relates that The Other Magpie, seeking revenge for the death of her brother at the hands of the Lakota, joined the Crow scouting contingent. Women often accompanied such expeditions, though not necessarily to fight. At the Rosebud battle Native Americans fought on both sides, and it was during one of these Indian-against-Indian fights that The Other Magpie and a Crow homosexual by the name of Osh-Tisch (Finds-Them-and-Kills-Them) saved the life of Bullsnake when he was badly wounded by Lakota fire. Osh-Tisch rode to his protection, firing at the Lakota even as they charged. The Other Magpie, armed only with a short belt knife and willow coup stick, rode toward the Lakota, breaking their charge. During this attack she counted coup on a Lakota warrior. Osh-Tisch shot and killed this Lakota, and The Other Magpie took his scalp. Upon return to her village she took part in the traditional Scalp Dance and danced her "coup" with the men. Pretty Shield told Linderman that all the women could tell this story but that the men would not talk of it.

The Other Magpie is one of the most enigmatic historical members of the Crow Nation today. Carson Walks over Ice, archivist at the Little Big Horn College on the Crow Reservation, stated in a 1997 interview that The Other Magpie was not counted in the 1886 census of the nation, or at least her name did not appear on the rolls. He feels certain that she married, changed her name before the census (a not-uncommon occurrence), and died childless. If she had children, there would be a family history to keep her memory alive in the oral tradition. In spite of years of searching, no other records about her have been found.

However, articles by James S. Brust in both *American Heritage* and *Montana, the Magazine of Western History*, tell of a western frontier photographer who may have captured in one single photograph both Osh-Tisch and The Other Magpie. Brust and Mike Cowdrey both hypothesized that the Crow *berdache* (homosexual) labeled as "Squaw Jim" in the picture was in fact Osh-Tisch, and that the figure seated next to him was The Other Magpie. Osh-Tisch has been positively identified by Crow historians. However, the lack of any records or family makes it next to impossible to conclusively identify the second person as The Other Magpie.

Bibliography: Brust, "Into the Face of History," 1992, and "John H. Fouch," 1994; Carson Walks Over Ice, telephone interview with author (14 April 1997); Ewers, "Deadlier than the Male," 1965; Linderman, *Pretty Shield*, 1972; Vaughn, *With Crook at the Rosebud*, 1956.

—Rodney G. Thomas

P

PAN, first name unknown (born ?; died ca. 510, China). An occasional cavalry fighter during the Northern Wei dynasty (386–534).

According to the biography of her husband Yang Dayan (died ca. 517), Pan was good at archery on horseback and often hunted and even fought by his side. When her husband was the regional inspector of Jingzhou, Pan visited his camp. Whenever there was a mission, he would ask her to dress in full armor, and they would fight together. After returning from combat or a hunt, both chatted with his subordinates in the tent; Dayan often jokingly referred to her as "General Pan." They had three sons. In 507 Dayan was demoted to the Ying Prefecture because of a failed military maneuver at Zhongli. Pan stayed in Luoyang during this time, and it was said that her conduct was far from exemplary. In 510, when Dayan was promoted to administrator of Zhongshan, he heard of Pan's reported infidelity and killed her.

Bibliography: Chen Menglei, *Gujin tushu jicheng (Qin ding)*, 1993; Li Yanshou, *Bei shi*, 1974; Wei Shou, *Wei shu*, 1974.

—Sherry J. Mou

PARAGUAYAN WOMEN IN THE WAR OF THE TRIPLE ALLIANCE, 1864–1870.

Paraguayan women played a significant role in Latin America's most devastating conflict, the War of the Triple Alliance, also known as the Paraguayan War. A number of Paraguayan women actively sought involvement in the struggle against the allied armies of the Triple Alliance (Brazil, Argentina, and Uruguay). A much larger number of women, especially rural women of the

Paraguayan Women in the War of the Triple Alliance

A Paraguayan mother bearing her dead child to the grave. (Reproduced from "The War in Paraguay," *Harper's New Monthly Magazine,* April 1870: 633–647)

lower classes, found themselves caught up in the struggle and had no choice but to become involved, especially in the military campaigns, once the foreign troops had invaded their homeland. Still others, numbering in the thousands and including women of every social class, were among the victims and casualties of that conflict.

From the beginning of the war rural women were attached to all divisions of the Paraguayan army as camp followers. Many went along as wives, mistresses, or companions of the common soldiers. Camp women lived in their own barracks or row of huts far to the rear of the camps. Although they were not an official part of the army, camp women named sergeants and captains among themselves who were responsible for all the women. As volunteers, they also received neither rations nor wages, but had to survive on what food soldiers could spare. Yet their role was crucial to the war effort, since they moved munitions wagons, fed the soldiers, carried supplies, dug trenches, kept the camps in order, nursed the wounded, and buried the dead. Camp followers also served as messengers, since they were always coming and going from the military camps to the capital, carrying the latest news about what was happening at the front. In performing all of their duties, these women were exposed to the same dangers as the rest of the army. Indeed, a few were taken prisoner by the Brazilian army. The Brazilian commander, the Marqués de Caxias, for example, noted in his war diary in July 1868 that among the 1,327 prisoners taken, there were 99 officers and 3 women. At the Battle of Avay in early December 1868 the Brazilian army found more than 300 women and children on the battlefield following the Paraguayan defeat. In the Battle of Itá Ybaté (Lomas Valentinas), 21–28 December 1868, camp followers did all the burying of the dead and moved munitions wagons, besides nursing the wounded. The U.S. minister to Paraguay, Martin T. McMahon, observed thousands of women at the front, many of whom were killed or wounded. According to contemporary Brazilian sources, some 1,500 Paraguayan women and children contended with 21,000 allied troops, mainly Brazilians. Without rifles the women militants fought with bottles, sand, glass, and any other objects they could get their hands on. The Paraguayan defenders proved no match for the Brazilian infantry. Several hundred women and children died on the battlefields. Nimia Candía was one of these combatants. Having suffered wounds six months earlier at the Battle of Itá Ybaté, she was lanced in the Battle of Piribebuy and later died in the local military hospital when it was burned down by the allies. Today the names of at least fifteen of the women who fought in the Battle of Piribebuy in August 1869 are commemorated in this village.

Bibliography: Ganson, "Following Their Children into Battle," 1990; Massare de Kostianovsky, *La mujer paraguaya*, 1970; Potthast-Jutkeit, *Paradies Mohammeds oder Land der Frauen?*, 1994.

—Barbara Ganson

PARENT, MARIE-BARBE IGNACE JOSEPHE (born 4 December 1772; died August 1829). Soldier, France, revolutionary armies.

Marie-Barbe Parent became a distinguished soldier in the French revolutionary armies (1792–1795) in spite of her parents' efforts to keep her from military service. In 1792, at the age of twenty, she joined the military disguised as a man in order to escape from the watchfulness of her parents. It is unknown what name she used for her enlistment. Enrolled in the 9th Battalion of the Fédérés du Nord, she served under General Dumouriez in Belgium until her parents found her and forced her to return home.

A year later she enrolled anew in the 139th Demi-brigade of the Army of the Rhine and of the Moselle, where she served with a distinguished record until the fall of 1794. At that time she was discharged because she was female, to the delight of her parents, who, according to one historian, "despaired at having reared a woman soldier." Later she married and became the mother of two children.

In the spring of 1795 Parent was brought before the National Convention of France, where she was awarded 600 livres for "her zealousness, her bravery,

A camp woman accompanying the Paraguayan Army through the pass at Ascurra. (Reproduced from "Paraguay and Her Enemies," *Harper's New Monthly Magazine,* February 1870: 42–49)

and the decency of her behavior" while on campaign. She was allowed to accept the award in the uniform of the French army.

Bibliography: Brice, *La femme et les armées*, n.d.; Cère, *Mme Sans-Gêne*, 1894; Dossier XR49, "Parent," Service historique de l'Armée, Vincennes, France; France, *Procès-verbal de la Convention nationale*, 7 fructidor III, 156.

—S.P. Conner

PARIS COMMUNE, WOMEN'S MILITARY ROLES. Insurrection in Paris, France, during which women for the first time in French history fought in armed units officially supported by a governing body.

Illustration from the late nineteenth-century pamphlet of the Paris Commune. (Library of Congress)

The Paris Commune (18 March–28 May 1871) came into being after France was defeated by Prussia in the Franco-Prussian War. During the German siege of Paris (September 1870–January 1871) the Parisian National Guard defended the city, operating independently of the government, since the army was unable to protect Paris. During the siege a "Women's Committee of the rue d'Arras" sought to form an "armed legion to gather the wounded on the battlefield, care for them in fixed or mobile hospitals, and if necessary replace men at the ramparts." Another proposal envisioned ten battalions of "Amazones de la Seine" to guard barricades and nurse the wounded, in the hope that shared hardship and sacrifice would prove that women merited "emancipation and civil equality," and 1,500 "Amazones" reportedly signed up; however, the proposal was rejected by military authorities.

After peace was made with Germany, a conflict developed when the Parisian National Guard refused to give up its cannon to the regular army. The citizens of Paris elected a council that proclaimed Paris autonomous, with a stated goal of re-creating France as a confederation of communes, and a series of social reforms was initiated. In the meantime, the French army was attacking Paris to regain control; the Communards organized defensive efforts. Women were now allowed to join the defenders, mainly due to the *Union des femmes pour le défense de Paris et les soins aux blessés* [The Union of Women for the Defense of Paris and Care for the Wounded]. Founded after an abortive march on the government at Versailles in imitation of the women of 1789, the *Union des femmes* organized women "to fight and to triumph or die for the defense of our Communal rights." The *Union des femmes* asked for and received material and financial aid from the insurrectionary Communal Council "so that the necessary number of women will be ready to serve in the first aid stations and, if needed, on the barricades." Ten of the twenty Parisian districts are known to have witnessed efforts to organize groups of barricade fighters.

According to newspaper accounts, on 14 May a member of the Commune distributed rifles to about a hundred women, and on 21 May a French government spy reported that "troops of armed *citoyennes* [citizenesses], preceded by a brass band," were marching about the city. One such group was the Légion des fédérées, founded by Union des femmes leaders with the support of local officials in the twelfth ward. While it and the Union des femmes organized women to fight on barricades should government forces enter the city, policing the internal enemies of the Commune was a more immediate daily concern; Légion members Catherine Rogissart and Julie Magot were among the many women later convicted for having arrested men who refused to fight for the Commune. In addition, many individual women, including **Louise Michel**, served with National Guard units as nurses and canteen workers, some of whom also engaged in combat.

After the army breached the walls of Paris on 21 May 1871, the Union des femmes mustered women to defend barricades at the place Pigalle, the place Blanche, the place de Clichy, and the boulevard Magenta, among others. A women's unit defended another barricade on the place du Château-d'Eau, where, according to a newspaper account, "they came forward on the double and began to fire, crying 'Long live the Commune.' They were armed with Snider carbines, and shot admirably. They fought like devils." Perhaps a hundred women died in the barricade fighting or were massacred by French troops after the end of hostilities. Fires ignited during the fighting gave rise to the legend of *pétroleuses*, or women arsonists; there were no organized bands of women arsonists, but a few women lit defensive fires. In the final "Bloody Week" (21–28 May 1871) of barricade fighting, during which government troops massacred perhaps 20,000 Parisians, including at least 100

Woman Soldier Is Honored by France—Fought at Polish Revolution and Franco-German War of 1871; Twice Honored

PARIS—Antoinette Lix was born in Colmar, Alsace, May 31, 1839; she was a crack shot, fencer and rider at ten, was lieutenant of Polish Uhlans at 25, lieutenant of French sharpshooters in 1870, died in 1909 in the convent of Notre Dame de Sion, Paris. Her birthplace at Colmar was marked with a plaque recently.

Daughter of an ex-soldier and only child, she was brought up like a boy. She was an assiduous student and at 17 was graduated as a teacher and went to Poland as governess to the children of Princess Lubienska. Revolution broke out there in 1863. The Lubiensky castle was surrounded and the Prince, fearing capture, fled one night after entrusting his family to Antoinette Lix.

Learning that the revolutionaries were preparing an attack on Gen. Boncza, she donned a soldier's uniform, leaped on a horse and arrived at the Polish camp. She was just in time. Galvanizing the men, she led them into action. Thus Antoinette Lix, a maid from the Vosges, became known as "Dark Michael," lieutenant of Uhlans.

Her native country refused to take Antoinette in the army, but made her postmistress at Lamarche, Vosges, a job she held at the outbreak of the Franco-Prussian War. Once more refused by the army, she organized a corps of sharpshooters and fought the victorious Germans with such bravery and resourcefulness that she was honored, for the second time in her life, with the grade of lieutenant. Broken-hearted because of the loss to the enemy of her native province and defeat of her fatherland, she retired after the war to the Parisian convent, where she died in 1909.

—*World Echo,* 6 January 1934.

women, the insurrectionary Communal Council openly encouraged military action by women: "Let even the women join their brothers, their fathers, their husbands! Those who have no weapons can tend the wounded, and can haul paving-stones up into their rooms to crush the invader." The *Légion des fédérées* and the *Union des femmes* were vital in laying the foundation for such an appeal, for in them French women attained demands made since 1792 and 1848 for some of the military rights and duties of citizenship.

Representative Participants

Lefebvre, Blanche. Founding member of the *Union des femmes*. She died fighting behind a barricade.

Lemel, Nathalie. Union des femmes leader who called upon women to "take up arms for the Commune and fight to the last drop of blood." She commanded women fighting at a barricade, an eyewitness reporting that "she was the only older woman amid a group of young girls, all armed with rifles and wearing ambulance nurses' arm-bands as well as red scarfs."

Neckbecker, Louise. "Capitaine" of the *Légion des fédérées*. She was a lace maker who was said by police to be a prostitute (although that term was used rather freely in police reports of the time).

Rogissart, Catherine. She was said by neighbors to have been the flag bearer of the *Légion* and the vice-president of the revolutionary club held in the Saint-Eloi Church, where the Légion met.

Valentin, Adélaïde. A working-class woman, she was a founding member of the Union des femmes and the "Colonelle" of the *Légion des fédérées*. Valentin called upon "all *citoyennes* to make themselves useful to the cause that we defend" and "guard positions in Paris while the men go to fight" at the front.

Bibliography: Gullickson, *Unruly Women of Paris,* 1996; Johnson, "Citizenship and Gender," 1994; Schulkind, "Socialist Women during the 1871 Paris Commune," 1985; Thomas, *The Women Incendiaries,* 1966.

—Martin P. Johnson

PAVLICHENKO, LIUDMILA MIKHAILOVNA (born 12 July 1916, Belaya Tserkov' [now Kiev Region, Ukraine]; died 27 October 1974, USSR). Sniper, Second Company, Second Battalion, 54th Razin Regiment, 25th "V.I. Chapayev" Division, Independent Marine Army. Major (ret.). Soviet Union, Second World War.

Liudmila Mikhailovna Pavlichenko is the top-scoring female military sniper in history, with 309 kills logged (36 of them enemy snipers), which she accumulated in less than one year in combat. Celebrated in the Soviet wartime press as an exemplary sniper and model for Soviet women, in 1942 Pavlichenko also served as an unofficial ambassador to the United States and Canada to raise American support for the Soviet war effort.

Pavlichenko earned a civilian sharpshooter's badge after small-arms training in a military club at the Arsenal Factory in Kiev and sniper training in the **Osoaviakhim**. She volunteered for combat duty at her local Recruitment Center in June 1941 at the age of twenty-five and was assigned to a sniper platoon in the 2nd Company of the famous 25th Chapaev Division stationed near Odessa. Pavlichenko soon demonstrated her skill in August 1941 by shooting two Romanians as "trials" and her first German sniper after three days of effort. During one engagement in the defense of Odessa she assumed command of the company after the commander and his replacement had been hit. By mid-October and the evacuation of Soviet troops from Odessa, Pavlichenko had scored 187 kills; during the defense of Sevastopol, where she fought in several sectors, she raised her score to 309. According to Pavlichenko, the fight at Sevastopol was particularly difficult since not only did the Germans have all sniper positions marked and under fire, "they knew the snipers by name." Wounded four times (twice seriously), she was medically evacuated from Sevastopol in June 1942. After recuperating, she was sent to the United States in August 1942 at the invitation of Eleanor Roosevelt—the first Soviet citizen to visit the White House. Pavlichenko toured forty-three American cities before visiting Canada and Great Britain and spoke in support of Western-Soviet cooperation against Germany.

On her return to Russia, Pavlichenko became a master sniper instructor; she personally trained eighty snipers who accounted for more than 2,000 kills. In 1943 Pavlichenko graduated from the Vystrel courses for officers and was promoted to the rank of major in the coastal services. From 1945 to 1953 she worked as a research officer in the Naval Staff Headquarters. Awarded the **Hero of the Soviet Union** on 25 October 1943, Pavlichenko was also decorated with two Orders of Lenin and other medals. She died at the age of fifty-eight and was buried in Novodevich'e Cemetery.

Liudmila Pavlickhenko, a twenty-six-year-old Soviet sniper, tries out a rifle belonging to a member of the British Home Guard, November 1942. (Hulton-Deutsch Collection/CORBIS)

Bibliography: Cottam, *Women in War and Resistance,* 1998; Erickson, "Soviet Women at War," 1993; Pavlichenko, *Geroicheskaia byl,* 1960, and "I Was a Sniper," 1977; Shkadov, *Geroi Sovetskogo Soiuza,* 1988; Toropov, *Geroini,* 1969.

—Mary Allen

PETITJEAN, MADELEINE (born 1746, Metz; died ?). Cannoneer, 4th Company of the Battalion of the Sorbonne, French Revolutionary Wars, 1793.

Madeleine Petitjean joined the military at the age of forty-six after having lost fifteen of seventeen children and having been widowed three times. For nearly a year during the period after her third husband's death Petitjean served in the military, where she found steady employment and a subsistence wage. During service in the Vendée in western France she was captured by the enemy and tortured. Later released, she served until September 1793, when she was seriously injured (a knife wound in the right thigh and a musket shot to the right foot), leaving her partially incapacitated.

Like many other women, Petitjean was expelled from the military after the 1793 enactment of a law restricting women from serving in the French military (see **French Revolution and Napoleonic Era, women soldiers in**). Petitjean petitioned for assistance on the basis of her disability. When her final petition reached the National Convention in 1794, the deputies noted that Petitjean's first two husbands had died in naval service, and her third at the Bastille, yet she had not strayed from service to her nation. She was, according to their recognition, "a true republican... who has neither a place to live nor any means of subsistence." She was accorded thirty sous per day, which at the time was a bare minimum. Later in 1794, based on the record of her third husband and in recognition of her own military service and injuries, Petitjean was granted 636 livres per year as a pension.

Bibliography: Cère, *Mme Sans-Gêne,* 1894; Hennet, Léon, "Madeleine Petitjean," *Carnet de la Sabretache* 5 (31 August 1907): 533–536; Dossier XR48, "Petitjean," Service historique de l'Armée, Vincennes, France; France, *Procès-verbal de la Convention nationale,* 25 prairial II, 254–255.

—S.P. Conner

PHILIPPA OF HAINAULT (born 1314; died 1369). Queen of England.

The daughter of William III the Good, count of Hainault, and Jeanne de Valois, Philippa married King Edward III of England around 1328. She was described by Tout as "a paragon of English queens," and the unity of the royal family and the state owed much to her. The mother of twelve children, including the Black Prince, she was the patron of Froissart and Chaucer as well as of the Queen's College, Oxford, which is named in her honor.

Froissart's works describe a queen who took an active military role in the defense of her kingdom, but this romantic view is not supported by other evidence. According to Froissart, Queen Philippa personally led troops in battle against the Scottish invaders who had attacked England in 1346 at the request of Philip VI of Valois, whose forces were being besieged at Calais in France by Philippa's husband. Having assembled all the forces she could, according to Froissart, Philippa then marched them to Newcastle (although other, more reliable evidence suggests that they were probably marched to Durham) and led the makeshift English army at the Battle of Neville's Cross, where she captured King David of Scotland. However, contemporary evidence does not support Froissart's account of the battle: the English army was raised by William Zouch, archbishop of York, and Philippa did not go to battle herself; she instead spent the day of the battle in Durham praying for the success of the English army and only went to the battlefield after her army had defeated the Scots.

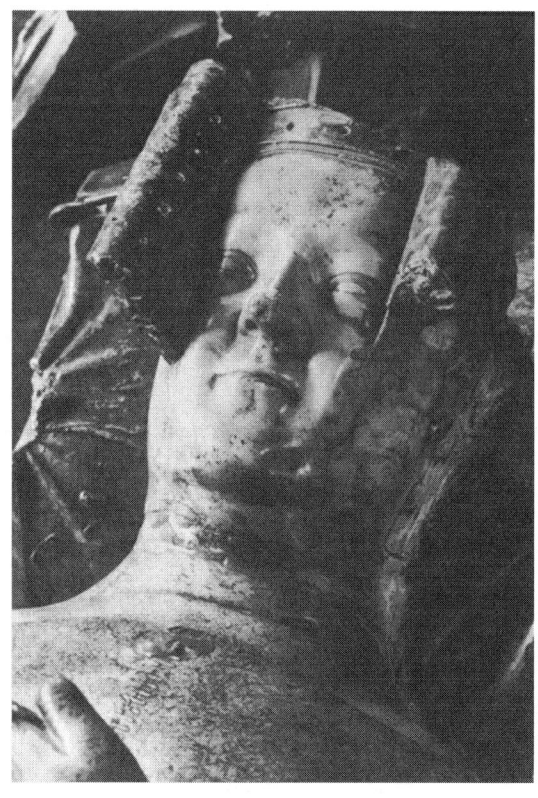

Detail of the tomb of Philippa of Hainault by Jean de Liege, Westminster Abbey. (Perry-Castañeda Library)

When Calais finally fell after a long siege, Edward III ordered the six leading burgesses of the city to be taken from the city in chains and then beheaded. According to Froissart, the pregnant Queen Philippa fell to her knees in tears to beseech her husband, as proof of his love for her, to show mercy upon them. Edward released the six burgesses to her care and ceased further military action against the people of the conquered city of Calais. There is no real evidence in support of Froissart's tale, although it has the ring of truth: sixteen years earlier Philippa had intervened in a similar way on behalf of the people of Westminster.

Bibliography: Froissart, *The Chronicle of Froissart*, 1967; Green, "The Household of Edward, the Black Prince," 1997; Hardy, *Philippa of Hainault and Her Times*, 1910; Harvey, *The Black Prince and His Age*, 1976; McKisack, *The Fourteenth Century*, 1959; Nicholson, *Edward III and the Scots*, 1965.

—J.M.B. Porter

PHIPPS, ANITA EVANS (born 1886, Augusta Arsenal, Maine, United States; died 18 July 1953, Nonquitt, Massachusetts, United States). Director of women's relations, United States Army, 1921–1931.

Anita Phipps wrote the first complete and workable plan for a women's army corps in the years between the two world wars. In 1920 the Nineteenth Amendment to the U.S. Constitution was ratified, giving women the right to vote. The growing political power of women voters, many of whom became active in antimilitary and pacifist movements, was viewed with alarm by the army. To persuade women voters that a strong military was a necessity, the position of director of women's relations, United States Army, was established. The director's job was to maintain liaison between the War Department and women's organizations to promote the idea of a strong military. After the first appointee resigned, the thirty-five-year-old Anita Phipps was appointed director. Phipps, the daughter of an army general, had served as the director of the Motor Corps Service of the Pennsylvania-Delaware Division of the Red Cross during the First World War and had experience in dealing with the military.

Phipps encountered problems from the start. The War Department's failure to give her military status and support for her position caused her to lose credibility with women's organizations. Despite repeated pleas, her job duties were never clearly defined and her work was dismissed by the War Department. Phipps decided to define her position herself and began planning for what she called a Women's Service Corps, presumably with her as the head. She collected information on the utilization of women by different organizations during the First World War, including the British army and the United States Navy. She concluded that the policy of enlisting women during the First World War, but giving them no military training, housing, discipline, or courtesies, was a mistake. She proposed that her women's corps should serve in, not auxiliary to, the army, be fully trained, and serve under the command of women officers. By surveying various army corps, commanders, and branch chiefs, Phipps thought that about 170,000 women could be utilized in wartime.

> One new indulgence was to go out evenings alone. This I worked out carefully in my mind, as not only a right but a duty. Why should a woman be deprived of her only free time, the time allotted to recreation? Why must she be dependent on some man, and thus forced to please him if she wished to go anywhere at night?
>
> A stalwart man once sharply contested my claim to this freedom to go alone. "Any true man," he said with fervor, "is ready to go with a woman at night. He is her natural protector." "Against what?" I inquired. As a matter of fact, the thing a woman is most afraid to meet on a dark street is her natural protector.
>
> —Charlotte Perkins Gilman, *The Living of Charlotte Perkins Gilman: An Autobiography*, reprint of the 1935 ed. (Madison: University of Wisconsin Press, 1991), 72.

The plan was rejected by the General Staff in 1926, which cited difficulties in the cost of housing, transportation, and toilet facilities and in the personnel policy involved. Privately, officers in the War Department were fearful that Phipps's plan would establish a powerful organization of women within the army. Finally, in 1931, Phipps asked the secretary of war to either define her duties and authority or abolish her position. Nothing was done, and in 1931 the new chief of staff, General Douglas MacArthur, informed the War Department that Phipps's duties were of no military value and the position was eliminated. Although all her efforts seemed to have been in vain, Phipps's study helped lay the groundwork for the creation of the **WAC** in 1942. Phipps later held the position of head of American Women's Voluntary Services in the District of Columbia during the Second World War.

Bibliography: Holm, *Women in the Military*, 1992; Morden, *The Women's Army Corps*, 1990; Treadwell, *The Women's Army Corps*, 1954.

—Vicki L. Friedl

PINGYANG, PRINCESS (died 623, China), posthumously titled "Luminous." Commander at the end of the Sui dynasty (581–618).

Princess Pingyang commanded an army of more than 70,000 in the last years of the Sui dynasty and helped her father found the Tang dynasty (618–907). The third daughter of Gaozu, who would become the first emperor of the Tang dynasty (r. 618–626), Pingyang married Chai Shao (died 638), the swordsman guard to the heir apparent of the Sui emperor. When her father led a revolt against the government in Taiyuan in 617, he secretly sent to Chai Shao for help. Chai Shao and the princess were in the capital Chang'an (present Xi'an in Shaanxi Province); the princess encouraged her husband to join her father, though he did not want to leave her behind.

After Chai Shao left, Princess Pingyang went to Hou Prefecture (present Huxian in Shanxi Province) and used all the family assets to hire soldiers. She attracted several hundred dissidents and bandits from nearby mountains and formed a regiment; she disciplined her army with clear instructions and regulations against plunder. As a result, many came to her from near and far, and soon she led more than 70,000 soldiers. Though the Sui government often sent troops against her, she rebuffed the attacks and notified her father, who was greatly pleased with the situation.

After her father crossed the Hui River, he sent Chai Shao with several hundred cavaliers to welcome the princess, who led more than 10,000 crack troops to converge with her father's army. The armies of Pingyang and her husband encircled the capital; her troops were called "women's troops" by all other camps. After the capital was captured, she was installed as Princess Pingyang, and for her military merits her share of royal gifts often differed from others.

When Pingyang died in 623, her father Emperor Gaozu added various decorations and soldiers to the funeral procession, including a military band. The chamberlain for ceremonials petitioned the throne that according to the codes of propriety, a woman's funeral should not be accompanied by a military band. Emperor Gaozu disagreed, declaring: "A military band plays military music; since the princess raised and commanded armies in the past in response to the righteous call of dynastic change, she earned military merits. . . . The Princess's achievements matched those of a minister, and she should not be compared to ordinary women. How could her funeral have no military band!" So he increased the size of the band to demonstrate her extraordinary achievements and ordered the appropriate department to confer upon her the posthumous title "Luminous."

Bibliography: Liu Xu, *Jiu Tang shu*, 1975.

—Sherry J. Mou

PLATER, EMILIA (born 13 November 1806, Wilno [Vilnius], in the Russian-held area of partitioned Poland; died 23 December 1831, Justianów, near the present-day border of Poland and Lithuania). Served as a captain in the Polish infantry during the November Uprising (1830–1831) against Russia.

One of several women to take an active part in the unsuccessful Polish insurrection of 1831, Emilia Plater fought in a number of skirmishes and battles against Russian forces in Lithuania during the first half of 1831. Her life and activities have been immortalized in Polish art and literature. She ranks as the best-known woman in the national insurrectionary tradition and is recognized as a symbol of Polish patriotism.

Plater was raised by her mother and received a classical education at home, where she also learned to ride and shoot. In addition, she took a keen interest in national history, modeling her behavior on the mythical heroines of Poland's distant past. During a visit to Warsaw in 1829, unlike most young women, she ignored social events, preferring to tour the remains of the city's seventeenth-century defenses. More than one acquaintance suggested that she was aspiring to emulate the exploits of **Joan of Arc**.

The opportunity to do so soon presented itself. In the spring of 1831 all of Lithuania was ablaze with patriotic manifestations in opposition to Russian rule. In late March Plater announced her desire to fight against the tsarist government; in a letter to a relative she claimed that she had known from her earliest childhood that she would one day go off to war. Days later, wearing a saber and dressed in a partisan uniform, Plater addressed a crowd of local inhabitants, calling on them to join the struggle. One source claims that she was able to assemble a unit of more than 300 soldiers. She aimed to approach the fortress at Dyneburg (Daugavpils) in the hope of making contact with sympathizers from the city's cadet academy. En route, her partisans surprised and defeated a small Russian unit that was guarding a postal station, acquiring sixty horses in the process. On 2 April Plater's forces repelled a second Russian detachment. Two days later, however, her cousin, Cezary Plater, took charge of the unit. Nevertheless, it was Emilia who acted in the name of the formation at the town of Jeziorosy, where she signed the act of uprising in accordance with the traditions of the old Polish-Lithuanian Commonwealth.

It is difficult to separate fact from legend regarding Plater's military activities. Several sources note that she served in a number of partisan formations with distinction and honor. One insurgent, for example, recalled that she "courageously withstood the heaviest fire" during the Polish defeat at Prystowiany. On the other hand, other documents hint that she sometimes became a liability in battle, losing consciousness or falling from her horse. Yet it is certain that she fought in a number of armed actions against the Russians, including the victory at Wiłkomierz on 17 May.

As the uprising spread, insurgent units were organized into regular army formations. While the Polish commander in Lithuania, General Dezydery Chłapowski, applauded her efforts, he initially recommended that Plater return home. The story goes that upon meeting with the general she declared, "I decided to be a soldier and to fight for the liberation of the homeland, until Poland regains complete freedom." Upon observing her determination and desire to fight, Chłapowski gave Plater the rank of captain and named her commander of the First Company of the First Lithuanian Infantry Regiment (later the 25th Infantry Regiment).

In this role she fought in the Battle of Kowno (25 June), during which the Poles were defeated and her regiment was decimated. In one of the last actions of the conflict in Lithuania, the Poles tried to attack the garrison of Szawel (6 July). This attempt also failed, although Plater distinguished herself by saving a portion of the Polish supply column. In the face of increasing Russian strength, however, the Polish cause was lost. The Polish high command in Lithuania gave up the struggle as hopeless and ordered its troops into Prussian territory. In July, while serving in the corps of General Antoni Giełgud, she learned of the general's decision to give up the fight and retire across the Prussian border. Plater rejected this action as "shameful." As the defeated troops moved toward the border, legend has it that she declared, "As for me—as long as I feel the spark of life in my breast—I will fight for the Homeland." She intended instead to make her way to Warsaw to join other Polish units. She fell ill on the way, however, and at age twenty-five died—legend has it—upon hearing of the uprising's defeat.

Adam Mickiewicz glorified Plater's life in 1832 in the poem "The Death of the Colonel," describing her as the ideal leader, loved by her troops and the common people. Plater also became the subject of a painting, songs, and several written works. In interwar Poland her portrait was used on twenty-złoty banknotes, and the Twenty-second Infantry Regiment was named after her. She was also the patron of the **Emilia Plater Independent Women's Battalion** of the First Kosciuszko Division (associated with the Red Army) in the Second World War. Today one of the main streets of Warsaw bears her name.

Bibliography: Ciepieńko-Zielińska, Donata, *Emilia Plater* (Warsaw: 1966); *Encyklopedja Wojskowa*, s.v. "Platerówna, Emilja"; Król, Stefan, *101 kobiet polskich* (Warsaw: 1988); *Polski Słownik Biograficzny*, s.v. "Plater, Emilia"; Wawrzykowska-Wierciochowa, Dioniza, *Sercem i orężem ojczyźnie służyły* (Warsaw: 1982).

—Robert Ponichtera

PLSK (POMOCNICZA LOTNICZA SŁUŻBA KOBIET, WOMEN'S AUXILIARY AIR FORCE). Formed in 1943, disbanded in 1945, Great Britain. Auxiliary military formation, Great Britain, Second World War.

PLSK (Women's Auxiliary Air Force)

The PLSK was founded by order of the minister of defense of the Polish government in exile in mid-December 1942 on the model of the British Women's Auxiliary Air Force (WAAF) to cooperate with the Polish air force and replace Polish servicemen in service and support roles on Polish air bases in the United Kingdom. Initially, thirty-six women candidates were sent in May 1943 for basic training to Falkirk, Scotland. After undergoing successively higher levels of specialized training, the group qualified as instructors. Twelve women graduated from an officers' course and the remaining from an NCO course. In October 1943 they were awarded both Polish and British ranks. In November 1943 the general recruitment to the PLSK began. There are discrepancies in the total number of officers and NCOs given. A recent Polish source cites the total strength of the PLSK as being 1,436, constituting 10 percent of the Polish Air Force strength in the West, with 52 officers and 110 NCOs, while an earlier Polish source cites a total of 1,653, including 52 officers and 163 NCOs.

Apart from those from Great Britain, where the PLSK recruited most of its members, the PLSK also attracted Polish women from Canada, the United States, France, Argentina, Switzerland, China, and Japan. A large number of volunteers came from military units organized in the Soviet Union and subsequently evacuated to the Middle East. The women served in twenty-six units of the Polish Air Force. The PLSK recruits were trained in forty-five specialties to which they were directed on the basis of their abilities and preferences. For example, the women were employed in communications, laboratory work, and deciphering of aerial photographs. They served as clerks, cooks, mechanics, radio and telephone operators, physicians, dentists, sentries, and parachute folders. Women officers served in administrative, logistical, intelligence, and educational capacities.

Bibliography: Mleczak, Eugeniusz, "Dziewczęta z tamtych lat: Odsłonięcie tablicy w Katedrze Polowej WP," *Polska Zbrojna* 109. 921 V (7 June 1994): 2; "Polki na wojnie," *Skrzydlata Polska* 10 (3 October 1985); 12.

— Kazimiera J. Cottam

Q

QIN LIANG-YÜ (born 1574, Zhongzhou, Sichuan Province, China; died 1648, Shizhu, Sichuan, China), byname Zhensu, also known as Qin Taibao. Native chieftain of Shizhu with the title of Pacification Commissioner. Assistant commissioner in chief, regional commander of Sichuan.

As a general of the late Ming dynasty (1368–1644), Qin Liangyu fought with her troops against the invading Manchus in northeastern China and against local rebellions and bandits in the southwest. She is the only woman who ever reached the highest post in the military command system of the Ming dynasty, that of regional commander. Her bravery, loyalty, and military skill attracted the attention of several Chinese historiographers, and her biography, which appears in the section on military officials, is the only biography of a woman in a section other than the biographies of women in the official history of the Ming dynasty. Because Qin Liang-yü fought to put down peasant rebellions in late Ming times, writers in Communist China have been ambivalent in their treatment of her. In one historical novel of the 1970s her role was even changed to acting as a partisan of the peasant class.

In her youth Qin's father gave her the same military training as her brothers, including horseback riding and archery. Her special field of interest is said to have been military strategy. After she married Ma Qiancheng, a native Shizhu chieftain in eastern Sichuan, she joined him in training the local troops. According to some sources, she even drilled a corps of women, but there is no confirmation of that in the primary sources.

> Perhaps women were once so dangerous that they had to have their feet bound.
>
> —Maxine Hong Kingston, *The Woman Warrior: Memoirs of a Girlhood among Ghosts* (New York: Vintage Books, 1977), 23.

In 1599–1600 Qin accompanied her husband in the suppression of a local rebellion, the Bozhou campaign in the southwest of China, with an elite corps of 500 soldiers under her personal command, which is her first recorded combat experience. After Ma's death in 1613 she took over his title and office as well as the command of the local troops, in accordance with hereditary leadership rights under the Ming court.

When Liaodong in northeastern China was threatened by the Manchus in 1620, Emperor Guangzong (r. 1620) ordered her to send a detachment. She sent two of her brothers with several thousand recruits, and after they suffered an overwhelming defeat, she went personally in 1621 with 3,000 elite soldiers. After victories at Shanhai Pass she returned to Sichuan on imperial orders to raise more troops for the support of the northeast. Back home, she immediately helped suppress the riots of local bandits, for which she was given the post of regional commander of Sichuan and was promoted to assistant commissioner in chief. After the Manchus had taken several cities near Beijing in 1630, Emperor Sizong (r. 1628–1644) again summoned Qin to send an army to strengthen the defense at the capital. Her encampment in Beijing was known as Sichuan Camp until recent times.

When in the mid-1630s Sichuan was again shaken by local uprisings and invading bandits, Qin was entrusted with exterminating them. Her troops initially won several victories, but in 1640 they were defeated, mostly owing to errors on the part of Qin's superiors in administrative civil offices. She tried in vain to prevent one of the main leaders of the peasant rebellions, Zhang Xianzhong (1605–1647), from conquering the province, but did succeed in protecting Shizhu from devastation.

Emperor Yongli of the Southern Ming (r. 1645–1646) bestowed on her the honorary title of "Marquis of Zhongzhen" and announced her "Grand Guardian of the Heir Apparent," the highest honor an official could get in imperial China. After her death in 1648 Qin was buried at the Huilong Mountain several kilometers east of Shizhu. Her tomb there underwent restoration in 1987.

Bibliography: Chen Menglei, *Gujin tushu jicheng (Qin ding)*, 1993; Hummel, *Eminent Chinese of the Ch'ing Period*, 1943–1944; Qin Liangyu shi yanjiu bianji weiyuanhui, *Qin Liangyu shiliao jicheng*, 1987; Zhang, *Ming shi*, 1974; Wang Huailing, ed., *Buji Shizhu ting xinzhi* (China: Woodblock print of 1843).

—Doris Kehry-Kurz

QUATSAULT, ANNE (born 1775, Loiret, France; died ?), also known as "Quatre-Sous" by her soldier comrades. Guardian of horses, driver in artillery transport, and infantry soldier during the French revolutionary era, 1791–1794.

According to stories, thirteen-year-old Anne Quatsault attempted to join the French volunteers from her hometown in 1788. Although she was in male disguise, she was denied entry into the military because she was deemed too young and too short. Undeterred, Quatsault continued her efforts, and in 1791 at either Fontainebleau or Versailles (sources differ) she managed to obtain a position in artillery transport, again in male disguise.

Quatsault served in the French military until April 1794, when she could not escape the requirements of the 1793 French law that excluded women from military service (see **French Revolution and Napoleonic Era, women soldiers in**). Although she had served in disguise the entire term of her military employment, she was apparently recognized and forced from service along with most other women. By that time she had served in the Vendée and in the sieges of Liège, Aix-la-Chapelle, Namur, Maastricht, and Dunkirk and had had two horses shot out from under her at Hondschoote, resulting in a debilitating injury. When she petitioned the government for a pension, based on her military record, her petition request was seconded by representatives of the government who congratulated her for her valor, her meritorious conduct "although she was a woman," and her successful disguise.

For her military achievements, which were well documented, Quatsault received 300 livres per year in recompense, to be augmented by 200 livres when she married. In accordance with the 1793 law, the French national government also provided the sum of 150 livres to assist her in obtaining women's clothing for her return to civilian life. Although the pension was supposedly guaranteed for life, Quatsault was forced numerous times to remind the French government of her service, and in 1817, under the reign of Louis XVIII, she sold her belongings in order to subsist while awaiting the reinstatement of her pension. Nothing further is known about her later years in France.

Bibliography: Brice, *La Femme et les armées*, n.d.; Cère, *Mme Sans-Gêne*, 1894; Dossier XR49, "Quatsault," Service historique de l'Armée, Vincennes, France; France, *Procès-verbal de la Convention nationale*, 3 floreal an II, 60.

—S.P. Conner

QUITÉRIA DE JESUS, MARIA (born 1792; died 21 August 1853, Salvador, Bahia, Brazil). Soldier and First Cadet, Exército Pacificador (Brazilian independence army).

The daughter of a petty rancher from the interior of the province of Bahia, Brazil, Maria Quitéria—known then and today by her first names, rather than her devotional surname— left her father's home in September 1822 to

Maria Quitéria de Jesus, 1823. (Reproduced from Maria Graham, *Journal of a Voyage to Brazil and Residence There during Part of the Years 1821, 1822, 1823,* London: Longman, Hurst, Rees, Orme, Brown, and Green, 1824)

Quitéria de Jesus, Maria

The monument to Maria Quitéra de Jesus, erected in the 1950s, in Salvador, Brazil. (Courtesy of Hendrik Kraay)

volunteer for the patriot forces then being organized to expel the Portuguese army from the provincial capital of Salvador. She claimed to be her brother-in-law's son and enlisted in an infantry battalion. Though her sex was discovered when her father applied to have her discharged a few days later, she remained in the ranks, modifying her uniform by adding a tartan kilt. She fought in three skirmishes, once in chest-deep water to repel a Portuguese landing, and another time in an enemy trench where she took prisoners. To reward her for this feat, the commander of Brazilian forces named her a first cadet, an honorific rank reserved for sons of nobles and senior officers. Discharged after the war, Maria Quitéria was received in Rio de Janeiro by the new Brazilian emperor, who dubbed her a knight of the Imperial Order of the Cross (cavaleiro da Ordem Imperial do Cruzeiro) in 1823 and commissioned her an *alferes* (second lieutenant), which entitled her to an army pension.

Bibliography: Graham, *Journal of a Voyage to Brazil*, 1824; Reis Júnior, *Maria Quitéria*, 1953; Souza, "Maria Quitéria," 1979.

—Hendrik Kraay